AN ANALYSIS OF
THE FINITE ELEMENT METHOD

Prentice-Hall
Series in Automatic Computation

George Forsythe, editor

MARTIN AND NORMAN, *The Computerized Society*

MATHISON AND WALKER, *Computers and Telecommunications: Issues in Public Policy*

MCKEEMAN, et. al., *A Compiler Generator*

MEYERS, *Time-Sharing Computation in the Social Sciences*

MINSKY, *Computation: Finite and Infinite Machines*

NIEVERGELT, et al., *Computer Approaches to Mathematical Problems*

PLANE AND MCMILLAN, *Discrete Optimization: Integer Programming and Network Analysis for Management Decisions*

PRITSKER AND KIVIAT, *Simulation with GASP II: a FORTRAN-Based Simulation Language*

PYLYSHYN, editor, *Perspectives on the Computer Revolution*

RICH, *Internal Sorting Methods: Illustrated with PL/1 Program*

RUSTIN, editor, *Algorithm Specification*

RUSTIN, editor, *Computer Networks*

RUSTIN, editor, *Data Base Systems*

RUSTIN, editor, *Debugging Techniques in Large Systems*

RUSTIN, editor, *Design and Optimization of Compilers*

RUSTIN, editor, *Formal Semantics of Programming Languages*

SACKMAN AND CITRENBAUM, editors, *On-Line Planning: Towards Creative Problem-Solving*

SALTON, editor, *The SMART Retrieval System: Experiments in Automatic Document Processing*

SAMMET, *Programming Languages: History and Fundamentals*

SCHAEFER, *A Mathematical Theory of Global Program Optimization*

SCHULTZ, *Spline Analysis*

SCHWARZ, et al., *Numerical Analysis of Symmetric Matrices*

SHERMAN, *Techniques in Computer Programming*

SIMON AND SIKLOSSY, *Representation and Meaning: Experiments with Information Processing Systems*

STERBENZ, *Floating-Point Computation*

STERLING AND POLLACK, *Introduction to Statistical Data Processing*

STOUTEMYER, *PL/1 Programming for Engineering and Science*

STRANG AND FIX, *An Analysis of the Finite Element Method*

STROUD, *Approximate Calculation of Multiple Integrals*

TAVISS, editor, *The Computer Impact*

TRAUB, *Iterative Methods for the Solution of Polynomial Equations*

UHR, *Pattern Recognition, Learning, and Thought*

VAN TASSEL, *Computer Security Management*

VARGA, *Matrix Iterative Analysis*

WAITE, *Implementing Software for Non-Numeric Application*

WILKINSON, *Rounding Errors in Algebraic Processes*

WIRTH, *Systematic Programming: An Introduction*

AN ANALYSIS OF THE FINITE ELEMENT METHOD

GILBERT STRANG

Massachusetts Institute of Technology

GEORGE J. FIX

University of Maryland

PRENTICE-HALL, INC.

ENGLEWOOD CLIFFS, N.J.

Library of Congress Cataloging in Publication Data

STRANG, GILBERT,
    An analysis of the finite element method.

    (Prentice-Hall series in automatic computation)
    Bibliography: p.
    1. Finite element method. I. Fix, George J.,
    joint author. II. Title.
TA335.S77   515'.624   72-12642
ISBN 0-13-032946-0

10 9 8 7 6

Printed in the United States of America.

PRENTICE-HALL INTERNATIONAL, INC., *London*
PRENTICE-HALL OF AUSTRALIA, PTY. LTD., *Sydney*
PRENTICE-HALL OF CANADA, LTD., *Toronto*
PRENTICE-HALL OF INDIA PRIVATE LIMITED, *New Delhi*
PRENTICE-HALL OF JAPAN, INC., *Tokyo*

To Jill and Linda

cherchez la f.e.m.

PREFACE

The finite element method has been an astonishing success. It was created to solve the complicated equations of elasticity and structural mechanics, and for those problems it has essentially superseded the method of finite differences. Now other applications are rapidly developing. Whenever flexibility in the geometry is important—and the power of the computer is needed not only to *solve* a system of equations, but also to *formulate* and *assemble* the discrete approximation in the first place—the finite element method has something to contribute.

From a mathematical point of view, the method is an extension of the Rayleigh–Ritz–Galerkin technique. It therefore applies to a wide class of partial differential equations. The Ritz technique does not, however, operate directly with the differential equation; instead, the continuous problem is put into an equivalent variational form, and the approximate solution is assumed to be a combination $\sum q_j \varphi_j$ of given trial functions $\varphi_j(x)$. This is the method of weighted residuals, and the weights q_j are computed from the underlying variational principle. It is this discrete problem which the computer actually solves.

So far the idea is an old one. What is new is the choice of trial functions: in the finite element method they are *piecewise polynomials*. That choice is responsible for the method's success. Each function φ_j is zero over most of the domain, and enters the computation only in the neighborhood of a particular node. In that neighborhood φ_j is pieced together from polynomials of low degree, and the computations are as simple as possible. It is remarkable that simultaneously, and quite independently, piecewise polynomials have become preeminent in the mathematical theory of approximation of functions. Apparently it was the right idea at the right time.

Because the mathematical foundations are sound, it is possible to understand why the method works. This is the real reason for our book. Its purpose is to explain the effect of each of the approximations that are essential for the finite element technique to be computationally efficient. We list here some of these approximations:

(1) interpolation of the original physical data
(2) choice of a finite number of polynomial trial functions
(3) simplification of the geometry of the domain
(4) modification of the boundary conditions
(5) numerical integration of the underlying functional in the variational principle
(6) roundoff error in the solution of the discrete system.

These questions are fundamentally mathematical, and so are the authors. Nevertheless this book is *absolutely not* intended for the exclusive use of specialists in numerical analysis. On the contrary, we hope it may help establish closer communication between the mathematical engineer and the mathematical analyst. It seems to us that the finite element method provides a special opportunity for this communication: the theory is attractive, the applications are growing, and best of all, the method is so new that the gap between theory and application ought not yet to be insurmountable.

Of course we recognize that there are obstacles which cannot be made to disappear. One of them is the language itself; we have kept the mathematical notations to a minimum, and indexed them (with definitions) at the end of the book. We also know that, even after a norm has been interpreted as a natural measure of strain energy, and a Hilbert space identified with the class of admissible functions in a physically derived variational principle, there still remains the hardest problem: to become comfortable with these ideas, and to make them one's own. This requires genuine patience and tolerance on both sides, as well as effort. Perhaps this book at least exhibits the kind of problems which a mathematician is trained to solve, and those for which he is useless.

In the last few years a great many numerical analysts have turned to finite elements, and we are very much in their debt. This is acknowledged explicitly throughout the book, and implicitly in the bibliography, even though we have by no means attempted a formal history. Here, before the book begins, we want to thank two others—engineers rather than mathematicians—for help that was the most important of all. One is Isaac Fried, whose influence led us to abandon an earlier (and completed) "Fourier Analysis of the Finite Element Method," and to study instead the real thing. The other is Bruce Irons, whose remarkable intuitions are described (and proved correct, as often as we can) in the book itself.

Chapter 1 is very much longer than the others, and was used by the first author as the text in an introductory course at M.I.T. The only homework was to go out and program some finite elements. Where such programs are already available, students could be asked to combine computational experiments with a theoretical seminar based on the book.

Chapters 2 to 5 were also written by the first author. The last three chapters were drafted by the second author, and then revised and "homogenized" by the first. And the whole was typed by Mrs. Ingrid Naaman, who has gracefully allowed us to believe that she enjoyed it; thank you.

GILBERT STRANG

GEORGE J. FIX

Cambridge, Massachusetts

CONTENTS

1 AN INTRODUCTION TO THE THEORY

1.1. THE BASIC IDEAS

The finite element method can be described in a few words. Suppose that the problem to be solved is in variational form—it may be required to find the function u which minimizes a given expression of potential energy. This minimizing property leads to a differential equation for u (the Euler equation), but normally an exact solution is impossible and some approximation is necessary. The Rayleigh–Ritz–Galerkin idea is to choose a finite number of trial functions $\varphi_1, \ldots, \varphi_N$, and among all their linear combinations $\sum q_j \varphi_j$ to find the one which is minimizing. This is the Ritz approximation. The unknown weights q_j are determined, not by a differential equation, but by a system of N discrete algebraic equations which the computer can handle. The theoretical justification for this method is simple, and compelling: *The minimizing process automatically seeks out the combination which is closest to u.* Therefore, the goal is to choose trial functions φ_j which are convenient enough for the potential energy to be computed and minimized, and at the same time general enough to approximate closely the unknown solution u.

The real difficulty is the first one, to achieve convenience and computability. In theory there always exists a set of trial functions which is complete—their linear combinations fill the space of all possible solutions as $N \to \infty$, and therefore the Ritz approximations converge—but to be able to compute with them is another matter. This is what finite elements have accomplished.

The underlying idea is simple. It starts by a subdivision of the structure, or the region of physical interest, into smaller pieces. These pieces must be easy for the computer to record and identify; they may be triangles or rectangles. Then within each piece the trial functions are given an extremely

1

simple form—normally they are polynomials, of at most the third or fifth degree. Boundary conditions are infinitely easier to impose locally, along the edge of a triangle or rectangle, than globally along a more complicated boundary. The accuracy of the approximation can be increased, if that is necessary, but not by the classical Ritz method of including more and more complex trial functions. Instead, the same polynomials are retained, and the subdivision is refined. The computer follows a nearly identical set of instructions, and just takes longer to finish. In fact, a large-scale finite element system can use the power of the computer, for the *formulation* of approximate equations as well as their solution, to a degree never before achieved in complicated physical problems.

Unhappily none of the credit for this idea goes to numerical analysts. The method was created by structural engineers, and it was not recognized at the start as an instance of the Rayleigh–Ritz principle. The subdivision into simpler pieces, and the equations of equilibrium and compatibility between the pieces, were initially constructed on the basis of physical reasoning. The later development of more accurate elements happened in a similar way; it was recognized that increasing the degree of the polynomials would greatly improve the accuracy, *but the unknowns q_j computed in the discrete approximation have always retained a physical significance.* In this respect the computer output is much easier to interpret than the weights produced by the classical method.

The whole procedure became mathematically respectable at the moment when the unknowns were identified as the coefficients in a Ritz approximation $u \approx \sum q_j \varphi_j$, and the discrete equations were seen to be exactly the conditions for minimizing the potential energy. Surely Argyris in Germany and England, and Martin and Clough in America, were among those responsible; we dare not guess who was first. The effect was instantly to provide a sound theoretical basis for the method. As the techniques of constructing more refined elements have matured, the underlying theory has also begun to take shape.

The fundamental problem is to discover how closely piecewise polynomials can approximate an unknown solution u. In other words, we must determine how well finite elements—which were developed on the basis of computational simplicity—satisfy the second requirement of good trial functions, to be effective in approximation. Intuitively, any reasonable function u can be approached to arbitrary accuracy by piecewise linear functions. The mathematical task is to estimate the error as closely as possible and to determine how rapidly the error decreases as the number of pieces (or the degree of the polynomial within each piece) is increased. Of course, the finite element method can proceed without the support of precise mathematical theorems; it got on pretty well for more than 10 years. But we believe it will be useful, especially in the future development of the method, to understand and consolidate what has already been done.

We have attempted a fairly complete analysis of *linear problems* and the *displacement method*. A comparable theory for fully nonlinear equations does not yet exist, although it would certainly be possible to treat semilinear equations—in which the difficulties are confined to lower-order terms. We make a few preliminary comments on nonlinear equations, but this remains an outstanding problem for the future. In our choice of the displacement method over the alternative variational formulations described in Chapter 2, we have opted to side with the majority. This is the most commonly used version of the finite element method. Of course, the approximation theory would be the same for all formulations, and the duality which is so rampant throughout the whole subject makes the conversion between displacement methods and force methods nearly automatic.

Our goal in this chapter is to illustrate the basic steps in the finite element method:

1. The variational formulation of the problem.
2. The construction of piecewise polynomial trial functions.
3. The computation of the stiffness matrix and solution of the discrete system.
4. The estimation of accuracy in the final Ritz approximation.

We take the opportunity, when stating the problem variationally, to insert some of the key mathematical ideas needed for a precise theory—the Hilbert spaces $\mathcal{3C}^s$ and their norms, the estimates for the solution in terms of the data, and the energy inner product which is naturally associated with the specific problem. With these tools, the convergence of finite elements can be proved even for a very complicated geometry. In fact, the simplicity of variational arguments permits an analysis which already goes beyond what has been achieved for finite differences.

1.2. A TWO-POINT BOUNDARY-VALUE PROBLEM

Our plan is to introduce the finite element method, and the mathematics which lies behind it, in terms of a specific and familiar example. It makes sense to choose a one-dimensional problem, in order that the construction of elements shall be simple and natural, and also in order that the mathematical manipulations shall be straightforward—requiring integration by parts rather than some general Green's formula. Therefore, our choice falls on the equation

$$(1) \qquad\qquad -\frac{d}{dx}\left(p(x)\frac{du}{dx}\right) + q(x)u = f(x).$$

With suitable boundary conditions at the endpoints $x = 0$ and $x = \pi$, this is a classical Sturm–Liouville problem. It represents a number of different physical processes—the distribution of temperature along a rod, for example, or the displacement of a rotating string. Mathematically, the first point to emphasize is that the equation and boundary conditions arise from a *steady-state problem*, and not one which unfolds in time from initial conditions of displacement and velocity. It would correspond in more space dimensions to an *elliptic boundary-value problem*, governed for example by Laplace's equation.

In order to illustrate the treatment of different types of boundary conditions, especially in the variational statement of the problem, we fix the left-hand end of the string and let the other be free. Thus at the end $x = 0$ there is an *essential* (or *kinematic*, or *restrained*, or *geometric*) boundary condition, in other words, one of *Dirichlet* type:

$$u(0) = 0.$$

At the right-hand end $x = \pi$, the string is not constrained, and it assumes a *natural* (or *dynamic*, or *stress*) boundary condition, in other words, one of *Neumann* type:

$$u'(\pi) = 0.$$

We propose to consider this model problem from four different points of view, in the following order:

1. Pure mathematics.
2. Applied mathematics.
3. Numerical approximation by finite differences.
4. Numerical approximation by finite elements.

It is essential to recognize the common features of these four approaches to the same problem; the tools which are useful to the pure mathematician in proving the existence and uniqueness of the solution, and to the applied mathematician in understanding its qualitative behavior, ought to be applied also to the study of the numerical algorithms.

We begin with the pure mathematician, who combines the differential equation and boundary conditions into a single equation,

$$Lu = f.$$

L is a linear operator, acting on a certain class of functions—those which in some sense satisfy the boundary conditions and can be differentiated twice. Mathematically, the fundamental question is precisely this: *to match such a space of functions u with a class of inhomogeneous terms f*, in such a way that *to each f there corresponds one and only one solution u*. Once this correspondence between f and u has been established, the problem $Lu = f$ is in an abstract sense "solved." Of course there is still a little way to go in actually discovering which solution u corresponds to a given f. That step is the real subject of this book. But we believe it is worthwhile, and not just useless fussiness, to try first to get these function spaces right. In fact, it is of special importance for the variational principles and their approximation to know exactly which space of functions is admissible. (The references to "spaces" of functions carry the implication that if u_1 and u_2 are admissible, then so is $c_1 u_1 + c_2 u_2$; this superposition is a natural property in linear problems.)

We want to consider one specific choice, the one which is perhaps most important to the theory, for the space of inhomogeneous data: Those f are admitted which have *finite energy*. This means that

$$(2) \qquad \int_0^\pi (f(x))^2 \, dx < \infty.$$

Any piecewise smooth function f is thereby included, but the Dirac δ-function is not; we shall return later to this case of a "point load." The space of functions satisfying (2) is often denoted by L_2; we prefer the notation $\mathfrak{IC}^0$, indicating by the superscript how many derivatives of f are required to have finite energy (in this case it is only f itself).

For the simplest Sturm–Liouville equation $-u'' = f$, it is not hard to guess the corresponding space of solutions. This solution space is denoted by $\mathfrak{IC}_B^2$—the subscript B refers to the boundary conditions $u(0) = u'(\pi) = 0$, and the superscript 2 requires that the second derivative of u has finite energy.†
The role of the pure mathematician is then to show, under the assumptions $p(x) \geq p_{min} > 0$ and $q(x) \geq 0$, that $\mathfrak{IC}_B^2$ is still the solution space for the more general equation $-(pu')' + qu = f$. In fact, his final theorem can be stated in the following way:

The operator L is a one-to-one transformation from $\mathfrak{IC}_B^2$ onto $\mathfrak{IC}^0$, so that for each f in $\mathfrak{IC}^0$ the differential equation (1) has a unique solution u in $\mathfrak{IC}_B^2$. Furthermore, the solution depends continuously on the data: If f is small, then so is u.

The last sentence requires further explanation; we need *norms* in which to measure the size of f and u. The two norms will be different, since the data space and solution space are different. Fortunately, there is a natural choice for the norms in terms of the energy, or rather its square root:

$$\| f \|_0 = \left[\int (f(x))^2 \, dx \right]^{1/2},$$

$$\| u \|_2 = \left[\int \left((u''(x))^2 + (u'(x))^2 + (u(x))^2 \right) dx \right]^{1/2}.$$

---

†These spaces are defined again in the index of notations at the end of the book.

With these definitions, the continuous dependence of the solution on the data can be expressed in a quantitative form: There exists a constant C such that

$$(3) \qquad\qquad \|u\|_2 \le C\|f\|_0.$$

The uniqueness of the solution follows immediately from this estimate: If $f = 0$, then necessarily $u = 0$. In fact, it is such estimates which lie at the very center of the modern theory of partial differential equations. A general technique for proving (3), which applies also to boundary-value problems in several space dimensions, has been created only in the last generation. In this book, we shall accept such estimates as proved: For elliptic equations of order $2m$, this means that

$$(4) \qquad\qquad \|u\|_{2m} \le C\|f\|_0.$$

We move now to a more applied question, the actual construction of the solution. If the coefficients p and q are constant, then this can be carried out in terms of an infinite series. The key lies in knowing the eigenvalues and eigenfunctions of L:

$$(5) \qquad u_n(x) = \sqrt{\frac{\pi}{2}} \sin(n - \tfrac{1}{2})x, \qquad \lambda_n = p(n - \tfrac{1}{2})^2 + q.$$

It is immediate to check that $Lu_n = -pu_n'' + qu_n = \lambda_n u_n$, that the functions u_n satisfy the boundary conditions and therefore lie in $\mathcal{H}_B^2$, and that they are orthonormal:

$$\int_0^\pi u_n(x)u_m(x)\,dx = \delta_{nm}.$$

Suppose the inhomogeneous term is expanded in a series of eigenfunctions:

$$(6) \qquad\qquad f(x) = \sum_{n=1}^\infty a_n \sqrt{\frac{\pi}{2}} \sin(n - \tfrac{1}{2})x.$$

Then integrating formally, the orthogonality of the u_n gives

$$\|f\|_0^2 = \int_0^\pi f^2\,dx = \sum_{n=1}^\infty a_n^2.$$

The functions f in $\mathcal{H}^0$ are exactly those which admit a harmonic expansion of the form (6), with coefficients satisfying $\sum a_n^2 < \infty$. Actually, this ought to seem a little paradoxical, since apparently every f of the form (6) will satisfy $f(0) = 0$, $f'(\pi) = 0$, whereas no boundary conditions were meant to be imposed on f: The elements of $\mathcal{H}^0$ are required only to have finite energy, $\int f^2 < \infty$. The paradox is resolved by the completeness of the eigenfunctions u_n in $\mathcal{K}^0$. Whether f satisfies these spurious boundary conditions or not, its expansion is valid in the mean-square sense,

$$\int_0^\pi \left(f(x) - \sum_1^N a_n \sqrt{\frac{\pi}{2}} \sin(n - \tfrac{1}{2})x \right)^2 dx \longrightarrow 0 \qquad \text{as } N \to \infty.$$

The boundary conditions on f are thus unstable and disappear in the limit as $N \to \infty$. Figure 1.1 shows how a sequence of functions f_N, all lying in $\mathcal{K}_B^2$, could still converge to a function f outside that space.

The Sturm-Liouville differential equation $Lu = f$ is now ready to be solved: if $f = \sum a_n u_n$, then u has the expansion

$$(7) \qquad u = \sum \frac{a_n}{\lambda_n} u_n = \sqrt{\frac{\pi}{2}} \sum_1^\infty \frac{a_n \sin(n - \tfrac{1}{2})x}{p(n - \tfrac{1}{2})^2 + q}.$$

With this explicit construction, the estimate $\|u\|_2 \leq C\|f\|_0$ and the matching of data space $\mathcal{K}^0$ with solution space $\mathcal{K}_B^2$ can be verified directly.

The question of the boundary conditions is more subtle and deserves further comment. We have seen already that even though f can be expanded in terms of the u_n, which do satisfy the boundary conditions, still f has absolutely nothing to do with these conditions. Therefore the question is: What is different about u? Why does u satisfy the boundary conditions? The answer is that the series expansion for u converges in a much stronger sense than the expansion for f: not only does $\sum a_n u_n / \lambda_n$ converge to u in the mean-square sense, but so do its first and second derivatives. More precisely,

$$\left\| u - \sum_1^N \frac{a_n}{\lambda_n} u_n \right\|_2 \longrightarrow 0 \qquad \text{as } N \to \infty.$$

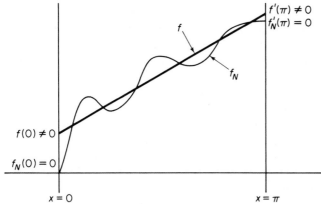

Fig. 1.1 f_N in $\mathcal{K}_B^2$ approximating a general function f.

The point is that when even the second derivatives converge, the boundary conditions are *stable*; the limit function u is compelled to satisfy the boundary conditions. (Note that in Figure 1.1, the second derivatives of f_N did not converge to those of f; therefore the limit f did not have to satisfy the boundary conditions, and was outside the space $\mathcal{3C}_B^2$. This is what will not happen for u.)

The general rule is this: *boundary conditions which involve only derivatives below order s will make sense in the $\mathcal{3C}^s$ norm*; those involving derivatives of order s or higher will be unstable and will not apply to functions in the space $\mathcal{3C}^s$. We shall see that this is the rule which distinguishes between essential boundary conditions, which stay, and natural boundary conditions, which go. The distinction becomes crucial in the variational problem, whose statement is in terms of *first derivatives*, that is, the $\mathcal{3C}^1$ norm. The finite element approximations will be required to satisfy all boundary conditions below order 1—that means the condition $u(0) = 0$—but they will not be required to satisfy the condition on the first derivative. This leniency at $x = \pi$ will not prevent the finite element approximations from converging in the $\mathcal{3C}^1$ norm to a solution u which does satisfy $u'(\pi) = 0$. This is the key to the following section, which extends the "pure mathematics" standpoint to the equivalent variational problem.

1.3.   THE VARIATIONAL FORM OF THE PROBLEM

The linear equation $Lu = f$ is related to the quadratic functional

$$I(v) = (Lv, v) - 2(f, v)$$

in the following way: $I(v)$ is minimized at $v = u$ only if its derivative (or first variation) vanishes there, and the condition for the vanishing of this derivative is exactly the *Euler equation* $Lu = f$. The problems of inverting L and minimizing I are equivalent; they produce the same solution u. Therefore, such problems can be investigated either in an *operational form*, in terms of the linear operator L, or in *variational form*, in terms of the quadratic I. The goal in this section is to find the exact variational equivalent of our two-point boundary-value problem.

This equivalence of differential equations with variational problems is basic also to the choice of a computational scheme. The differential equation may be approximated by a discrete system, using finite differences, or the variational integral can be minimized over a discrete class of functions, as in the finite element method. In many applications—particularly in steady-state rather than transient problems—the variational statement is the primary physical principle, and the differential equation only a secondary consequence. Therefore, it is not surprising to find in such applications a strong movement toward the minimization of the quadratic functional as the fundamental problem to be approximated.

The relationship between the linear and quadratic problems is transparent if L, v, and f are just real numbers. $I(v) = Lv^2 - 2fv$ is a parabola and, *provided that L is positive*, its minimum is attained at the point u where

$$\frac{dI}{dv}\bigg|_{v=u} = 2(Lu - f) = 0.$$

If L were not positive, the problem of minimization would break down: either the minimum is $-\infty$ or, if $L = 0$, the parabola degenerates into a straight line.

A more interesting case is when v and f are n-dimensional vectors and L is a symmetric positive definite matrix of order n. The notation $(\ ,\ )$ then stands for the usual inner product (or dot product) of two vectors, so

$$I(v) = \sum_{j,k} L_{jk} v_k v_j - 2 \sum_j f_j v_j.$$

Applying the symmetry $L_{jk} = L_{kj}$, the Euler equation is

$$\frac{\partial I}{\partial v_m}\bigg|_{v=u} = 2[\sum_k L_{mk} u_k - f_m] = 0, \qquad m = 1, \ldots, n.$$

These n simultaneous equations make up the vector equation $Lu = f$. The minimum value of I, attained at $u = L^{-1}f$, is

$$I(L^{-1}f) = (f, L^{-1}f) - 2(f, L^{-1}f) = -(f, L^{-1}f).$$

Since L is positive definite, so is L^{-1}, and this minimum is negative (or zero, when $f = 0$). Geometrically $I(v)$ is represented by a convex surface, a paraboloid opening upward, when L is positive definite.

A vector derivation of the minimizing equation $Lu = f$ comes by varying all n components at once. If I has a minimum at u, then for all v and ϵ,

$$I(u) \le I(u + \epsilon v) = I(u) + 2\epsilon[(Lu, v) - (f, v)] + \epsilon^2(Lv, v).$$

Since ϵ can be arbitrarily small and of either sign, its coefficient must vanish:

(8) $\qquad\qquad (Lu, v) = (f, v) \qquad$ for every v.

This forces $Lu = f$. We call the equation (8) the *weak form*, or *Galerkin form*, of the problem. It no longer requires L to be positive definite or even symmetric, since it deals not necessarily with a minimum but only with a stationary point. In other words, it states only that the first variation vanishes. In this form the problem leads to Galerkin's approximation process.

It is worth mentioning that $(Lv, v) - 2(f, v)$ is not the only quadratic whose minimum occurs at the point where $Lu = f$. It is obvious that the least-squares functional $Q(v) = (Lv - f, Lv - f)$ has its minimum (zero) at the same point. There is one significant difference, however: the Euler equation, which arises by equating the derivatives $\partial Q/\partial v_m$ to zero, is not $Lu = f$, but $L^T Lu = L^T f$. The two are theoretically equivalent, assuming that L is invertible, but in practice the appearance of $L^T L$ is a disadvantage.

Consider now the differential equation of the previous section:

$$Lu = \left[-\frac{d}{dx}\left(p(x)\frac{d}{dx} \right) + q(x) \right] u = f,$$

$$u(0) = u'(\pi) = 0.$$

We want to construct $I(v) = (Lv, v) - 2(f, v)$. These inner products now involve functions defined on the interval $0 \le x \le \pi$ rather than vectors with n components, but their definition is completely analogous to the vector case:

$$(f, v) = \int_0^\pi f(x)v(x)\, dx.$$

Note that our data f and functions v are assumed to be *real*, as in almost all applications. This is a convenient assumption, and the modifications to be made in the complex case are well known; in the integrand above, one of the factors should be conjugated.

The computation of I involves an integration by parts (indeed the whole theory of differential operators rests on this rule):

(9)
$$(Lv, v) = \int_0^\pi [-(pv')' + qv]v \, dx$$
$$= \int_0^\pi [p(v')^2 + qv^2] \, dx - pv'v \Big|_0^\pi.$$

If v satisfies the boundary conditions $v(0) = v'(\pi) = 0$, then the integrated term vanishes and the quadratic functional is

$$I(v) = \int_0^\pi [p(x)(v'(x))^2 + q(x)(v(x))^2 - 2f(x)v(x)] \, dx.$$

This is the functional to be minimized.

The solution of the differential problem $Lu = f$ is expected to coincide with the function u that minimizes I. But within what class of functions shall we search for a minimum? Since we are looking for a solution u in $\mathcal{H}_B^2$, then certainly we ought to admit as candidates at least all members of $\mathcal{H}_B^2$. This leads to a correct and equivalent variational problem; the minimum of $I(v)$ over all functions v in $\mathcal{H}_B^2$ occurs at the desired point $v = u$. However, one is struck by the fact that *the expression for I involves no second derivatives*; because of the integration by parts, it involves only v and v'. It follows that $I(v)$ will be well defined if only the first derivative of v, rather than the second, is required to have finite energy. Therefore, the possibility arises of enlarging the class of functions which are admissible in the minimization problem to a space bigger than $\mathcal{H}_B^2$.

Our guiding principle will be this: Any function v will be admitted, provided it can be obtained as the limit of a sequence v_N in $\mathcal{H}_B^2$, where by the word "limit" we mean that the quadratic terms in the potential-energy functional $I(v)$ converge:

$$(10) \qquad \int_0^\pi p(v' - v_N')^2 + q(v - v_N)^2 \to 0 \qquad \text{as } N \to \infty.$$

Notice that *such an enlargement of our space cannot actually lower the minimum value of I*; each new value $I(v)$ is the limit of old values $I(v_N)$. Thus if the mimimum of I was already attained for some u in $\mathcal{H}_B^2$, that remains the minimizing function. This is exactly what we know to be the case. However, we now have the enormous advantage, while searching for this minimum, of being permitted to try functions v which were outside the original class $\mathcal{H}_B^2$. In practice, this means that we can now try functions which are continuous but only piecewise linear—they are easy to construct, and their first derivative has finite energy, but they do not lie in $\mathcal{H}_B^2$.

Our problem is now to describe this new and larger admissible space. In other words, we want to discover the properties possessed by a function v which is the limit in the sense of (10)—this is effectively the $\mathcal{H}^1$ norm—of functions v_N which have two derivatives and satisfy all the boundary conditions.

There are two properties to be determined, the smoothness required of the admissible functions and the boundary conditions they must satisfy. The first is comparatively easy: since the requirement (10) notices only convergence of the first derivative, the limiting function v need only lie in $\mathcal{H}^1$. This means that the norm

$$\| v \|_1 = \left[\int_0^\pi \left(v^2 + (v')^2 \right) dx \right]^{1/2}$$

must be finite.

Again the boundary conditions pose a more subtle problem: Since any sequence from $\mathcal{H}_B^2$ will satisfy $v_N(0) = 0$ and $v_N'(\pi) = 0$, will its limit v inherit both these properties? It turns out that *the first condition is preserved, and the other is lost*. To see that the limit need not satisfy the Neumann condition $v'(\pi) = 0$, suppose for example that $v(x) = x$ and consider the sequence v_N in Fig. 1.2. Since $v - v_N$ is zero except over the small interval at the end, and there $0 \le v' - v_N' \le 1$, the requirement (10) is obviously satisfied. Thus

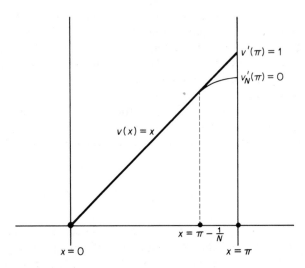

Fig. 1.2   Convergence in $\mathfrak{K}^1$, with $v'_N(\pi) = 0$ but $v'(\pi) \neq 0$.

$v(x) = x$ lies in the limiting space of admissible functions, even though $v'(\pi) \neq 0$.

On the other hand, the condition $v(0) = 0$ continues to hold in the limit. In fact, v_N will converge at each individual point x, since by the Schwarz inequality

$$|v_N(x) - v_M(x)|^2 = \left| \int_0^x \left(v'_N(y) - v'_M(y) \right) dy \right|^2$$

$$\leq \int_0^x 1^2 \, dy \int_0^x (v'_N - v'_M)^2 \, dy \longrightarrow 0.$$

By standard results in analysis, the limiting function v is continuous and the convergence of v_N to v is uniform in x. In particular, at the point $x = 0$, $v(0) = \lim v_N(0) = 0$. On the other hand, the first derivatives converge only in mean square, giving no assurance that $v'(\pi) = 0$.

The space of admissible functions v in the minimization is therefore $\mathfrak{K}^1_E$; its members have first derivatives with finite energy and satisfy the essential boundary condition $v(0) = 0$ indicated by the subscript E. The natural boundary condition $v'(\pi) = 0$ is not imposed. Notice that if our mathematics is consistent, then the function u in $\mathfrak{K}^1_E$ which minimizes I should *automatically* satisfy $u'(\pi) = 0$. This is easy to confirm, since for any ϵ and any v in $\mathfrak{K}^1_E$,

$$I(u) \leq I(u + \epsilon v)$$

$$= I(u) + 2\epsilon \int_0^\pi pu'v' + quv - fv + \epsilon^2 \int_0^\pi p(v')^2 + qv^2.$$

Since this holds for ϵ on both sides of zero, the linear term (first variation) must vanish:

(11)
$$0 = \int_0^\pi pu'v' + quv - fv$$
$$= \int_0^\pi [-(pu')' + qu - f]v + p(\pi)u'(\pi)v(\pi).$$

If the minimizing u has two derivatives, permitting the last integration by parts, this expression will be zero for all v in $\mathcal{3C}_E^1$ only if both $-(pu')' + qu = f$ in the interval, and the natural condition $u'(\pi) = 0$ holds at the boundary. Also $u(0) = 0$, because like every other function in $\mathcal{3C}_E^1$, u satisfies the essential boundary condition. This completes the cycle: The minimization of I over $\mathcal{3C}_E^1$ is equivalent to the solution of $Lu = f$, and the computation of u can be approached from either direction.

Geometrically, the process of completion from $\mathcal{3C}_B^2$ to $\mathcal{3C}_E^1$ has a simple interpretation. The quadratic I is represented by a convex surface, a paraboloid in infinite dimensions. At first, when I was defined only for v in $\mathcal{3C}_B^2$, there were "holes" in this surface; all we have done is to fill them in. The surface has changed neither its shape nor its minimum value; it is just that the pinpricks, corresponding to functions v lying in $\mathcal{3C}_E^1$ but not in $\mathcal{3C}_B^2$, no longer exist.

To conclude this section, we remark on two singular problems which are of great importance in applications. It is noteworthy that in both cases the variational form which was just established remains completely valid; $I(v)$ is still to be minimized over the admissible space $\mathcal{3C}_E^1$, and the minimizing function u is the desired solution. In contrast, the operational form $Lu = f$ becomes more complex, special conditions enter at the singularity x_0, and *the solution is no longer in* $\mathcal{3C}_B^2$. From the viewpoint of applied mathematics, this special behavior cannot be ignored; it is probably the very point to be understood. From the viewpoint of finite element approximations, however, it is significant that the algorithm can proceed without complete information on the singularity. Such information will be extremely valuable in speeding the convergence of the approximations, as in Chapter 8 on singularities, but the algorithm will not break down without it.

Remark 1. The coefficient $p(x)$ may be discontinuous at a point x_0 where the elastic property of the string (or the diffusivity of the medium in heat flow) changes abruptly. At such a point there appears an *internal boundary*; the solution u will no longer have two derivatives. To find the "jump condition" at x_0, we depend on the variational form of the problem; the first variation $\int pu'v' + quv - fv$ must still vanish for all v if u is minimizing. Assuming no other points of difficulty, integration by parts over the separate intervals $(0, x_0)$ and (x_0, π) yields

$$
0 = \int_0^{x_0} [-(pu')' + qu - f]v + p_- u'_- v_-
$$
$$
+ \int_{x_0}^{\pi} [-(pu')' + qu - f]v + p(\pi)u'(\pi)v(\pi) - p_+ u'_+ v_+.
$$

The subscripts $-$ and $+$ indicate the limiting values as x approaches x_0 from the left and right, respectively. Recall that $v_- = v_+$ for any v in $\mathcal{K}_E^1$, since v is continuous, and in particular that $u_- = u_+$. Varying v, it follows that the differential equation holds in each interval, that $u'(\pi) = 0$ at the far end, and that

$$
p_- u'_- = p_+ u'_+.
$$

This is the natural boundary condition at x_0, and is a direct consequence of the variational form: u' has a jump, but *the combination pu' remains continuous.*

Since there is a jump in u', the solution lies in $\mathcal{K}_E^1$ but not in $\mathcal{K}_B^2$. This is a case in which *one of the holes in the surface $I(v)$ was actually at the bottom.* The minimum value of $I(v)$ would have been the same over the original space $\mathcal{K}_B^2$, but within that space no function attained the minimum. The surface came arbitrarily near to the hole, but it was filled in only by the function satisfying the jump condition.

The standard error estimates for finite elements, which rest on an assumed smoothness of u, will degenerate at the discontinuity x_0. These estimates can be saved, however, if by placing a node at x_0 we permit the trial functions in the approximation to copy the jump condition. Since the condition is natural rather than essential, not every trial function has to satisfy it; as long as the trial functions are not forced to have a continuous derivative, and thereby violate the jump condition, the approximation will be good.

Remark 2. Up to now the inhomogeneous term f has been required to come from $\mathcal{K}^0$, thereby excluding the δ-function. Physically this would be an interesting choice, representing a point load or point source, and mathematically the corresponding u is the *fundamental solution.* Therefore we try harder to include it, by reconsidering the functional

$$
I(v) = \int_0^{\pi} p(v')^2 + qv^2 - 2fv,
$$

which is still to be minimized over $\mathcal{K}_E^1$.

Suppose, for the sake of example, that $p \equiv 1$ and $q \equiv 0$. If f has finite energy, then the integral I is finite and its minimization is straightforward.

But also for $f = \delta(x_0)$, $0 < x_0 < 1$, the integral is finite. In this case

$$I(v) = \int_0^\pi (v')^2\, dx - 2v(x_0),$$

and the minimum occurs at the *ramp function* $v = u$ shown in Fig. 1.3. Again the solution is not in $\mathcal{K}_B^2$, and the hole corresponding to this ramp function happens to fall at the bottom of the infinite-dimensional paraboloid $I(v)$.

One more possibility: Suppose the point load is at the end $x_0 = \pi$, so that $f = \delta(\pi)$. Then the solution is $u(x) = x$, and *the ramp function never levels off.* In this case *the solution violates the natural boundary condition* $u'(\pi) = 0$. Looking back at (11) this is perfectly consistent; with $p = 1$, $q = 0$, and $u(x) = x$, the first variation is

$$\int_0^\pi (u'v' - fv)\, dx = \int_0^\pi \left(v' - \delta(\pi)v \right) dx \equiv 0$$

for any v in $\mathcal{K}_E^1$. Thus the first variation vanishes at $u(x) = x$, which is indeed minimizing. The only remarkable point is that the integration by parts in (11), and therefore the subsequent derivation of the natural boundary condition $u'(\pi) = 0$, falls through. Thus if f is allowed to be singular at $x = \pi$, there may no longer be a natural boundary condition at that point.

Once a δ-function has been accepted as a possible choice of f, there arises the general question: Which class of inhomogeneous terms can be permitted? More precisely, which space of data matches the solution space $\mathcal{K}_E^1$? Roughly speaking, as long as $I(v)$ remains finite for all v in $\mathcal{K}_E^1$, the minimization can proceed. This rule accepts δ-functions and their linear combinations but not

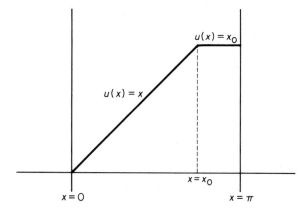

Fig. 1.3 Fundamental solution for a point load; f and u not in $\mathcal{K}^0$ and $\mathcal{K}^2$.

their derivatives. For example, the *dipole function* $f = \delta'(x_0)$ would give

$$\int_0^\pi fv = -\int_0^\pi v'\delta = -v'(x_0),$$

and this may not be finite; v' can have an unbounded peak at x_0 and still have finite energy. The dipole is too singular to be allowed.

The data f which are now accepted come from the space denoted by $\mathcal{H}^{-1}$; f's derivative of order -1, that is, its indefinite integral, is in $\mathcal{H}^0$. The second-order operator L takes the space $\mathcal{H}_E^1$ into $\mathcal{H}^{-1}$, just as it took $\mathcal{H}_B^2$ into $\mathcal{H}^0$. A suitable norm in $\mathcal{H}^{-1}$ is

$$(12) \qquad \|f\|_{-1} = \max_{v \text{ in } \mathcal{H}^1} \frac{\left|\int f(x)v(x)\,dx\right|}{\|v\|_1}.$$

There is, however, a peculiar result if f happens to be a δ-function concentrated at the origin, namely that $\int fv = 0$ for every v in the admissible space $\mathcal{H}_E^1$. Such a phenomenon means that the variational principle will treat this f as zero, and the solution will be $u \equiv 0$. (This is just the ramp function of Fig. 1.3, with $x_0 = 0$.) This means that the solution space $\mathcal{H}_E^1$ matches the data space $\mathcal{H}^{-1}$ only under the following proviso: There is to be no distinction between data f_1 and f_2, if they differ by a multiple of the δ-function at the origin.

Finally, the solution should depend continuously on f, using the solution-space norm for u and the data-space norm for f. The proof depends on the vanishing of the first variation for every v in $\mathcal{H}_E^1$, in particular for $v = u$:

$$\int_0^\pi p(u')^2 + qu^2 = \int_0^\pi fu.$$

The right side, from the definition of the data norm $\|f\|_{-1}$, is bounded by $\|f\|_{-1}\|u\|_1$. The left side is obviously larger than $p_{\min}\|u'\|_0^2$, and this is easily shown to be larger than $\sigma\|u\|_1^2$ for some positive σ. Therefore,

$$\sigma\|u\|_1^2 \leq \|f\|_{-1}\|u\|_1 \quad \text{or} \quad \|u\|_1 \leq \frac{1}{\sigma}\|f\|_{-1}.$$

This expresses the continuous dependence of u on f.

1.4. FINITE DIFFERENCE APPROXIMATIONS

It is almost an article of faith in numerical analysis that whatever can be solved abstractly can also be solved by concrete numerical computations.

"Every continuous mapping admits a convergent series of discrete approximations." The previous sections established two continuous mappings from f to u, one for the differential equation ($\|u\|_2 \le C\|f\|_0$) and the other for its variational equivalent ($\|u\|_1 \le \sigma^{-1}\|f\|_{-1}$). Therefore, both these problems should be ripe for numerical solution.

We begin with the differential equation $Lu = f$ and replace the derivatives by difference quotients. The result is a finite linear system $L^h U^h = f^h$—a discrete operational form. There are two key steps in the theoretical analysis of this difference equation:

1. To compute the local truncation error, or discretization error, by means of a Taylor series expansion.

2. To establish that the system is globally stable, in other words, that U^h depends continuously on f^h as the mesh size h approaches zero.

Together, these two steps establish the rate of convergence of U^h to the true solution u as $h \to 0$. The point of our discussion is really to be able to contrast the convergence theory for difference equations with the techniques used in the next section, and throughout the rest of the book, to prove convergence in the variational problem.† It is remarkable how differently the two steps above would be approached in the variational framework: the local truncation error is abandoned in favor of verifying the approximation properties (or completeness) of the trial functions, and stability needs no special proof—it is automatically present, for finite elements.

As usual, the interval $[0, \pi]$ is divided into equal pieces of length $h = \pi/N$ by the points $x_i = ih$, $i = 0, 1, \ldots, N$. Then the derivatives in the equation $-(pu')' + qu = f$ are removed in favor of centered difference quotients:

$$u'(x) \longrightarrow \Delta^h u(x) = \frac{u(x + h/2) - u(x - h/2)}{h}$$

The result is a discrete equation $-\Delta^h(p\Delta^h U) + qU = f$, which we require to hold at the interior mesh points x_i:

(13)
$$\frac{1}{h^2}\left[-p\left(x_i + \frac{h}{2}\right)(U^h_{i+1} - U^h_i) + p\left(x_i - \frac{h}{2}\right)(U^h_i - U^h_{i-1})\right]$$
$$+ q(x_i)U^h_i = f(x_i).$$

Since this is a second-order equation, it requires a boundary condition at each end of the interval. At the left end, $U^h_0 = 0$ is the obvious choice. At the

---

†This section is a digression from our main theme, but the finite element and finite difference methods are so closely related that we need to be in a position to compare them. The reader may discern our own preference and, if he shares it, skip this section.

other end, there is no unique difference replacement for $u'(\pi) = 0$, and we shall consider two alternatives: a one-sided difference

(14a)
$$\frac{U_N^h - U_{N-1}^h}{h} = 0$$

and a centered difference

(14b)
$$\frac{U_{N+1}^h - U_{N-1}^h}{2h} = 0.$$

In the first case the difference equation (13) holds for $0 < i < N$, which together with the two boundary conditions produces $N + 1$ equations in $N + 1$ unknowns. In the second case the difference equation applies also at $x_N = \pi$, to compensate for the extra unknown U_{N+1}^h. These boundary conditions are easy to compare when $p = 1$ and $q = 0$; after eliminating the unknowns at the extreme ends, the difference equations are, respectively,

(15a)
$$L^h U^h = \frac{1}{h^2} \begin{pmatrix} 2 & -1 & 0 & \cdot \\ -1 & 2 & \cdot & 0 \\ 0 & \cdot & 2 & -1 \\ \cdot & 0 & -1 & 1 \end{pmatrix} \begin{pmatrix} U_1 \\ \cdot \\ \cdot \\ U_{N-1} \end{pmatrix} = \begin{pmatrix} f_1 \\ \cdot \\ \cdot \\ f_{N-1} \end{pmatrix}$$

and

(15b)
$$L^h U^h = \frac{1}{h^2} \begin{pmatrix} 2 & -1 & 0 & \cdot \\ -1 & 2 & \cdot & 0 \\ 0 & \cdot & 2 & -1 \\ \cdot & 0 & -2 & 2 \end{pmatrix} \begin{pmatrix} U_1 \\ \cdot \\ \cdot \\ U_N \end{pmatrix} = \begin{pmatrix} f_1 \\ \cdot \\ \cdot \\ f_N \end{pmatrix}$$

To analyze the difference equation with variable p and q, we consider the *local truncation error* $\tau^h(x)$ which arises when the true solution u is substituted for U^h:

$$-\Delta^h(p\Delta^h u) + pu - f = \tau^h.$$

This is a completely formal computation, in which $u(x_i \pm h)$ and $p(x_i \pm h/2)$ are expanded in a Taylor series around the central point x_i. Because u satisfies the differential equation, the zero-order term vanishes; this cancellation expresses the *consistency* of the difference and differential equations. The terms which remain are

$$\tau^h = -\frac{h^2}{24}[(pu')''' + (pu''')'] + O(h^4).$$

The same process applies at the boundaries. For the one-sided difference operator,

$$\frac{u(\pi) - u(\pi - h)}{h} = u'(\pi) - \frac{h}{2}u''(\pi) + \cdots$$

$$= -\frac{h}{2}u''(\pi) + O(h^3).$$

Again, consistency with the true condition $u'(\pi) = 0$ cancelled the zero-order term. The centered difference is of course more accurate:

$$\frac{u(\pi + h) - u(\pi - h)}{2h} = \frac{h^2}{6}u'''(\pi) + O(h^4).$$

If the difference equation and boundary condition are treated together, rather than separately, the truncation errors look quite different. We take as an example the matrices L^h displayed above, in which the boundary conditions have been used to eliminate the last unknown. Expanding the final row by Taylor series, the one-sided and centered conditions yield, respectively,

$$\frac{u(\pi - h) - u(\pi - 2h)}{h^2} - f(\pi - h) = -\frac{1}{2}u''(\pi) + O(h),$$

$$\frac{2u(\pi) - 2u(\pi - h)}{h^2} - f(\pi) = -\frac{h}{3}u'''(\pi) + O(h^2).$$

In this form, it would appear that the overall errors are $O(1)$ and $O(h)$, which is completely wrong.

For an estimate of the error $E^h = U^h - u$, we look at the difference equation which it satisfies:

$$-\Delta^h(p\Delta E^h) + qE^h = \tau^h.$$

With the centered difference at $x = \pi$, the boundary conditions in this error equation are

$$E_0^h = 0, \qquad \frac{E_{N+1}^h - E_{N-1}^h}{2h} = \frac{h^2}{6}u'''(\pi) + O(h^4).$$

Notice that this difference equation is analogous to the original differential problem, except that *the inhomogeneous terms now come from the local truncation error*. We therefore expect the leading term in E^h, say $h^2 e_2$, to arise from the terms which are of order h^2 in the local error:

$$-(pe_2')' + qe_2 = -\tfrac{1}{24}[(pu')''' + (pu''')'],$$

$$e_2(0) = 0, \qquad e_2'(\pi) = \tfrac{1}{6}u'''(\pi).$$

This solution $e_2(x)$ is the *principal error function*. To compute the next term in the error, we substitute $u + h^2 e_2$ into the difference problem. This yields a truncation error which starts at h^4, and the coefficient of this term is the right-hand side in the equation for e_4.

In short, we may recursively determine an expansion

$$(16) \qquad U^h = u + h^2 e_2 + h^4 e_4 + \ldots = \sum h^n e_n(x).$$

The calculation of the error terms e_n is entirely mechanical; it has to be stopped only when there is a break in the smoothness of the solution or when the boundary conditions no longer permit such an expansion. One virtue of (16) is that the error varies with x in the proper way; we have more than just an unrealistic maximum bound over the whole interval. Furthermore, this expansion justifies *Richardson's extrapolation to $h = 0$*; one computes with two or more choices of h, and chooses a combination of the results so as to cancel the leading terms in the expansion. With the centered boundary condition in our example, the linear term $h e_1$ vanishes identically, and the combination of U^h and U^{2h} which increases the accuracy from second to fourth order is

$$\tfrac{4}{3} U^h - \tfrac{1}{3} U^{2h} = u + O(h^4).$$

This extrapolation technique has been discussed and verified numerically any number of times but has not yet been widely adopted in practice. The accuracy of the boundary conditions is its most severe limitation, particularly in more dimensions when the boundary of the region intersects the mesh in a completely erratic way.

Applying the same ideas to the one-sided boundary condition, the first term $h e_1$ arises from the $O(h)$ truncation error at the boundary:

$$-(p e_1')' + q e_1 = 0,$$
$$e_1(0) = 0, \qquad e_1'(\pi) = -\tfrac{1}{2} u''(\pi).$$

Naturally the centered difference is to be preferred.

This first-order accuracy appears also if the difference equation is linked with a variational problem. As always, the positive-definite symmetric system $L^h U^h = f^h$ is the Euler equation for the vanishing of the first derivative, at the minimizing U^h, of

$$I_\Delta(V^h) = (L^h V^h, V^h) - 2(f^h, V^h).$$

With the one-sided boundary condition, this functional is

$$I_\Delta = \sum_{i=1}^{N-1} \left[p_{i-1/2} \left(\frac{V_i^h - V_{i-1}^h}{h} \right)^2 + q_i (V_i^h)^2 - 2 f_i V_i^h \right].$$

Obviously I_Δ is a finite difference analogue of the true quadratic functional

$$I(v) = \int_0^\pi p(v')^2 + qv^2 - 2fv.$$

And, almost as obviously, I_Δ is accurate only to first order. Instead of using the centered *trapezoidal rule*, which is of second order, the integral I has been replaced by sums biased to one side.

This hints at a technique which lies in between finite differences and finite elements—the integral $I(v)$ is approximated by a sum I_Δ involving difference quotients, and then minimized. It is an easy way to derive approximations of a low order of accuracy, and deserves more analysis than it has received. Its advantages tend to disappear, however, when higher accuracy is demanded.

Up to this point our analysis of the difference equation has been completely formal, leading to the error expansion $\sum h^n e_n$. There is now a second step to be taken: to prove that U^h converges to u and that the expansion is asymptotically valid. [The expansion itself cannot converge for a finite h, since this would require that U^h is an analytic function of h, and that the problem is well posed even for complex h; all one hopes to prove is that $U^h - \sum_0^M h^n e_n$ $= O(h^{M+1})$ as h decreases to zero.] This second step demands the same kind of estimate as in the differential problem: the solution U^h must depend continuously on the data f^h.

We return to the difference equation and ask the first question: Does there exist a unique solution U^h for every f^h? Equivalently, is the matrix L^h nonsingular? One of the most effective proofs of the invertibility of L^h leads at the same time to a discrete maximum principle, as follows.

Suppose $L^h U^h = 0$. Let the largest component $|U_i^h|$ be the nth, and choose the sign of U^h to make $U_n^h \geq 0$. Then the difference equation at x_n is

$$p_{n+1/2}(U_n^h - U_{n+1}^h) + p_{n-1/2}(U_n^h - U_{n-1}^h) + h^2 q_n U_n^h = 0.$$

Since each term is nonnegative, all three must vanish. If the zero-order coefficient q_n is positive, it follows immediately that $U_n^h = 0$. In any case, the other terms lead to $U_{n+1}^h = U_n^h = U_{n-1}^h$. Thus these components are also maximal, and the whole argument can be repeated with $n - 1$ or $n + 1$ in place of n. Ultimately, after enough repetitions, it follows that $U_n^h = U_0^h = 0$. Thus $L^h U^h = 0$ holds only if U^h is the zero vector, and L^h *must be invertible*. Allowing inhomogeneous boundary conditions, the same argument would show that no component U_i^h can be larger than both U_0^h and U_N^h. This is a *discrete maximum principle*, from which it follows that *the discrete Green's function $(L^h)^{-1}$ is a nonnegative matrix.*

By the way, a similar proof leads to Gerschgorin's theorem in matrix theory: Every eigenvalue λ of a matrix A lies in at least one of the circles

$$|\lambda - A_{ii}| \leq \sum_{j \neq i} |A_{ij}|.$$

Choosing A to be either of the matrices $h^2 L^h$ displayed in (15), all eigenvalues satisfy $|\lambda - 2| \leq 2$. Thus the theorem does not rule out the possibility that $\lambda = 0$ is an eigenvalue, that is, that L^h is singular; it was at this point that the repetitions of the argument at $i = n - 1, n - 2, \ldots, 1$ were needed.

Gerschgorin's theorem becomes perfectly useless in a fourth-order problem. The leading coefficients in the simplest case are $A_{ii} = 6$, $A_{i,i\pm 1} = -4$, and $A_{i,i\pm 2} = 1$, and the Gerschgorin circles are $|\lambda - 6| \leq 10$. Since $\lambda = 0$ lies inside these circles, the Gerschgorin argument fails to prove even that A is semidefinite. This difficulty simply reflects the absence of a maximum principle for fourth-order problems. If we compare $u'' = 0$ with $u^{(iv)} = 0$, for example, it is obvious that straight lines attain their extrema at the ends of the interval, whereas cubics need not.

The maximum principle, when it holds, can be made to yield a simple proof of convergence. We prefer instead to preserve a close analogy between differential and difference equations, by discussing the discrete inequality which corresponds to $\|u\|_2 \leq C\|f\|_0$. The statement of such an inequality requires, first of all, a redefinition of the norm to make it apply to grid functions. An obvious choice for the discrete energy is

$$\|f^h\|_0 = (\sum h |f^h_j|^2)^{1/2}.$$

For the square of the 2-norm, we introduce the energy in the function and its first and second forward-difference quotients:

$$\|U^h\|_2^2 = \|U^h\|_0^2 + \|\Delta_+ U^h\|_0^2 + \|\Delta_+^2 U^h\|_0^2.$$

These sums extend only over the grid points at which they are defined; the forward-difference quotient $\Delta_+ f_i = (f_{i+1} - f_i)/h$ makes no sense at the last grid point.

There is one other new feature to be introduced, inhomogeneous boundary conditions. For the two-point boundary-value problem, the continuous dependence of u on f and on the boundary data is expressed by

$$(17) \qquad \|u\|_2 \leq C(\|f\|_0 + |u(0)| + |u'(\pi)|).$$

For the finite difference equation, in which Δ_π denotes whichever boundary operator is applied at the right-hand end, the corresponding inequality is

$$(18) \qquad \|U^h\|_2 \leq C(\|f^h\|_0 + |U^h_0| + |\Delta_\pi U^h|).$$

For each choice of difference equation, this is the basic estimate to be proved. It is not an automatic consequence of (17), although it implies (17): *The continuous inequality is necessary but not sufficient for the discrete inequality.* In this sense the theory of difference equations is the more difficult; there is an enormous variety of possible difference schemes, and each requires a more-or-less new proof of (18). As in the continuous problem, we shall accept the inequality as true and concentrate on its implications; the technique of proof in one-dimensional problems is summarized by Kreiss [K7]. The inequality (18) asserts the *stability of the difference equation*: The discrete solution U^h depends continuously on the discrete data f^h, uniformly in h.

To establish that U^h converges to u, we need the most celebrated theorem of numerical analysis: *Consistency and stability imply convergence.* This theorem is proved in two steps:

1. The error E^h satisfies the same difference equation (13) as U^h, and therefore by stability it depends continuously on *its* data, which is nothing but the local truncation error:

$$\| E^h \|_2 \leq C(\| \tau^h \|_0 + | E^h(0) | + | \Delta_n E^h |).$$

2. By consistency, which was reflected in the Taylor expansions of τ^h, $E^h(0)$, and $\Delta_n E^h$, the right side approaches zero as $h \to 0$. Therefore, convergence is proved.

We call attention to one more point. Because τ^h involves the fourth derivative of u, the error estimate in the centered difference case is effectively $\| E^h \|_2 \leq Ch^2 \| u \|_4$. With a little extra patience this can be reduced to

$$\| E^h \|_0 \leq C'h^2 \| u \|_2.$$

Thus the convergence is of order h^2 whenever u is in $\mathcal{H}^2$, or in other words whenever f is in $\mathcal{H}^0$. This is the same rate of convergence as in the simplest finite element method given in Section 1.6, but the proof there is very much easier.

Altogether, it appears that a satisfactory theory for one-dimensional difference equations is possible but not trivial. In multidimensional problems, that is, for partial differential equations, almost every convergence proof in the numerical analysis literature has depended on a maximum principle. Without this principle there have been a few ad hoc arguments, adequate for special problems, but the general theory now being developed is so delicate that it looks extremely difficult to apply. The problem is that a single differential equation allows an enormous variety of difference approximations, above all at a curved boundary. In contrast, the variational methods are governed by stricter rules, and it is just these restrictions which permit a more complete theory.

We shall now concentrate exclusively on this theory—the construction and convergence of finite elements.

1.5.   THE RITZ METHOD AND LINEAR ELEMENTS

In this section we begin on the finite element method itself. The general framework is already established, that there is a choice of approximating either the individual terms in the differential equation or the underlying variational principle. The finite element method chooses the latter. At the same time, the discrete equations which arise in the variational approximation are effectively difference equations.

In variational form, the problem is to minimize the quadratic functional

$$I(v) = \int_0^\pi [p(x)(v'(x))^2 + q(x)(v(x))^2 - 2f(x)v(x)]\, dx$$

over the infinite-dimensional space $\mathcal{JC}_E^1$. *The Ritz method is to replace $\mathcal{JC}_E^1$ in this variational problem by a finite-dimensional subspace S, or more precisely by a sequence of finite-dimensional subspaces S^h contained in $\mathcal{JC}_E^1$.* The elements v^h of S^h are called *trial functions*. Because they belong to $\mathcal{JC}_E^1$, they satisfy the essential boundary condition $v^h(0) = 0$. Over each space S^h the minimization of I leads to the solution of a system of simultaneous linear equations; the number of equations coincides with the dimension of S^h. Then *the Ritz approximation is the function u^h which minimizes I over the space S^h*:

$$I(u^h) \leq I(v^h) \qquad \text{for all } v^h \text{ in } S^h.$$

The fundamental problems are (1) to determine u^h, and (2) to estimate the distance between u^h and the true solution u. This section is devoted to problem 1 and the next section to problem 2.

We begin with two examples which illustrate the classical Ritz method. The subspaces S^h will not have the special form associated with the finite element method; instead, each subspace in the sequence will contain the preceding one.

Suppose in the first example that the coefficients p and q are constant, and let S^h be the subspace spanned by the first $N = 1/h$ eigenfunctions of the continuous problem. The trial functions—the members of S^h—are then the linear combinations

$$v^h(x) = \sum_1^N q_j \varphi_j(x) = \sum_1^N q_j \sqrt{\frac{\pi}{2}} \sin\left(j - \tfrac{1}{2}\right)x.$$

Clearly these functions lie in $\mathcal{3C}_E^1$; they satisfy even the natural boundary condition $(v^h)'(\pi) = 0$, which is not necessary.

The weights q_j are to be determined so as to minimize I. From the orthogonality property of the eigenfunctions, the integral has the special form

$$I(v^h) = \sum_1^N \left[q_j^2 \lambda_j - 2 \int_0^\pi f q_j \varphi_j \, dx \right],$$

where $\lambda_j = p(j - \frac{1}{2})^2 + q$. The minimizing equation $\partial I / \partial q_j = 0$ leads directly to the optimal values

$$Q_j = \frac{1}{\lambda_j} \int_0^\pi f \varphi_j \, dx = \frac{(f, \varphi_j)}{\lambda_j}, \qquad j = 1, \ldots, N.$$

Therefore, the Ritz approximation is

$$u^h = \sum_1^N \frac{(f, \varphi_j)\varphi_j}{\lambda_j}.$$

In this case the system of linear equations $\partial I / \partial q_j = 0$ determining the optimal coordinates Q_j was trivial to solve; its coefficient matrix is diagonal, because the eigenfunctions are orthogonal. Furthermore, u^h *is the projection of the true solution*

$$u = \sum_1^\infty \frac{(f, \varphi_j)\varphi_j}{\lambda_j}$$

onto the first N eigenfunctions.

In this example it is easy to compute how much is gained at each step in the Ritz sequence of approximations. By including the Nth trial function φ_N, u^h is improved by the term $\lambda_N^{-1}(f, \varphi_N)\varphi_N$. For smooth f these corrections approach zero very rapidly as $N \longrightarrow \infty$, and even for an arbitrary f the correction is at most $\lambda_N^{-1} \| f \|_0$. In a more realistic problem the exact eigenfunctions will not be known, but if the geometry remains simple, then the use of sines and cosines as trial functions is still of great importance. (Orszag and others have shown how the fast Fourier transform keeps the computational effort within reasonable bounds.) For more complicated geometries we prefer finite elements.

In the second example we keep p and q constant, and choose trial functions which are polynomials:

$$v^h(x) = q_1 x + q_2 x^2 + \ldots + q_N x^N.$$

Again these satisfy the essential boundary condition $v^h(0) = 0$, but this time not the natural boundary condition. In this case

$$I(v^h) = \int_0^\pi [p(\sum q_j j x^{j-1})^2 + q(\sum q_j x^j)^2 - 2f \sum q_j x^j] \, dx.$$

Differentiating I with respect to the parameters q_j, we find a system of N linear equations for the optimal parameters $Q_1, \ldots, Q_N$:

$$KQ = F.$$

The unknown vector is $Q = (Q_1, \ldots, Q_N)$, the components of F are the "moments" $F_j = \int f x^j \, dx$, and the coefficient matrix K is a mess:

$$K_{ij} = \int_0^\pi [p(ix^{i-1})(jx^{j-1}) + qx^i x^j] \, dx$$

$$= \frac{pij\pi^{i+j-1}}{i+j-1} + \frac{q\pi^{i+j+1}}{i+j+1}.$$

The matrix whose entries are $(i + j + 1)^{-1}$ is the notorious Hilbert matrix, with which computations are virtually impossible. The eigenvalues of this matrix are so out of scale—the matrix is so ill-conditioned—that normally even for $N \approx 6$ roundoff error destroys all information in the data f. With the other terms of K included, the situation is even worse. The difficulty is that the powers x^j are nearly linearly dependent; all of them have essentially all their weight in the neighborhood of $x = \pi$. Numerical stability will depend on choosing a more strongly independent basis for the subspace.

The remedy is to cope with the excessive sensitivity of the problem *before* introducing the data, by orthogonalizing the original trial functions $\varphi_j = x^j$. In most cases this means introducing Legendre or Chebyshev polynomials as a new basis for the polynomial space S^h. Such a basis is useful on an interval, and remains so in more dimensions provided the geometry is very simple. For a general domain, however, these orthogonal polynomials again become unworkable.

We turn now to the construction of a finite element subspace S^h. The domain—in this case the interval $[0, \pi]$—is divided into pieces, and on each piece the trial functions v^h are polynomials. Some degree of continuity is imposed at the boundaries between pieces, but ordinarily this continuity is no more than is required in order that

1. The functions v^h will be admissible in the variational principle.
2. The quantities of physical interest, usually displacements, stresses, or moments, can be conveniently recovered from the approximate solution u^h.

It is very difficult to construct piecewise polynomials if too much continuity is required at the boundaries between the pieces.

In our example the admissible space is $\mathcal{3C}_E^1$, whose members are continuous. This rules out piecewise constant functions. Therefore, the simplest choice for S^h is *the space of functions which are linear over each interval* $[(j-1)h, jh]$, *continuous at the nodes* $x = jh$, *and zero at* $x = 0$. The derivative of such a function is piecewise constant and obviously has finite energy; thus S^h is a subspace of $\mathcal{3C}_E^1$. We shall refer to these trial functions as *linear elements*.

For $j = 1, \ldots, N$, let φ_j^h be the function in S^h which equals one at the particular node $x = jh$, and vanishes at all the others (Fig. 1.4). These *roof functions* constitute a basis for the subspace, since every member of S^h can be written as a combination

$$v^h(x) = \sum_1^N q_j \varphi_j^h(x).$$

Notice the following important fact about the coefficient q_j: It coincides with the value of v at the jth node $x = jh$. Since the coordinates q_j are nothing but the nodal values of the function, the optimal coordinates Q_j will have direct physical significance; they are the Ritz approximations to the displacement of the string at the nodes. Notice also that the φ_j^h form a *local basis*, or patch basis, because each φ_j^h is identically zero except in a region of diameter $2h$. In fact, it is obvious that φ_{j-1}^h *is orthogonal to* φ_{j+1}^h, since whenever one of these functions doesn't vanish, the other does. Thus although the basis is not quite orthogonal, it is only adjacent elements which will be coupled.

With coefficients normalized to $p = q = 1$, the problem is to minimize

$$I(v^h) = \int_0^\pi [((v^h)')^2 + (v^h)^2 - 2fv^h]\, dx.$$

With $v^h = \sum q^j \varphi_j^h$, this integral is a quadratic function of the coordinates $q_1, \ldots, q_N$, and it can be computed over one subinterval at a time. On the

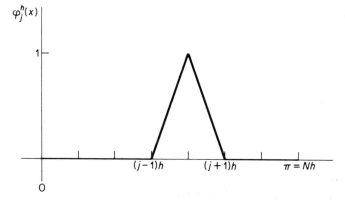

Fig. 1.4 Piecewise linear basis functions.

jth subinterval, the function v^h goes linearly from q_{j-1} to q_j and $(v^h)' = (q_j - q_{j-1})/h$. (By convention $q_0 = 0$.) Therefore,

$$\int_{(j-1)h}^{jh} ((v^h)')^2 \, dx = \frac{(q_j - q_{j-1})^2}{h}.$$

A slightly longer computation gives

$$\int_{(j-1)h}^{jh} (v^h)^2 \, dx = \frac{h}{3} (q_j^2 + q_j q_{j-1} + q_{j-1}^2).$$

These terms correspond to a single piece of string, with linearly varying displacement. For the whole string, the second-degree term in $I(v^h)$ is the sum

$$\int_0^\pi p((v^h)')^2 + q(v^h)^2 = \sum_1^N \frac{p(q_j - q_{j-1})^2}{h} + \frac{qh(q_j^2 + q_j q_{j-1} + q_{j-1}^2)}{3}.$$

This is not a particularly convenient form for the result. We would prefer to have it in the matrix form $q^T K q$, in other words (Kq, q), because it is the matrix K which we shall eventually need. The reason is this: The expression $I(v^h)$ is quadratic in the parameters $q = (q_1, \ldots, q_N)$, of the form

$$I(v^h) = q^T K q - 2F^T q.$$

The minimum of such an expression occurs (as we know from the matrix case, Section 1.3, where we set $\partial I / \partial q_m = 0$) at the vector $Q = (Q_1, \ldots, Q_N)$ determined by

$$KQ = F.$$

This is the system we shall have to solve, and therefore all we need to know is the matrix K and the vector F.

The best plan is to find the contribution to K from each "element," that is, each piece of the string. Therefore, we go back to

$$\int_{(j-1)h}^{jh} ((v^h)')^2 = \frac{(q_j - q_{j-1})^2}{h},$$

and record the right side (the value of the integral) in the matrix form

$$(q_{j-1} q_j) \frac{1}{h} \begin{pmatrix} 1 & -1 \\ -1 & 1 \end{pmatrix} \begin{pmatrix} q_{j-1} \\ q_j \end{pmatrix} = (q_{j-1} q_j) k_1 \begin{pmatrix} q_{j-1} \\ q_j \end{pmatrix}.$$

The matrix k_1 is an *element stiffness matrix*. It represents a computation which needs to be done only once, since it is independent of the particular differential equation. Similarly, the calculation of the zero-order term $\int (v_h)^2 \, dx$ over a single element is recorded once and for all in terms of the *element mass matrix*:

$$\frac{h}{3}(q_{j-1}^2 + q_{j-1}q_j + q_j^2) = (q_{j-1}q_j)\frac{h}{6}\begin{pmatrix} 2 & 1 \\ 1 & 2 \end{pmatrix}\begin{pmatrix} q_{j-1} \\ q_j \end{pmatrix}$$

$$= (q_{j-1}q_j)k_0\begin{pmatrix} q_{j-1} \\ q_j \end{pmatrix}.$$

Now the summation over the elements $j = 1, \ldots, N$ is replaced by *an assembly of the global stiffness matrix K*. This means that the element matrices are added together, after being placed into their proper positions in the global array. The matrix associated with $\int_0^\pi ((v^h)')^2\, dx$, recalling that the unknown q_0 is to be discarded because of the essential boundary condition, is

$$K_1 = \frac{1}{h}\begin{pmatrix} 1 & & & \\ & & & \\ & & & \\ & & & \end{pmatrix} + \frac{1}{h}\begin{pmatrix} 1 & -1 & \\ -1 & 1 & \\ & & \end{pmatrix} + \cdots + \frac{1}{h}\begin{pmatrix} & & \\ & 1 & -1 \\ & -1 & 1 \end{pmatrix}$$

$$= \frac{1}{h}\begin{pmatrix} 2 & -1 & 0 & 0 & \cdot \\ -1 & 2 & -1 & \cdot & 0 \\ 0 & -1 & \cdot & -1 & 0 \\ 0 & \cdot & -1 & 2 & -1 \\ \cdot & 0 & 0 & -1 & 1 \end{pmatrix}.$$

Again the relationship between the matrix and the integral is that

$$\int_0^\pi ((v^h)')^2\, dx = (q_1 \cdots q_N)K_1\begin{pmatrix} q_1 \\ \cdot \\ \cdot \\ q_N \end{pmatrix} = q^T K_1 q.$$

The integral of the undifferentiated term $(v^h)^2$ is given by $q^T K_0 q$, where the mass matrix K_0 (later denoted by M) is formed by the same assembling process:

$$K_0 = \frac{h}{6}\begin{pmatrix} 4 & 1 & 0 & 0 & \cdot \\ 1 & 4 & 1 & \cdot & 0 \\ 0 & 1 & \cdot & 1 & 0 \\ 0 & \cdot & 1 & 4 & 1 \\ \cdot & 0 & 0 & 1 & 2 \end{pmatrix}.$$

The required matrix K is now the sum $K_1 + K_0$. These matrices need not be assembled all at one time and stored; instead the entries can be computed when they are required in the process of solving the final system, $KQ = F$.

It remains to compute the term $\int fv^h$, by which the inhomogeneous load f enters the approximation. This integral is linear in the coordinates q_j:

$$\int_0^\pi fv^h\, dx = \sum_1^N q_j \int_0^\pi f\varphi_j^h\, dx = F^T q,$$

if we define the load vector F to have components

$$F_j = \int_0^\pi f\varphi_j^h\, dx.$$

In practice these numbers are computed, just like the stiffness and mass matrices, by integrating over one element at a time. Suppose that over the jth interval

$$\int_{(j-1)h}^{jh} fv^h = \alpha_j q_{j-1} + \beta_j q_j.$$

When these integrals are summed, the coefficient of a given q_k is $F_k = \beta_k + \alpha_{k+1}$. One sees here in the most rudimentary form the bookkeeping which must be carried out by the computer. A given node kh enters as the right endpoint of the kth subinterval, leading to a term $\beta_k q_k$ in the integral, and also as the left endpoint of the next subinterval, leading to $\alpha_{k+1}q_k$. The assembly subroutine must be aware that both these subintervals are incident on the kth node, and combine the results. (There seems to be no doubt that, starting from scratch, it takes longer to program finite elements than finite differences for a two-point boundary-value problem; again the real benefit is felt first in more than one dimension.)

For arbitrary f these integrals cannot be computed exactly, and some numerical quadrature will be necessary. One possibility is *to approximate f by linear interpolation at the nodes*. In other words, f is replaced by its piecewise linear interpolate $f_I = \sum f_k \varphi_k^h(x)$, where f_k is the value of f at the node $x = kh$. Then the integral $\int f_I v^h$ involves *the same inner products $\int \varphi_k^h \varphi_j^h$ which were computed earlier in forming the mass matrices k_0*. Over the jth interval

$$\int_{(j-1)h}^{jh} f_I v^h\, dx = (f_{j-1} \quad f_j) k_0 \begin{pmatrix} q_{j-1} \\ q_j \end{pmatrix},$$

since the only difference between this and the integration of $(v_h)^2$ is that one pair of coefficients q_{j-1} and q_j is replaced by the two nodal values of f.

Again the computer will sum these results from $j = 1$ to $j = N$ by assembling the mass matrix:

$$\int_0^\pi f_I v^h \, dx = (f_0 \cdots f_N)\tilde{K}_0 \begin{pmatrix} q_1 \\ \cdot \\ \cdot \\ \cdot \\ q_N \end{pmatrix}.$$

$\tilde{K}_0$ differs from K_0 only in that the zeroth row has to be retained, since in general $f(0) \neq 0$.

This result approximates the true linear term $F^T q$, that is, the one which would be obtained by exact integrations. We denote the approximation by $\tilde{F}^T q$; the mass matrix K_0 gives the coefficients in the approximate load vector $\tilde{F}$:

$$\tilde{F}_j = \frac{h}{6}[f((j-1)h) + 4f(jh) + f((j+1)h)].$$

The difference between $\tilde{F}_j$ and the true value $F_j = \int f \varphi_j^h \, dx$ is not difficult to estimate. If f is linear, then of course the two agree, since the interpolate f_I coincides with f itself. For a quadratic f, however, a discrepancy appears. If $f(x) = x^2$, then

$$F_j = \int_{(j-1)h}^{jh} x^2\left(-j + 1 + \frac{x}{h}\right) dx + \int_{jh}^{(j+1)h} x^2\left(j + 1 - \frac{x}{h}\right) = j^2 h^3 + \frac{h^3}{6},$$

whereas

$$\tilde{F}_j = \frac{h^3}{6}[(j-1)^2 + 4j^2 + (j+1)^2] = j^2 h^3 + \frac{h^3}{3}.$$

Therefore, the error in numerical integration is $h^3/6 = h^3 f''(jh)/12$. For an arbitrary smooth f, this will be the leading term in $\tilde{F}_j - F_j$.

There does exist an integration formula, with the same simplicity as $\tilde{F}_j$, which gets also the quadratic (and even cubic) terms correct. It is given by $\bar{F}_j = h(f_{j-1} + 10f_j + f_{j+1})/12$ and is known as *Collatz's Mehrstellenverfahren* [H3]. It can be derived by a *quadratic* interpolation of f over the double interval $[(j-1)h, (j+1)h]$, followed by exact integration of $\bar{F}_j = \int f_I \varphi_j^h \, dx$. This derivation is unnatural, however, for finite element programs. Instead of computing once and for all an integral over $[jh, (j+1)h]$, we have to go over each such interval twice, once to evaluate $\bar{F}_j$ using the quadratic interpolate at the nodes x_{j-1}, x_j, x_{j+1}, and then again to evaluate $\bar{F}_{j+1}$ using the interpolate at x_j, x_{j+1}, x_{j+2}. This is a typical instance in which the most efficient formula in a certain class is not found by the finite element assembly process; given complete freedom to find the best formula in each special case, finite differences can win. But the essential point for complicated problems is that the finite element formula is not far from optimal, and it is generated in a systematic and painless way by the computer.

In practice, the replacement of f by its interpolate f_I has been superseded by *direct numerical integration*. Over each subinterval, fv^h is integrated by a standard quadrature formula

$$\int fv^h = \sum w_i f(\xi_i) v^h(\xi_i).$$

The most common choice, and the most efficient, is Gaussian quadrature, for example with equal weights w_i at the two symmetrically placed evaluation points $\xi_i = (j + 1/2 \pm 1/\sqrt{3})h$. This is again exactly correct for cubic f, and therefore it will automatically attain the same accuracy as the Collatz formula described above. The *one-point* Gauss rule, with ξ_i at the midpoint of each interval, is already sufficient to maintain the accuracy inherent in linear trial functions: It yields $\tilde{F}_j = h(f_{j-1/2} + f_{j+1/2})/2$. [The trapezoidal rule, with equal intervals, gives exactly the right side $\tilde{F}_j = hf(jh)$ of the simple three-point difference equation of Section 1.4.]

The result of any numerical integration is to replace the true linear terms $F^T q$ by some approximate expression $\tilde{F}^T q$, still linear in the unknowns $q_1, \ldots, q_N$.

The same ideas apply to the quadratic terms, if the coefficients $p(x)$ and $q(x)$ in the differential equation actually depend on x. The integrals of $p(x)((v^h)')^2$ and $q(x)(v^h)^2$ are again computed on each subinterval by a numerical quadrature. The assembled results are stored as approximate stiffness matrices, whose quadratic forms are close to the true integrals $q^T K_1 q$ and $q^T K_0 q$. We shall assume for the present that all these integrals are computed exactly, and study in Section 4.3 the effect of errors in numerical integration.

Now we try to assemble our own ideas. Writing K for the sum $K_1 + K_0$, the computations so far have produced the formula

$$I(v^h) = I(\sum q_j \varphi_j^h) = q^T K q - 2F^T q.$$

This is the discrete expression to be minimized in the Ritz method. Observe that it is in exactly the standard variational form; the minimizing vector Q is determined by the linear equation

$$KQ = F.$$

We refer to this as the *finite element equation*. Its solution is the central step in the numerical computations, and if h is small, it will be a large system. The matrix K is guaranteed to be positive definite, and therefore invertible; since

$p > 0$,

$$q^T K q = \int p(x)(\sum q_j \varphi'_j)^2 + q(x)(\sum q_j \varphi_j)^2$$

can be zero only if $\sum q_j \varphi'_j$ is identically zero, and this happens only if every $q_j = 0$.

In fact, the global stiffness matrix K is remarkably like the finite difference matrix L^h, or rather hL^h, of the previous section. With constant coefficients, the leading terms are identical, both being proportional to the second difference with weights $-1, 2, -1$. The zero-order term qu entered only the diagonal entries of L^h. In K it appears also in the coupling between adjacent unknowns, and is "smoothed" with the weights $1, 4, 1$, reminiscent of Simpson's rule. We reemphasize that *once the approximating subspace S^h is chosen, the discrete forms of different terms in the equation are all determined.* The Ritz method is a "package deal," and neither requires nor permits the user to make independent decisions about different parts of the problem.

In particular, the treatment of boundary conditions is fixed, and recalling the alternatives in the difference equation, it is natural to wonder which condition, of which order of accuracy, is chosen "automatically" by the finite element equation. Looking at the last row of K, the equation at the boundary $x = \pi$ in the case $p = q = 1$ is

$$\frac{-q_{N-1} + q_N}{h} + \frac{h}{6}(q_{N-1} + 2q_N) = \int f\varphi_N^h = \int_{\pi-h}^{\pi} f(y)\left(1 + \frac{y - \pi}{h}\right) dx.$$

Substituting the true value $u(x_j)$ for q_j, and expanding u and f in Taylor series about $x = \pi$, the truncation error is

$$u' - \frac{h}{2}u'' + \frac{h^2}{6}u''' - \frac{h^3}{24}u^{iv} + \frac{h}{6}\left(3u - hu' + \frac{h^2}{2}u''\right)$$

$$- \frac{h}{2}f + \frac{h^2}{6}f' - \frac{h^3}{24}f'' + \cdots \approx \frac{h^3}{24}u''(\pi) + \cdots$$

using the differential equation $-u'' + u = f$ and boundary condition $u'(\pi) = 0$. In terms of difference equations, this means that the boundary condition happens to be third-order accurate.

The final step in computing the finite element approximation u^h is to solve the linear system $KQ = F$. We propose to discuss only direct elimination methods, which are preferred to iterative techniques in an overwhelming majority of finite element programs. (It is fascinating to reflect on the rise and fall of iterative methods during the last generation. A tremendous amount of hard work and good mathematics went into the development of overrelaxation and alternating direction methods; this was a dominant theme in numerical analysis. Now these methods are increasingly squeezed between elimination and devices like the fast Fourier transform; the latter is unquestionably more efficient when the geometry and the equation are appropriate, and otherwise—especially when the same linear system is to be solved for many right-hand sides, as in design problems—elimination is convenient and straightforward.)

We want to discuss very briefly the theory of Gaussian elimination. In applying this familiar algorithm to a general coefficient matrix K, the first unknown Q_1 is eliminated from the last $N - 1$ equations, then Q_2 is eliminated from the last $N - 2$ equations, and finally Q_{N-1} is eliminated from the last equation. The system $KQ = F$ is transformed in this way to an equivalent system

$$
UQ = \begin{pmatrix} U_{11} & \cdot & \cdot & \cdot & U_{1N} \\ & U_{22} & \cdot & & U_{2N} \\ & & \cdot & \cdot & \cdot \\ & & & & U_{NN} \end{pmatrix} \begin{pmatrix} Q_1 \\ \cdot \\ \cdot \\ \cdot \\ Q_N \end{pmatrix} = F'.
$$

The unknowns Q_j are now determined by *back substitution*, solving the last equation for Q_N, the next to last for Q_{N-1}, and so on.

It is important to understand in matrix terms what has happened. Suppose we begin to undo the elimination process by adding back to the last equation the multiple $l_{N,N-1}$ of the $(N - 1)$st equation, which was subtracted off in the process of eliminating Q_{N-1}. Next we add back to the last two equations the multiples $l_{N-1,N-2}$ and $l_{N,N-2}$ of equation $N - 2$, which were subtracted in the elimination of Q_{N-2}. Finally, we recover the original system $KQ = F$, having added back to each equation i the multiples l_{ij} of the previous equations, $j = 1$ through $j = i - 1$, which were subtracted in the elimination of Q_1 through Q_{i-1}. In matrix terms *the system $KQ = F$ has been recovered by multiplying the transformed system $UQ = F'$ by the matrix*

$$
L = \begin{pmatrix} 1 \\ l_{21} & 1 \\ \cdot & \cdot & 1 \\ \cdot & & \cdot & l_{N-1,\,N-2} & 1 \\ l_{N1} & \cdot & l_{N,\,N-2} & l_{N,\,N-1} & 1 \end{pmatrix}.
$$

This means that $LUQ = LF'$ is precisely the same as $KQ = F$; *Gaussian elimination is nothing but the factorization of K into a product*

$$
K = LU
$$

of a lower triangular matrix times an upper triangular matrix. Thus $K^{-1}F$, which is the solution Q we want to compute, is identical with $U^{-1}L^{-1}F$, and the two triangular matrices are easy to invert. In fact, $L^{-1}F$ is F', the inhomogeneous term as it stands after elimination, and then Q is $U^{-1}F'$, the result after back substitution. (If there are many systems $KQ = F_n$ to be solved, with different data but the same stiffness matrix K, then the factors L and U should be stored.)

Now we consider what is special about the elimination process when K is known, as in our example, to be symmetric, positive-definite, and tridiagonal. The first point is that the process succeeds—the eliminations can be carried out, and the factorization $K = LU$ exists. It would not have succeeded with $K = \begin{pmatrix} 0 & 1 \\ 1 & 0 \end{pmatrix}$, since Q_1 could not be eliminated from the second equation by subtracting a multiple $l_{2,1}$ of the first. The condition for success is that each of the matrices in the upper left corner of K,

$$K^{(1)} = (K_{11}), \quad K^{(2)} = \begin{pmatrix} K_{11} & K_{12} \\ K_{21} & K_{22} \end{pmatrix}, \quad \cdots$$

should have a nonzero determinant. For a positive-definite matrix, these determinants are all positive, and therefore elimination can be carried out with no exchanges of rows. In fact, the determinant of $K^{(j)}$ equals the product $U_{11}U_{22} \cdots U_{jj}$, so that the pivot elements U_{jj}, lying on the main diagonal of U, are all positive.

Something more is required for the elimination algorithm to be numerically stable: The pivots U_{jj} must be not only nonzero but also sufficiently large. Otherwise, the information in the original coefficients K_{ij} will be destroyed as the algorithm progresses. We have only partial control over the size of these pivots, and therefore some roundoff error is unavoidable. The intrinsic sensitivity of K to small perturbations is measured by its *condition number*, roughly the ratio of its largest eigenvalue to its smallest, which is the subject of Chapter 5. This number will depend on the mesh size h and on the order of the differential equation. In some cases, however, numerical difficulties can spring not from an ill-conditioned K but from an ill-chosen algorithm. With a matrix like $K = \begin{pmatrix} \epsilon & 1 \\ 1 & 0 \end{pmatrix}$, for example, the two equations should first be exchanged, or pivoted. If we exchange rows at each stage so as to maximize the pivot element U_{jj}, then the Gaussian elimination algorithm becomes as stable as the condition number allows. The troublesome matrix K in this example is, of course, not positive-definite; it is more typical of the matrices which arise in the "mixed" method (Section 2.3).

The direct stiffness method—in which the unknowns are the displacements u, as in the greater part of this book—automatically yields a positive definite

K. In this case, *elimination without row exchanges is not only possible but also numerically stable*. To understand this, we first give the factorization $K = LU$ a more symmetric form, by dividing out the diagonal part of U:

$$D = \begin{pmatrix} U_{11} & & & \\ & \cdot & & \\ & & \cdot & \\ & & & \cdot & \\ & & & & U_{NN} \end{pmatrix}.$$

This leaves $K = LD(D^{-1}U)$, where all three factors are uniquely determined: L as a lower triangular matrix with unit diagonal, $D^{-1}U$ as an upper triangular matrix with unit diagonal, and D as a positive diagonal matrix. By symmetry, $D^{-1}U$ must be the transpose of L. Thus $K = LDL^T$, the symmetric form of the factorization. It is even possible to go one step further, by introducing a new lower triangular matrix $\tilde{L} = LD^{1/2}$; this yields the Cholesky factorization $K = \tilde{L}\tilde{L}^T$. Now we can explain why positive-definite symmetric matrices never require pivoting: The factor $\tilde{L}$ in this decomposition behaves like a square root of K and cannot get out of scale. In fact, the condition number of $\tilde{L}$, appropriately defined, is exactly the square root of that of K. This is to be compared with

$$\begin{pmatrix} \epsilon & 1 \\ 1 & 0 \end{pmatrix} = \begin{pmatrix} 1 & 0 \\ \epsilon^{-1} & 1 \end{pmatrix}\begin{pmatrix} \epsilon & 0 \\ 0 & -\epsilon^{-1} \end{pmatrix}\begin{pmatrix} 1 & \epsilon^{-1} \\ 0 & 1 \end{pmatrix},$$

where the factors on the right are large even though the matrix on the left is not.

Finally, the fact that K is tridiagonal means an enormous reduction in the number of computations. Since the first unknown Q_1 has only to be eliminated from the second equation—it does not appear in the others—all the multiples $l_{3,1}, \ldots, l_{N-1,1}$ of the first equation, normally required in eliminating Q_1, are zero. At the second stage, Q_2 has only to be eliminated from the third equation, and so on. Thus in the lower triangular matrix L, the only nonzero elements lie on the main diagonal and on the first subdiagonal; the factors of a tridiagonal matrix are bidiagonal.

Since the Cholesky factorization involves square roots of the pivots and presents some extra problems in avoiding operations with zeros, the most popular numerical algorithm is based on the decomposition $K = LDL^T$. The two factorizations are equivalent, both mathematically and from the point of view of numerical stability. (We assume no overflow.) For a tridiagonal matrix, the entries of L and D satisfy a simple recursion formula:

$$d_j = K_{j,j} - d_{j-1}l_{j,j-1}^2, \qquad d_0 = 0,$$

$$l_{j+1,j} = \frac{K_{j+1,j}}{d_j}.$$

Correspondingly, the vector $F' = L^{-1}F$ and the solution Q which results from back substitution are

$$F'_j = F_j - F'_{j-1}l_{j,j-1}, \qquad F_0 = 0,$$

$$Q_j = \frac{F'_j}{d_j} - Q_{j+1}l_{j+1,j}, \qquad Q_{N+1} = 0.$$

The important point is that the number of arithmetic operations involved is proportional only to N.

In many-dimensional problems, the stiffness matrix K will still be sparse and symmetric positive-definite. However, there will not be such an obvious ordering of the unknowns. In fact, *the nodal ordering becomes crucial,* and subroutines have been developed to yield an ordering that reduces the expense of Gaussian elimination.

The simplest and most popular criterion is the bandwidth of the matrix. Suppose it is known that only the first w subdiagonals of K, and the first w superdiagonals, are nonzero. (For tridiagonal matrices, the bandwidth by this definition is $w = 1$.) Then at each stage in the elimination process this information can be used to advantage. There are only w nonzero elements to be eliminated below each pivot, and furthermore each elimination—subtracting a multiple of one row from another—operates on rows known to have only $2w + 1$ nonzero entries. The band structure is preserved throughout elimination and is inherited by the triangular factors L and U. For a symmetric matrix, the number of operations is roughly $Nw^2/2$ rather than the $N^3/3$ required for a full matrix.

The simplest illustration of good and bad orderings, from the point of view of the bandwidth, is furnished in two dimensions by a rectangular array of nodes. If there are fewer nodes in the horizontal than in the vertical direction, the unknowns should be numbered consecutively along each row rather than each column. The bandwidth will be roughly the length of a row, since for a given node we must wait that long for the node above it (which is presumably linked to it by a nonzero entry in the stiffness matrix) to appear in the ordering. In general, finite element matrices are less systematic in structure than those from conventional difference equations, and an optimal ordering is far from self-evident.

There is a second criterion which takes into account the sparseness of a matrix; it is a little more precise and subtle than the bandwidth. It is based on the *profile*, or skyline, of the matrix. Consider the first nonzero element in the ith row. If this occurs in column j and that fact is known to the computer, it will not be necessary to subtract multiples of rows $1, 2, \ldots, j-1$ from row i. The multiplying factors $l_{i,1}, \ldots, l_{i,j-1}$ would all be zero anyway, since $Q_1, \ldots, Q_{j-1}$ do not require elimination from the ith equation; they are absent in the first place. The profile is formed by locating these first non-zero entries in each row, and like the band structure it is preserved throughout Gaussian elimination and inherited by the factor L. In case the profile goes significantly inside the band—so that a good many rows have length well below $2w + 1$ non-zero entries—it may be worthwhile for the computer to know the profile, and even to order the unknowns according to a "profile-reducing" algorithm.

We emphasize that the number of arithmetic operations is not the only criterion in choosing an algorithm; the storage requirements may be at least as important. For a band matrix, a standard procedure is to store successive diagonals of the matrix; this is probably near to optimal, and uses about Nw locations. For a linear or bilinear element on a 50×50 square mesh, $N = 2500$ and $w \approx 50$; at about this point, a typical large computer would have to store information outside of core, and the programming and data management become much more complex. Therefore considerable attention is being paid to algorithms which take account of, and maintain as far as possible, the sparseness of a matrix even within its band or profile. At the extreme, we may identify the location of every nonzero entry in A and order the unknowns by a "sparse matrix algorithm" so as to minimize the number of nonzero entries in the lower triangular L. It seems to us that for finite elements this is too expensive; it takes no account of the systematic structure of the matrices.

We prefer, if the problem is so large that the standard band or profile algorithm goes outside of core, to follow the analysis given in a series of papers by Alan George. His goal, for finite elements with N unknown parameters in the plane, is to achieve $O(N^{3/2})$ arithmetic operations in elimination and $O(N \log N)$ non-zero entries in L. This would be essentially optimal. (There are some special direct methods, analogous to the fast Fourier transform, which need only $O(N \log N)$ operations and $O(N)$ total storage locations; but they are restricted to simple problems on rectangles.) His goal is achieved by an ordering [G3] which is like the minimum degree algorithm: at each stage, the unknown to be eliminated should be the one currently connected to the fewest unknowns. This is very different from Irons' frontal method, and it appears to demand a large program overhead—probably too large. A more recent suggestion is given in George's paper "An efficient band-oriented scheme for solving n by n grid problems." He divides the domain into thin strips, and applies a band algorithm to the corresponding submatrices (the unknowns inside the strip having been numbered to reduce the bandwidth). Between two strips will come a line of unknowns, and in George's ordering they should appear *after the strips which they separate*. It is then the comparatively small number of unknowns on these separating lines which contribute to a filling out of the band; the bigger submatrices corresponding to the link between one strip and another are empty. With a number of strips proportional to $h^{-1/2}$, the storage requirement turns out to be $O(N^{5/4})$—not optimal, but better by the substantial factor $N^{1/4}$ than a straightforward band scheme using $Nw = N^{3/2}$ locations. The saving is the same in three dimensions.

Whatever the algorithm—this will obviously remain an area of active research—the eventual result of the elimination process is the solution of $KQ = F$. With this, the finite element approximation is determined.

1.6. THE ERROR WITH LINEAR ELEMENTS

How close is the Ritz approximation u^h to the true solution u? As close as possible, according to the following theorem, in the sense that *the energy in the error $u - u^h$ is a minimum*. The Ritz method is therefore optimal, assuming that the energy is measured in the natural way. This measure must be one which is associated with the particular problem at hand, and in fact it is specified by the functional $I(v)$ itself: The energy in v is the second-degree term in $I(v)$. (This definition differs by a factor $\frac{1}{2}$ from the one which is physically correct; it is convenient just to ignore this factor.) Thus if the functional is written in the form

$$(19) \qquad\qquad I(v) = a(v, v) - 2(f, v),$$

the energy in the function v is given by $a(v, v)$.

The energy therefore coincides with the term which up to now has arisen as (Lv, v) and been integrated by parts. This integration has produced a more symmetric expression, and this symmetry is emphasized by the notation $a(v, v)$. In particular, if (Lv, w) is integrated by parts, the result is the symmetric form

$$a(v, w) = \int_0^\pi \big(p(x)v'(x)w'(x) + q(x)v(x)w(x)\big)\, dx.$$

This is the *energy inner product*. It is defined for all v and w in the admissible space $\mathcal{K}_E^1$ and represents the inner product that is intrinsic to the given problem.

Our goal in this section is first to prove the theorem described above—that the energy in the error is minimized by the Ritz method—and then to apply this theorem in establishing bounds for the error with linear elements.

THEOREM 1.1

Suppose that u minimizes $I(v)$ over the full admissible space $\mathcal{K}_E^1$, and S^h is any closed subspace of $\mathcal{K}_E^1$. Then:

(a) *The minimum of $I(v^h)$ and the minimum of $a(u - v^h, u - v^h)$, as v^h ranges over the subspace S^h, are achieved by the same function u^h. Therefore*

$$(20) \qquad\qquad a(u - u^h, u - u^h) = \min_{v^h\ \text{in}\ S^h} a(u - v^h, u - v^h).$$

(b) *With respect to the energy inner product, u^h is the projection of u onto S^h. Equivalently, the error $u - u^h$ is orthogonal to S^h:*

$$(21) \qquad a(u - u^h, v^h) = 0 \qquad \text{for all } v^h \text{ in } S^h.$$

(c) *The minimizing function satisfies*

$$(22) \qquad a(u^h, v^h) = (f, v^h) \qquad \text{for all } v^h \text{ in } S^h.$$

In particular, if S^h is the whole space $\mathcal{3C}_E^1$, then

$$(23) \qquad a(u, v) = (f, v) \qquad \text{for all } v \text{ in } \mathcal{3C}_E^1.$$

COROLLARY

It follows from (21) that $a(u - u^h, u^h) = 0$, or $a(u, u^h) = a(u^h, u^h)$, and the Pythagorean theorem holds: The energy in the error equals the error in the energy,

$$a(u - u^h, u - u^h) = a(u, u) - a(u^h, u^h).$$

Furthermore, since the left side is necessarily positive, the strain energy in u^h always underestimates the strain energy in u:

$$(24) \qquad a(u^h, u^h) \leq a(u, u).$$

This whole theorem is fundamental to the Ritz theory, and its three parts are very much interdependent. We can deduce (b) immediately from (c): If (23) holds for all v, it holds for every v^h in S^h, and subtracting (22), the result is (21).

Also (b) follows from (a): In an inner product space, the function in a subspace S^h closest to a given u is always the projection of u onto S^h. In the opposite direction, to show that (b) implies (a), we compute

$$a(u - u^h - v^h, u - u^h - v^h) = a(u - u^h, u - u^h) - 2a(u - u^h, v^h) + a(v^h, v^h).$$

If (21) holds, then

$$a(u - u^h, u - u^h) \leq a(u - u^h - v^h, u - u^h - v^h).$$

Equality occurs only if $a(v^h, v^h) = 0$, in other words, only if $v^h = 0$. Thus u^h is the unique minimizing function in (20), and (a) is proved.

The problem is now to establish (c), since that will imply (b) and therefore (a). If u^h minimizes I over S^h, then for any ϵ and v^h,

$$I(u^h) \leq I(u^h + \epsilon v^h).$$

The right side is

$$a(u^h + \epsilon v^h, u^h + \epsilon v^h) - 2(f, u^h + \epsilon v^h) = I(u^h) + 2\epsilon[a(u^h, v^h) - (f, v^h)] \\ + \epsilon^2 a(v^h, v^h).$$

Therefore,

$$0 \le 2\epsilon[a(u^h, v^h) - (f, v^h)] + \epsilon^2 a(v^h, v^h).$$

Since this is true for small ϵ of either sign, it follows that $a(u^h, v^h) = (f, v^h)$. This equation expresses *the vanishing of the first variation of I at u^h, in the direction of v^h.* In particular, $a(u, v) = (f, v)$, so that at u the first variation vanishes in every direction v. This is the equation (11) derived earlier. Thus (c) is established; it is the *equation of virtual work*.

The choice $v = u$ in this equation leads to an interesting result, that *at the minimum, the strain energy is the negative of the potential energy:*

(25) $$I(u) = a(u, u) - 2(f, u) = -a(u, u).$$

Similarly, $I(u^h) = -a(u^h, u^h)$. In every case $I(u) \le I(u^h)$, since u is minimizing over a larger class of functions, and therefore a change of sign reproduces the result stated in the corollary, that the strain energy is always underestimated:

$$a(u^h, u^h) \le a(u, u).$$

The proof of the theorem is now complete, except for one point: Neither the existence nor the uniqueness of u^h (or, in case S^h is the whole space $\mathcal{H}_E^1$, of u itself) has been properly established. To a functional analyst, such an omission means that the proof has hardly begun. We shall try to mollify him by pointing to the key word in the hypothesis of the theorem: The subspace S^h is required to be *closed*. This means that the subspace must contain its own limiting functions. If there is a sequence v_N in S^h such that

$$a(v_N - v_M, v_N - v_M) \to 0 \qquad \text{as } N, M \to \infty,$$

then there must exist in S^h a limit v, such that

$$a(v_N - v, v_N - v) \to 0 \qquad \text{as } N \to \infty.$$

This will always be true if S^h is finite-dimensional, which is the case contemplated in the Ritz method. In general, one cannot guarantee that there exists a function u^h in S^h which is absolutely the closest to u, unless the subspace is closed. We cite the example $S^h = \mathcal{H}_B^2$, which is not a closed subspace. It contains functions arbitrarily close to $u(x) = x$, but there is no closest one; the "projection" breaks down.

To prove the existence of u, defined to be the minimizing function over the whole space $\mathcal{JC}_E^1$, we need to see that $\mathcal{JC}_E^1$ is itself closed. This is exactly what was achieved in the completion process, enlarging $\mathcal{JC}_B^2$ to the full admissible space $\mathcal{JC}_E^1$. In particular, it was in completing (or closing) the admissible space that the natural boundary condition $u'(\pi) = 0$ was dropped. There was one technical point in that process which we hurried by: the space was completed in the natural energy norm, $a(v - v_N, v - v_N) \to 0$ as in (10), and yet we have described the completed space in terms of the $\mathcal{JC}^1$ norm. This step is justified by the equivalence of the two norms: there exist constants σ and K such that

(26a) $$a(v, v) \leq K \| v \|_1^2,$$

(26b) $$a(v, v) \geq \sigma \| v \|_1^2.$$

The last inequality also yields the uniqueness of u and u^h, since it means that the energy is positive definite: $a(v, v) = 0$ if and only if $v = 0$. The surface $I(v)$ is strictly convex, and can have only one stationary point, at the minimum.

The first inequality is easy, since

$$\int [p(v')^2 + qv^2] \, dx \leq \max(p(x), q(x)) \int [(v')^2 + v^2] \, dx.$$

Therefore, K can be chosen as $\max(p, q)$.

The inequality (26b) in the other direction starts in the same way, since p is bounded below by a positive constant $p_{\min}$:

(27) $$\int p(v')^2 \, dx \geq p_{\min} \int (v')^2 \, dx.$$

The difficulty comes with the undifferentiated, or zero-order, terms, since q need not be bounded away from zero; in fact, we may have $q \equiv 0$. Therefore, we need an inequality of *Poincaré type*, bounding v in terms of v'. With the boundary condition $v(0) = 0$, the natural idea is to write

$$v(x_0) = \int_0^{x_0} v'(x) \, dx,$$

and apply the Schwarz inequality:

$$|v(x_0)|^2 \leq \left(\int_0^{x_0} 1^2 \right) \left(\int_0^{x_0} (v')^2 \right) \leq \pi \int_0^{x_0} (v')^2.$$

Integration over $0 \leq x_0 \leq \pi$ yields a Poincaré inequality:

$$\int_0^\pi v^2 \leq \pi^2 \int_0^\pi (v')^2.$$

Now by borrowing part of the right side of (27),

$$\int p(v')^2 + qv^2 \geq p_{\min} \int (v')^2 \geq \frac{1}{2\pi^2} p_{\min} \int (v')^2 + v^2.$$

This is the required inequality $a(v, v) \geq \sigma \|v\|_1^2$, which asserts that the problem is *elliptic*. Together with (26a), it implies that whether $\mathcal{3C}_B^2$ is completed in the standard norm $\|v\|_1$ or in the energy norm $\sqrt{a(v, v)}$, the completed space is the same, $\mathcal{3C}_E^1$.

Note that with a natural boundary condition at *both* ends, the situation is entirely different. The Poincaré inequality depended on pinning down $v(0) = 0$, and will be violated by every constant function. With $v = $ constant, and with $q \equiv 0$, the energy $a(v, v) = \int p(v')^2$ can be zero without implying that $v = 0$. Equivalently, the differential equation

$$-(pu')' = f,$$
$$u'(0) = u'(\pi) = 0$$

has no unique solution; u is determined only up to a constant function. Thus a pure Neumann problem—physically, a problem in which rigid body motions are possible—may lead to technical difficulties: the quadratic form $a(v, v)$ will be indefinite.

So much for the general theory. The goal is to apply it to the finite element example, with S^h composed of broken-line functions, and emerge with an estimate of the error $e^h = u - u^h$. The key lies in the minimizing property (20):

$$a(e^h, e^h) \leq a(u - v^h, u - v^h) \qquad \text{for all } v^h \text{ in } S^h.$$

Of course, the function u is not known; we can be certain only that if f is in $\mathcal{3C}^0$, then u is in $\mathcal{3C}_B^2$. Therefore, the question is: *How closely can an arbitrary function u in $\mathcal{3C}_B^2$ be approximated by members of S^h?* We need not attempt to work only with the Ritz approximation u^h; it will be sufficient to find in S^h a good approximation to u, since u^h will always be better. Thus the estimation of the error e^h becomes a straightforward problem in approximation theory: How far away are functions in $\mathcal{3C}_B^2$ from S^h, in the natural norm $\sqrt{a(v, v)}$?

The most convenient choice of a function in S^h which is close to u is its interpolate u_I. The two functions agree at every node $x = jh$, and u_I is linear in between. It can be expanded in terms of the *roof functions* as

$$u_I(x) = \sum_{1}^{N} u(jh) \varphi_j^h(x).$$

At any node, only the one corresponding basis function φ_j^h is nonzero.

We compare u with u_I, first on the basis of a simple Taylor-series argument, leading to a pointwise estimate of their difference.

THEOREM 1.2

If u'' is continuous, then

(28) $$\max| u(x) - u_I(x)| \leq \tfrac{1}{8} h^2 \max| u''(x)|$$

and

(29) $$\max| u'(x) - u'_I(x)| \leq h \max| u''(x)|.$$

Proof. Consider the difference $\Delta(x) = u(x) - u_I(x)$ over a typical interval $(j-1)h \leq x \leq jh$. Since Δ vanishes at both ends of the interval, there must be at least one point z where $\Delta'(z) = 0$. Then for any x,

$$\Delta'(x) = \int_z^x \Delta''(y)\, dy.$$

But $\Delta'' = u''$, since u_I is linear, and (29) follows immediately:

$$|\Delta'(x)| = \left| \int_z^x u''(y)\, dy \right| \leq h \max| u'' |.$$

The maximum of $|\Delta(x)|$ will occur at a point where the derivative vanishes, $\Delta'(z) = 0$. We look to see in which half of the interval z lies; suppose, for example, that it is closer to the right-hand endpoint, so that $jh - z \leq h/2$. Then expanding in a Taylor series about z,

$$\Delta(jh) = \Delta(z) + (jh - z)\Delta'(z) + \tfrac{1}{2}(jh - z)^2 \Delta''(w),$$

where $z < w < jh$. Since Δ vanishes at the endpoint jh, and $\Delta'' = u''$,

$$|\Delta(z)| = |\tfrac{1}{2}(jh - z)^2 \Delta''(w)| \leq \tfrac{1}{8} h^2 \max| u'' |.$$

This constant $\tfrac{1}{8}$ is the best possible, not only for the error in linear *interpolation*, but even if we allow an arbitrary piecewise linear approximation. The most difficult function to approximate has u'' alternating between $+1$ and -1 on successive intervals (Fig. 1.5). The best piecewise linear approximation in this extreme case is identically zero, and the error is $h^2/8$.

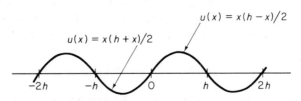

$u(x) = x(h + x)/2$

$u(x) = x(h - x)/2$

$-2h$ $-h$ 0 h $2h$

Fig. 1.5 The extreme case in piecewise linear approximation.

A more careful proof of (29) would improve it to $\max|\Delta'| \leq \frac{1}{2}h \max|u''|$, and the extreme function illustrated in Fig. 1.5 also shows this constant $\frac{1}{2}$ to be the best possible.

It follows from the theorem that if u'' is continuous, then

$$a(u - u_I, u - u_I) = \int_0^\pi p(\Delta')^2 + q\Delta^2 \leq Ch^2 \max|u''|^2.$$

Since the Ritz approximation u^h is at least as close as u_I, *the error in energy with linear elements satisfies*

$$a(u - u^h, u - u^h) \leq Ch^2 \max|u''|^2.$$

This is almost the result we want. The factor h^2 is perfectly correct, and reflects the observed rate of decrease of the error as the mesh is refined. The imperfection is in the other factor, $\max|u''|^2$. It is unsatisfactory to have to assume that u'' is continuous, or even that it is bounded, when the conclusion is an estimate of the error in the energy. It should be sufficient to assume only that u'' has finite energy in the $\mathfrak{JC}^0$ norm, in other words that $\int (u'')^2\, dx < \infty$. The proof of this sharper result, in the special instance of linear interpolation, will be based on Fourier series rather than Taylor series. The proper error estimate is given in (34) below.

THEOREM 1.3

If u'' lies in $\mathfrak{JC}^0$, then

(30)
$$\|u - u_I\|_0 \leq \frac{1}{\pi^2}h^2\|u''\|_0,$$

(31)
$$\|u' - u_I'\|_0 \leq \frac{1}{\pi}h\|u''\|_0,$$

(32)
$$a(u - u_I, u - u_I) \leq \left(\frac{h^2}{\pi^2}p_{max} + \frac{h^4}{\pi^4}q_{max}\right)\|u''\|_0^2.$$

Proof. Consider any subinterval of length h, for example the first one, $0 \leq x \leq h$. The difference $\Delta(x) = u(x) - u_I(x)$ vanishes at both endpoints and we represent it as a Fourier sine series:

$$\Delta(x) = \sum_1^\infty a_n \sin \frac{n\pi x}{h}.$$

By direct computation

$$\int_0^h (\Delta')^2\, dx = \frac{h}{2}\sum \left(\frac{n\pi}{h}\right)^2 a_n^2,$$

$$\int_0^h (\Delta'')^2\, dx = \frac{h}{2}\sum \left(\frac{n\pi}{h}\right)^4 a_n^2.$$

Since $n \geq 1$,

$$\left(\frac{n\pi}{h}\right)^2 a_n^2 \leq \frac{h^2}{\pi^2}\left(\frac{n\pi}{h}\right)^4 a_n^2.$$

Therefore, summing on n,

(33) $$\int_0^h (\Delta')^2 \leq \frac{h^2}{\pi^2}\int_0^h (\Delta'')^2 = \frac{h^2}{\pi^2}\int_0^h (u'')^2.$$

Here $\Delta'' = u''$ because u_I is linear. Equality holds if and only if every coefficient a_n after the first is zero; Δ must have the form $\sin \pi x/h$.

The conclusion (33) holds equally well over each subinterval, say from $(j-1)h$ to jh, and we can sum over all these subintervals:

$$\sum_1^N \int_{(j-1)h}^{jh} (\Delta')^2 \leq \frac{h^2}{\pi^2}\sum_1^N \int_{(j-1)h}^{jh} (u'')^2.$$

We want to simplify this to

$$\int_0^\pi (\Delta')^2 \leq \frac{h^2}{\pi^2}\int_0^\pi (u'')^2.$$

This step, which looks completely obvious, is justified only because there is no trouble at the points where the subintervals meet. Notice that if Δ'' were still on the right side, as in (33), the equality

$$\sum \int_{(j-1)h}^{jh} (\Delta'')^2 = \int_0^\pi (\Delta'')^2$$

would have been completely false, the right side being in reality infinite. (Δ'' is a δ-function at the nodes.) This point will recur when we consider the difference between "conforming" and "nonconforming" elements; when the trial functions are not smooth enough to lie in the admissible space, $I(v)$ cannot be computed element by element.

By a similar argument,

$$\int_0^\pi \Delta^2\, dx = \frac{h}{2}\sum a_n^2 \leq \frac{h^4}{\pi^4}\int_0^\pi (u'')^2.$$

Therefore,

$$a(\Delta, \Delta) = \int_0^\pi p(\Delta')^2 + q\Delta^2$$

$$\leq \left(\frac{h^2}{\pi^2}p_{\max} + \frac{h^4}{\pi^4}q_{\max}\right)\int_0^\pi (u'')^2 \leq C^2 h^2 \|u''\|_0^2.$$

This completes the proof.

As an alternative to Fourier series, the key inequality (33) could have been derived by solving a variational problem: Maximize $\int (\Delta')^2$, subject to $\Delta(0) = \Delta(h) = 0$, $\int (\Delta'')^2 = 1$. The stationary points are $\Delta = \sin n\pi x/h$, and the extremum falls naturally at $\sin \pi x/h$.

The interpolate u_I is not to be confused with the Ritz approximation u^h. Both are piecewise linear, but u^h is determined variationally while u_I is simply a convenient choice close to u. Theorem 1.1, which asserts that u^h is closer still, yields the first of the following inequalities.

COROLLARY

The error $e^h = u - u^h$ in the finite element method satisfies

$$(34) \qquad a(e^h, e^h) \leq C_1 h^2 \| u'' \|_0^2 \leq C_2 h^2 \| f \|_0^2.$$

The last inequality comes from (3), bounding the solution in terms of the data. The leading terms in the constants are $p_{\max}/\pi^2$ for C_1, and $p_{\max}/\pi^2 p_{\min}^2$ for C_2.

The final result is thus a bound of order h^2 for the error in energy. Computation shows this bound to be completely realistic, and in practice the error $a(u - u^h, u - u^h)$ is almost exactly *proportional* to h^2, beginning even at very crude meshes ($h = \frac{1}{2}$ or $\frac{1}{4}$?). Such regularity can be explained by establishing an asymptotic expansion for the error, of the kind introduced for difference equations in (16).

Our convergence proofs have so far assumed two derivatives for u, thereby excluding the case in which f is a δ-function and u is a piecewise-linear "ramp." An easy computation shows that the resulting error in energy is of order h, unless a node is placed right at the discontinuity in f. [It is also $O(h)$ for a line discontinuity in two dimensions.] In general, given only that u lies in $\mathcal{K}_E^1$, it is impossible to say anything about the *rate* of convergence; it may be arbitrarily slow as $h \to 0$. However, convergence itself is easy to prove.

THEOREM 1.4

For any solution u in $\mathcal{K}_E^1$—in other words for any data f in the matching data space—the finite element method converges in the energy norm:

$$a(e^h, e^h) \longrightarrow 0 \qquad \text{as } h \to 0.$$

Proof. Since $\mathcal{K}_E^1$ was constructed in the first place by completing $\mathcal{K}_B^2$, there is a sequence v_N in $\mathcal{K}_B^2$ converging in the energy norm to u. For each fixed N, the finite element approximations v_N^h converge to v_N as $h \to 0$, by Theorem 1.3. Therefore, choosing N large and then h small, there is a function v_N^h in S^h arbitrarily close to u. Since the projection u^h will be even closer, the sequence u^h must converge to u.

This argument applies unchanged to all such minimization problems and need not be repeated in every case. The necessary and sufficient condition for convergence in the Ritz method is obviously this, that *for every admissible* u, *the distance to the trial spaces* S^h (measured by the energy) *should approach zero as* $h \rightarrow 0$. The point of the previous theorem is that once this convergence has been verified for a "dense" subspace—one whose completion in the energy norm yields all admissible functions—then convergence is automatic for every u. Therefore, the interesting problem is to establish the *rate of convergence* in energy when u is sufficiently smooth.

It is equally interesting, and somewhat harder, to find this rate of convergence in a different norm. The result of the corollary is that the strains—the first derivatives $(u^h)'$—are in error by $O(h)$. We inquire next about *the error in the displacement*. How quickly does $e^h = u - u^h$ decrease when measured by $\|e^h\|_0$?

The crudest answer is to apply Poincaré's inequality $|e^h(x_0)| \leq \sqrt{\pi}\|e^h\|_1$, derived above. This bounds the error at every point x_0 uniformly by $O(h)$. One might expect, however, to improve this estimate to $O(h^2)$. Such an improvement appears almost obvious from (30), where the error in the interpolating element u_I is of second order. This at least proves that S^h contains a function which is within $O(h^2)$ of u in displacement. The difficulty is that *in the $\mathcal{K}^0$ norm the Ritz approximation u^h is not minimizing*. There is no assurance that u^h is as close to u as u_I is, and we shall later find fourth-order problems in which the error in displacement is no better than the error in slope.

In the present example, however, the displacement error is indeed $O(h^2)$. One possible proof is simply to forget the variational derivation of the finite element equations $KQ = F$, and to compute their truncation error as difference equations. (At the boundary $x = \pi$ this has already been done.) Applying the maximum principle, the result is actually a pointwise bound $|e^h(x)| = O(h^2)$, which is optimal. Nevertheless, this approach by way of finite differences is not entirely satisfactory, because its extension to irregular finite elements in two-dimensional problems becomes extremely difficult. Therefore, it is essential to look for an argument which will establish *variationally* the rate of convergence of the displacement error $\|e^h\|_0$.

The following trick is remarkably successful. Let z be the solution to the original variational problem over $\mathcal{K}_E^1$, when $e^h = u - u^h$ is chosen as the data. Then by the vanishing of the first variation,

$$(35) \qquad\qquad a(z, v) = (e^h, v) \qquad \text{for all } v \text{ in } \mathcal{K}_E^1.$$

In particular, we may choose $v = e^h$:

$$(36) \qquad\qquad\qquad a(z, e^h) = \|e^h\|_0^2.$$

On the other hand, Theorem 1.1 asserts that $a(v^h, e^h) = 0$ for all v^h in S^h. Subtracting from (36), this gives

$$(37) \qquad\qquad a(z - v^h, e^h) = \| e^h \|_0^2.$$

To the left side we apply the Schwarz inequality in the energy norm, that is,

$$| a(v, w)| \le \big(a(v, v)\big)^{1/2} \big(a(w, w)\big)^{1/2}$$

with $v = z - v^h$, $w = e^h$. By the corollary to Theorem 1.2,

$$\big(a(e^h, e^h)\big)^{1/2} \le Ch\| u'' \|_0.$$

Choosing v^h as the Ritz approximation to z, the same corollary gives

$$\big(a(z - v^h, z - v^h)\big)^{1/2} \le Ch\| z'' \|_0.$$

Thus the Schwarz inequality applied to (37) yields

$$\| e^h \|_0^2 \le C^2 h^2 \| u'' \|_0 \| z'' \|_0.$$

Finally, we can bound the solution z by its data e^h; according to (3),

$$\| z'' \|_0 \le \| z \|_2 \le \rho \| e^h \|_0.$$

Here is the key point; that to estimate the Ritz error e^h in the norm $\mathcal{H}^0$, which is variationally unnatural, one needs an equally unnatural bound for solutions in the norm of $\mathcal{H}^2$. The latter bound is entirely normal, however, from the viewpoint of differential equations; in fact, the central result of the theory was exactly this, to estimate the solution in $\mathcal{H}^2$ in terms of the data in $\mathcal{H}^0$. Substituting this bound into the previous inequality and cancelling the common factor $\| e^h \|_0$, this argument (known in the numerical analysis literature as *Nitsche's trick*) has established an h^2 estimate for the error in displacement:

THEOREM 1.5

The piecewise linear finite element approximation u^h, derived by the Ritz method, satisfies

$$(38) \qquad\qquad \| u - u^h \|_0 \le \rho C^2 h^2 \| u'' \|_0 \le \rho^2 C^2 h^2 \| f \|_0.$$

It is interesting that this bound was derived without appealing directly to the fact that approximation to order h^2 is possible in the $\mathcal{H}^0$ norm. This fact is therefore another consequence of the theorem: If S^h achieves $O(h)$ approximation in $\mathcal{H}^1$, then it can achieve $O(h^2)$ approximation in $\mathcal{H}^0$.

We have remarked that the rates of decrease predicted for the error—h^2 for displacement and h for strain—have been repeatedly confirmed by numerical experiment. Some experimenters have calculated only the errors at individual mesh points, instead of the mean-square errors over the interval, again with the same rates of convergence. [To predict these pointwise errors we must either return to the maximum principle, or assume more mean-square continuity of the data and refine the variational estimate. In a few important problems the Ritz solution is actually more accurate at the nodal points than elsewhere; for $-u'' = f$, $u(0) = u(\pi) = 0$, u^h agrees identically with u at the nodes, and the accuracy is infinite.] There have been a few cases, however, in which the expected convergence has *not* been confirmed, because of a simple flaw in the experiment: The computer has worked with the quantity

$$E^h = \max_j |e^h(jh)|.$$

This expression introduces, with every decrease in h, a new and larger set of mesh points. In particular, points nearer and nearer to the boundary, where the error is often greatest, will determine the computed value E^h. It is unreasonable to expect that this error, occurring at a point which varies with h, will still display the optimal h^2 rate of decrease.

Finally, there is the error which arises when the load f is replaced by its linear interpolate f_I. This change, which was introduced to simplify the integrations $\int f \varphi_j^h \, dx$, alters the load vector from F to $\tilde{F}$. It leads to approximate finite element solutions $\tilde{Q} = K^{-1}\tilde{F}$ and $\tilde{u}^h = \sum \tilde{Q}_j \varphi_j^h$, *which are the exact finite element approximations for the problem with data f_I.* Therefore, we have only to consider the change in the Ritz solution due to change in the data.

THEOREM 1.6

If f is replaced by its interpolate f_I, then the induced error $u^h - \tilde{u}^h$ in the finite element approximation satisfies

$$a(u^h - \tilde{u}^h, u^h - \tilde{u}^h) \leq \frac{K\rho^2}{\pi^4} h^4 \, \| f'' \|_0^2.$$

Proof. The exact solution $u - \tilde{u}$, corresponding to data $f - f_I$, is bounded by

(39)
$$\| u - \tilde{u} \|_2 \leq \rho \| f - f_I \|_0 \leq \frac{\rho}{\pi^2} h^2 \| f'' \|_0.$$

The last inequality gives the error in linear interpolation; it is copied from Theorem 1.3. Now it follows that

$$a(u - \tilde{u}, u - \tilde{u}) \leq K \| u - \tilde{u} \|_1^2 \leq K \| u - \tilde{u} \|_2^2 \geq \frac{K\rho^2}{\pi^4} h^4 \| f'' \|_0^2.$$

[We have not been able to use the strongest part of (39)—that even the second derivative of $u - \tilde{u}$ is $O(h^2)$.]

The proof is completed by applying the corollary to Theorem 1.1: $u^h - \tilde{u}^h$ is the projection onto S^h of $u - \tilde{u}$, and this projection cannot increase energy,

$$
\begin{aligned}
a(u^h - \tilde{u}^h, u^h - \tilde{u}^h) &= a(u - \tilde{u}, u - \tilde{u}) \\
&\quad - a((u - \tilde{u}) - (u^h - \tilde{u}^h), (u - \tilde{u}) - (u^h - \tilde{u}^h)) \\
&\leq a(u - \tilde{u}, u - \tilde{u}) \leq \frac{K\rho^2}{\pi^4} h^4 \| f'' \|_0^2.
\end{aligned}
$$

We conclude that the error due to interpolating the data is smaller (h^4 in the sense of energy) than the h^2 error inherent in basing the Ritz approximation on linear elements.

1.7. THE FINITE ELEMENT METHOD IN ONE DIMENSION

This section extends the previous ones in three directions—it introduces inhomogeneous boundary conditions, elements which are quadratic or cubic rather than linear, and differential equations of order 4 as well as 2. The error estimates for different finite elements will be stated rather than proved, since they later fall within the theory which is our main object to describe. The steps in the method itself are exactly the same as before: the variational statement of the problem, the construction of a piecewise polynomial subspace of the admissible space, and the assembly and solution of the linear equations $KQ = F$. The pattern will therefore be more or less complete, in one dimension.

We begin by retaining the same differential equation $-(pu')' + qu = f$, but with the more general boundary conditions

$$
u(0) = g, \qquad u'(\pi) + \alpha u(\pi) = b.
$$

The first of these conditions is again essential, and must be satisfied by every function v in the admissible space $\mathcal{H}_E^1$. Therefore, *the difference $v_0 = v_1 - v_2$ between any two admissible functions will satisfy the homogeneous condition* $v_0(0) = 0$. We denote by V_0 the space of these differences v_0; it was the admissible space when the essential conditions were homogeneous.

The boundary condition at the other end is of a new kind, involving both u' and u, and therefore the functional $I(v)$ must be recomputed. Physically, the system represents a string which is neither fixed nor completely free at $x = \pi$; instead it is connected to a spring. The new functional is

$$
I(v) = \int_0^\pi \left(p(v')^2 + qv^2 \right) dx + \alpha p(\pi) v^2(\pi) - 2 \int_0^\pi fv \, dx - 2bp(\pi)v(\pi).
$$

Thus the new condition has introduced a boundary term both in the linear part of the functional and in the energy

$$a(v, v) = \int \left(p(v')^2 + qv^2 \right) dx + \alpha p(\pi)v^2(\pi).$$

The last term represents the energy in the spring.

We now verify that the vanishing of the first variation in every direction v_0 leads to the same conditions on the minimizing u as the differential equation and boundary conditions. [Note that u is perturbed by functions v_0 in V_0, which assures that the essential condition $(u + \epsilon v_0)(0) = g$ is satisfied. The perturbations are *not* the functions v in $\mathcal{JC}_E^1$.] The coefficient of 2ϵ in $I(u + \epsilon v_0)$ is

$$\int [pu'v_0' + quv_0] + \alpha p(\pi)u(\pi)v_0(\pi) - \int fv_0 - bp(\pi)v_0(\pi)$$

$$= \int [-(pu')' + qu - f]v_0 + p(\pi)[u'(\pi) + \alpha u(\pi) - b]v_0(\pi).$$

This expression vanishes for all v_0 if and only if u satisfies the differential equation and the new boundary condition. *Therefore, this boundary condition is natural for the modified functional $I(v)$.*

In the general Ritz method, it no longer makes sense to ask that S^h be a subspace of $\mathcal{JC}_E^1$. $\mathcal{JC}_E^1$ itself is not a vector space; it has been shifted away from the origin. Therefore, we ask S^h to have the same form. *The trial functions v^h need not lie in the admissible space $\mathcal{JC}_E^1$, but the difference of any trial functions must be in the homogeneous space V_0.* These differences $v_0^h = v_1^h - v_2^h$ form a finite-dimensional space S_0^h, which is required to be a subspace of V_0.

For linear finite elements, the procedure is clear. There is no constraint at $x = \pi$, where the boundary condition is natural. The trial space S^h will therefore consist of *all piecewise linear functions which satisfy $v^h(0) = g$.* (This essential condition can be imposed exactly in one dimension, since it constrains v^h only at a point. In two or more dimensions, a boundary condition $v^h(x, y) = g(x, y)$ cannot be satisfied by a polynomial, and S^h is not contained in $\mathcal{JC}_E^1$.) S_0^h is the same piecewise linear trial space introduced in previous sections, vanishing at $x = 0$, and every v^h can be expanded as

$$v^h(x) = g\varphi_0^h(x) + \sum_1^N q_j\varphi_j^h(x).$$

Thus the coefficient of φ_0^h is fixed: $q_0 = g$.

The potential energy $I(v^h)$ is a quadratic in the unknowns $q_1, \ldots, q_N$, and its minimization leads again to a linear system $KQ = F$. In the interior of the interval—that is, for all but the first and last rows of the matrix—this system will be identical to the one in the previous sections. The first row of the matrix, corresponding to the left end of the interval, looks just like every other row except that $q_0 = g$.† Therefore the first equation in the system (with coefficients taken to be $p = q = 1$) is

$$\frac{-g + 2Q_1 - Q_2}{h} + \frac{h}{6}(g + 4Q_1 + Q_2) = \int_0^{2h} f\varphi_1^h \, dx.$$

Shifting the terms involving g to the other side, the first row of K is exactly as before. The inhomogeneous condition has altered only the first component F_1 of the load vector, by the amount $g/h - gh/6$.

At the other end the computation is almost as simple. The new terms in $I(v^h)$ are

$$\alpha p(\pi)\left(v^h(\pi)\right)^2 - 2bp(\pi)v^h(\pi) = \alpha p(\pi)q_N^2 - 2bp(\pi)q_N.$$

Therefore, in the last equation $\partial I/\partial q_N = 0$, after deleting the factor 2, there is an extra $bp(\pi)$ in the load component F_N and an extra $\alpha p(\pi)$ in the stiffness entry K_{NN}. Again the local nature of the basis was responsible for the simplicity of this change; the only basis function which is nonzero at $x = \pi$, and is therefore coupled to the boundary condition, is the last one.

The error estimates for this problem will be the same as in the previous section, namely

$$a(u - u^h, u - u^h) = O(h^2)$$

and

$$\|u - u^h\|_0 = O(h^2).$$

The first estimate again depends on the variational theorem: u is closer to u^h than it is to u_I. In turn, this depends on the presence of the linear interpolate u_I in the trial space S^h. Therefore, we may again appeal to the approximation Theorem 1.3 for the distance between a function and its interpolate.

The second extension of the method is the introduction of elements which are more "refined" than the piecewise linear functions. A given $u(x)$ is better approximated by quadratic or cubic than by linear interpolation, and there will be a corresponding improvement in the accuracy of u^h. Therefore, it is natural to construct trial spaces S^h which are composed of polynomials of higher degree.

To start, let S^h consist of all piecewise quadratic functions which are continuous at the nodes $x = jh$ and satisfy $v^h(0) = g$. Our first object is to

---

†This corresponds to the way in which essential boundary conditions are introduced in practice. They are ignored until the matrices are assembled, and only then is the unknown (in this case q_0) given the value which is prescribed by the boundary condition.

compute the dimension of S^h (the number of free parameters q_j) and to determine a basis. Note that as x passes a node, continuity imposes only one constraint on the parabola which begins at that node; two parameters of the parabola remain free. Therefore, the dimension must be twice the number of parabolas, or $2N$.

A basis can be constructed by introducing the midpoints $x = (j - \frac{1}{2})h$ as nodes, in addition to the endpoints $x = jh$. There are then $2N$ nodes, since $x = 0$ is excluded and $Nh = \pi$; we shall denote them by z_j, $j = 1, \ldots, 2N$. To each node there corresponds a continuous piecewise quadratic which equals one at z_j and zero at z_i, $i \neq j$;

(40) $$\varphi_j(z_i) = \delta_{ij}.$$

These functions are of two kinds, depending on whether z_j is an endpoint (Fig. 1.6a) or a midpoint (Fig. 1.6b). Notice that both are continuous and

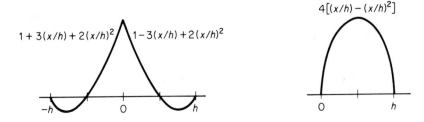

Fig. 1.6   The basis functions for piecewise quadratic elements.

therefore in $\mathcal{H}_E^1$. The function φ^2, which is nonzero only over one subinterval, is not intrinsically determined by the subspace; the internal node could have been chosen anywhere in the interval. The choice of the midpoint affected the basis but not the space itself.

The element stiffness matrix k_1, corresponding to the integral of $(v')^2$ over $0 \leq x \leq h$, is computed from the quadratic which equals q_0 at $x = 0$, $q_{1/2}$ at the midpoint $x = h/2$, and q_1 at $x = h$. This quadratic is

$$v^h(x) = q_0 + \frac{x}{h}(4q_{1/2} - q_1 - 3q_0) + \left(\frac{x}{h}\right)^2(2q_1 + 2q_0 - 4q_{1/2}).$$

To relate this to the basis functions, collect the coefficient of each q:

$$v^h(x) = q_0\left(1 - 3\frac{x}{h} + 2\frac{x^2}{h^2}\right) + q_{1/2}\left(4\frac{x}{h} - 4\frac{x^2}{h^2}\right) + q_1\left(-\frac{x}{h} + 2\frac{x^2}{h^2}\right).$$

These three coefficients are exactly the three parabolas in Fig. 1.6. The coefficient of q_0 is the right half of φ^1 in Fig. 1.6a—a parabola equal to 1 at $x = 0$ and to zero at $x = h/2$, $x = h$. The coefficient of $q_{1/2}$ is the parabola φ^2 in Fig. 1.6b, and the coefficient of q_1 is the first half of φ^1.

The element matrix k_1 is computed by integrating $(dv^h/dx)^2$ and writing the result as $(q_0 q_{1/2} q_1)^T k_1 (q_0 q_{1/2} q_1)$. Notice that k_1 is a 3×3 matrix, since three of the parameters q appear in any given interval; a parabola is determined by three conditions. In the calculation it is every man for himself, and we give only the final result:

$$k_1 = \frac{1}{3h} \begin{pmatrix} 7 & -8 & 1 \\ -8 & 16 & -8 \\ 1 & -8 & 7 \end{pmatrix}.$$

Notice that k_1 is singular; applied to the vector $(1, 1, 1)$ it yields zero. This vector $q_0 = 1, q_{1/2} = 1, q_1 = 1$ corresponds to a parabola v^h which is actually a horizontal line $v^h \equiv 1$, so its derivative is zero. This singularity of k_1 is a useful check; k_0 would not be singular.

These ideas extend directly to cubic elements. Continuity of the trial functions imposes one constraint at each of the nodes $x = jh$, leaving three free parameters in the cubic, and the dimension of S^h is $3N$. To construct a basis we place two nodes within each interval, say at a distance $h/3$ from the endpoints. Together with the endpoints this gives $3N$ nodes $z_j = jh/3$. The functions which satisfy $\varphi_j(z_i) = \delta_{ij}$ constitute a basis and are of three kinds (Fig. 1.7). The element matrices will be of order 4.

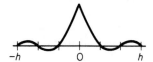

Fig. 1.7 Cubics which are only continuous at the nodes.

There is another cubic element which is better in almost every respect. It is constructed by imposing continuity not only on the function v^h but also on its first derivative. This means that the trial space is actually *a subspace of the preceding one*, with one new constraint at each of the $N - 1$ internal nodes $x = h, 2h, \ldots, \pi - h$. Therefore, the dimension of this new space will be $3N - (N - 1) = 2N + 1$, a reduction of one-third in the number of parameters to be calculated. The only way in which this new cubic space fails to improve on the old one is in the approximation of solutions u which do *not* have a continuous derivative. We saw in Section 1.3 that this is the case—u is in $\mathcal{JC}_E^1$ but not in $\mathcal{JC}_B^2$—for a point load or for a discontinuous coefficient $p(x)$ in the differential equation. In these cases of internal singularities it is essential not to require excess smoothness of the trial functions; the singularity x_0 should be chosen as a node, and across that node the cubic should only be continuous. This will preserve the order of convergence.

The placement of nodes for the smoother cubics becomes more interesting: There is a *double node* at each point $x = jh$. Instead of being determined by its values at the four distinct points 0, $h/3$, $2h/3$, and h, the cubic is now determined by its values and its first derivatives at the two endpoints, that is, by v_0, v_0', v_1, and v_1'. Both v_1 and v_1' will be shared by the cubic in the next subinterval, assuring continuity of v and v'. The basis functions are, therefore, of two kinds (Fig. 1.8). These basis functions have a double zero at the ends

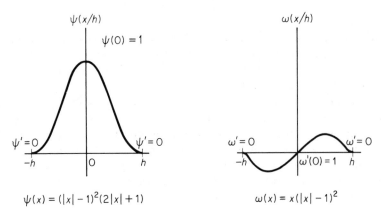

$$\psi(x) = (|x| - 1)^2 (2|x| + 1) \qquad \omega(x) = x(|x| - 1)^2$$

Fig. 1.8 Hermite cubics: v and v' continuous.

$(j \pm 1)h$. They are called *Hermite cubics*, since they interpolate both function value and derivative; the earlier examples might have been associated with the name of Lagrange.

The cubic polynomial on $0 \leq x \leq h$ which takes on the four prescribed values v_0, v_0', v_1, and v_1' is

$$(41) \quad v^h(x) = v_0 \psi\left(\frac{x}{h}\right) + v_0' \omega\left(\frac{x}{h}\right) + v_1 \psi\left(\frac{x-h}{h}\right) + v_1' \omega\left(\frac{x-h}{h}\right)$$

$$= v_0 + v_0' x + (3v_1 - 3v_0 - v_1' h - 2v_0' h)\frac{x^2}{h^2}$$

$$+ (2v_0 - 2v_1 + hv_1' + hv_0')\frac{x^3}{h^3}.$$

The element matrices, which are of order 4, are computed by the following integrations, where q is the column vector $(v_0, v_0', v_1, v_1')^T$:

Mass matrix k_0: $\displaystyle\int_0^h (v^h)^2 = q^T k_0 q,$

Stiffness matrix k_1: $\displaystyle\int_0^h ((v^h)')^2 = q^T k_1 q,$

Bending matrix k_2: $\displaystyle\int_0^h ((v^h)'')^2 = q^T k_2 q.$

Notice that because v^h is in $\mathcal{JC}^2$, it can be used in fourth-order problems; at that point the bending matrix will be needed.

There are several ways in which to organize the computation of the mass matrix k_0. One of the best is to form the matrix H which connects the four nodal parameters in the vector q to the four coefficients $A = (a_0, a_1, a_2, a_3)$ of the cubic polynomial v^h: $A = Hq$, or from (41),

$$
\begin{pmatrix} a_0 \\ a_1 \\ a_2 \\ a_3 \end{pmatrix}
=
\begin{pmatrix}
1 & 0 & 0 & 0 \\
0 & 1 & 0 & 0 \\
-\dfrac{3}{h^2} & -\dfrac{2}{h} & \dfrac{3}{h^2} & -\dfrac{1}{h} \\
\dfrac{2}{h^3} & \dfrac{1}{h^2} & -\dfrac{2}{h^3} & \dfrac{1}{h^2}
\end{pmatrix}
\begin{pmatrix} v_0 \\ v_0' \\ v_1 \\ v_1' \end{pmatrix}.
$$

The integration of $(v^h)^2 = (a_0 + a_1 x + a_2 x^2 + a_3 x^3)^2$ is completely trivial,

$$
\int_0^h (v^h)^2\, dx = (a_0 a_1 a_2 a_3)
\begin{pmatrix}
h & \dfrac{h^2}{2} & \dfrac{h^3}{3} & \dfrac{h^4}{4} \\
\dfrac{h^2}{2} & \dfrac{h^3}{3} & \dfrac{h^4}{4} & \dfrac{h^5}{5} \\
\dfrac{h^3}{3} & \dfrac{h^4}{4} & \dfrac{h^5}{5} & \dfrac{h^6}{6} \\
\dfrac{h^4}{4} & \dfrac{h^5}{5} & \dfrac{h^6}{6} & \dfrac{h^7}{7}
\end{pmatrix}
\begin{pmatrix} a_0 \\ a_1 \\ a_2 \\ a_3 \end{pmatrix}.
$$

Denoting this matrix by N_0, the result becomes

$$
\int_0^h (v^h)^2 = A^T N_0 A = q^T H^T N_0 H q.
$$

Therefore, the element mass matrix must be

$$
k_0 = H^T N_0 H.
$$

All this is easily programmed.

For the stiffness matrix, the link H between the nodal parameters q and the coefficient vector a is unchanged. The only difference now is that

$$\int_0^h ((v^h)')^2 \, dx = \int (a_1 + 2a_2 x + 3a_3 x^2)^2 = A^T N_1 A$$

$$= A^T \begin{pmatrix} 0 & 0 & 0 & 0 \\ 0 & h & h^2 & h^3 \\ 0 & h^2 & \dfrac{4h^3}{3} & \dfrac{3h^4}{2} \\ 0 & h^3 & \dfrac{3h^4}{2} & \dfrac{9h^5}{5} \end{pmatrix} A.$$

The element stiffness matrix is $k_1 = H^T N_1 H$. The results of these computations, and of a similar one for k_2, are the following matrices (which are to be completed by symmetry):

$$k_0 = \frac{h}{420} \begin{pmatrix} 156 & 22h & 54 & -13h \\ & 4h^2 & 13h & -3h^2 \\ & & 156 & -22h \\ & & & 4h^2 \end{pmatrix},$$

$$k_1 = \frac{1}{30h} \begin{pmatrix} 36 & 3h & -36 & 3h \\ & 4h^2 & -3h & -h^2 \\ & & 36 & -3h \\ & & & 4h^2 \end{pmatrix},$$

$$k_2 = \frac{1}{h^3} \begin{pmatrix} 12 & 6h & -12 & 6h \\ & 4h^2 & -6h & 2h^2 \\ & & 12 & -6h \\ & & & 4h^2 \end{pmatrix}.$$

The matrix k_0 is positive definite, whereas k_1 has a zero eigenvalue corresponding to the constant function $v^h \equiv 1$, that is, to $q = (1, 0, 1, 0)$. The matrix k_2 should be doubly singular, since every linear v^h will have $(v^h)'' \equiv 0$. The new null vector, corresponding to $v^h(x) \equiv x$, is $q = (0, 1, h, 1)$.

It is sometimes valuable to change k_1 and k_2 from the above *singular element matrices* to *natural element matrices*. This means that the matrices are made nonsingular by removing the *rigid body motions*, which are the cases $(v^h)' \equiv 0$ and $(v^h)'' \equiv 0$. The orders of the matrices are reduced to 3 and 2, respectively, and they now carry no redundant information; the fact that $(1, 0, 1, 0)$ is a null vector of k_1, which is automatic because it corresponds to a state with no strain, is taken for granted and not displayed. It appears that these "natural" matrices may make it simpler for a single program to accept a wide variety of elements, which is a tremendous asset; the proliferation of new elements, and of combinations between our present displacement method and force methods, has been overwhelming. In the book we keep the stiffness matrices in their singular form, because they display so clearly the role of all four nodal parameters v_0, v'_0, v_1, and v'_1.

For application to the problem $-(pu')' + qu = f$, these element matrices must be assembled into the global stiffness matrix K. Assuming constant coefficients, a typical row (or rather pair of rows, since there are two unknowns u_j and u'_j at the mesh point $x_j = jh$) of the assembled K is

$$\frac{p}{30h}(-36u_{j-1} - 3hu'_{j-1} + 72u_j - 36u_{j+1} + 3hu'_{j+1})$$

(42a)
$$+ \frac{qh}{420}(54u_{j-1} + 13hu'_{j-1} + 312u_j + 54u_{j-1} - 13hu'_{j-1})$$

$$= F_j = \int f(x)\psi\left(\frac{x}{h} - j\right),$$

$$\frac{p}{30}(3u_{j-1} - hu'_{j-1} + 8hu'_j - 3u_{j+1} - hu'_{j+1})$$

(42b)
$$+ \frac{qh^2}{420}(-13u_{j-1} - 3hu'_{j-1} + 8hu'_j + 13u_{j+1} - 3hu'_{j+1})$$

$$= F'_j = \int f(x)\omega\left(\frac{x}{h} - j\right).$$

It is interesting to make sense of this as a difference equation. Suppose $p = 1$, $q = 0$, and $f = 1$, so that the differential equation is $-u'' = 1$. The finite element equations above become

$$-\frac{6}{5}\frac{u_{j+1} - 2u_j + u_{j-1}}{h^2} + \frac{1}{5}\frac{u'_{j+1} - u'_{j-1}}{2h} = 1$$

$$-\frac{1}{5}\frac{u_{j+1} - u_{j-1}}{2h} + \frac{1}{5}u'_j - \frac{1}{30}(u'_{j+1} - 2u'_j + u'_{j-1}) = 0.$$

After Taylor expansions, the first equation is consistent with $-u'' = 1$, and the second with $-h^2u'''/15 = 0$, which can be derived by differentiating the first. *This is exactly the point at which finite element equations have contributed a new and valuable idea to the established finite difference technique.* Instead of operating only with the unknown u_j, so that there is always one equation per mesh point, the finite element difference equations may couple both displacements and slopes as unknowns, the equation for the slope being formally consistent with the original equation differentiated once. The effect is that high accuracy can be achieved, and derivatives of high order can be approximated, *without abandoning the strictly local nature of the difference equation.* The more irregular the mesh and the more curved the boundaries, the more important this innovation becomes. It should be taken seriously by the developers of difference schemes, since there—without being limited by having to derive discrete analogues from polynomial displacements and the Ritz method—even greater efficiency might be possible.

Suppose we apply Taylor expansions to test the order of accuracy of the Hermite difference equations (42). Starting with the assumptions

$$(43) \qquad v_j = u(jh) + \sum h^n e_n(jh), \qquad v'_j = u'(jh) + \sum h^n \epsilon_n(jh),$$

and expanding $v_{j\pm 1}$ and $v'_{j\pm 1}$ about the central point jh, it is possible to use the original differential equation and its differentiated forms to check that $\epsilon_n = e'_n$, and that these terms disappear for $n = 1, 2, 3$. In other words, the coupled difference equation is *fourth-order accurate*. This matches exactly the error estimates to be derived variationally, just as both the Taylor series and the variational bounds on $u - u_I$ were $O(h^2)$ in the linear case. However, an unanticipated and slightly distressing point emerged in [M8]: The asymptotic expansions (43) are spoiled by a mismatch at the boundaries, and the finite element errors seem to be less systematic as $h \longrightarrow 0$ than a simple power series in h. The errors are, however, definitely of order h^4.

There is still another important cubic space, formed from those functions for which even the *second* derivative is continuous at the nodes. A piecewise cubic which has continuous second derivatives is called a *cubic spline*. This space is again a subspace of the preceding one, the Hermite cubics, with one new constraint at each of the $N - 1$ internal nodes. Therefore, the dimension of the spline subspace is $3N - 2(N - 1) = N + 2$. This means that there is *one unknown per mesh point*, including the extra points $x_0 = 0$ (where $v_0 = 0$ but the slope v'_0 may be regarded as the free parameter) and $x_{N+1} = \pi + h$ (or we may prefer that the last parameter be v'_N). At internal mesh points the displacement v_j is the unknown, and the finite element equations will once more look exactly like conventional finite difference equations.

It is no longer obvious which four nodal parameters determine the shape of the cubic over a given subinterval, say $(j - 1)h \leq x \leq jh$. The nodes x_{j-1} and x_j, which belong to the interval, account for only two conditions, and the other two must come from outside. Therefore, *cubic splines cannot have a strictly local basis*, and the shape of v^h within an element must be affected by displacements outside the element. In fact, the spline which vanishes at all nodes except one (a cardinal spline) will be nonzero over *every* subinterval between nodes.

To compute with these splines, we need to construct the one which equals 1 at the origin and which drops to zero as rapidly as possible (Fig. 1.9). This function is known as a basic spline, or *B-spline*. It is fundamental to the whole theory of splines, and was among the many remarkable discoveries of I. J.

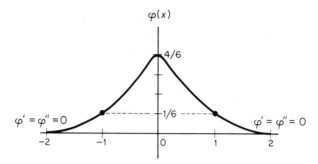

Fig. 1.9 The cubic B-spline: continuous second derivatives at the nodes.

Schoenberg. In particular, Schoenberg has proved that every cubic spline on $[0, \pi]$ can be written as a linear combination of B-splines:

$$v^h(x) = \sum_{-1}^{N+1} q_j \varphi_j^h(x).$$

These basis functions φ_j^h are formed from the B-spline in the figure by re-scaling the independent variable from x to x/h, and translating the origin to lie at the node jh:

$$\varphi_j^h(x) = \varphi\left(\frac{x}{h} - j\right).$$

In the expression for v^h, the constraint that it vanish at the origin has not been included. If φ^h and φ_{**}^h are combinations of φ_0^h and φ_1^h which satisfy this essential condition, the admissible cubic splines can be characterized exactly as

$$v^h(x) = q_* \varphi_*^h + q_{**} \varphi_{**}^h + \sum_2^{N+1} q_j \varphi_j^h(x).$$

In this form the $N + 2$ unknowns in the finite element system $KQ = F$ would be $q_*, \ldots, q_{N+1}$. The unknown q_{N+1} appears because the natural boundary condition imposes no constraint in the variational formulation; it would be interesting to see the effect on u^h of satisfying this condition and thereby removing the last unknown. Presumably this effect is exponentially small.

Two final remarks on splines:

1. It may be that their greatest importance comes in *approximation rather than minimization*. A given relationship between data is well and conveniently approximated by a spline, while the search for an *unknown* spline by minimization is less convenient.

2. With unequally spaced nodes they are still nonzero over four intervals (the minimum for a cubic spline), and if every node x_{2N-1} approaches x_{2N}, the B-splines degenerate into the Hermite basis functions ψ and ω.

It is easy to summarize the approximation properties of these more refined spaces. If the polynomials are of degree $k - 1$ ($k = 3$ for quadratics, $k = 4$ for cubics), then any smooth function u differs from its interpolate by

$$\|u - u_I\|_0 \leq Ch^k \|u^{(k)}\|_0.$$

The errors in the derivatives lose one power of h for every differentiation:

(44) $$\|u - u_I\|_s \leq C_s h^{k-s} \|u^{(k)}\|_0.$$

This estimate makes sense only if u_I is known to possess s derivatives, in other words, only if the piecewise polynomials lie in $\mathcal{H}^s$. Therefore, there is a limit $s \leq q$ on the inequality (44): $q = 1$ for the $\mathcal{C}^0$ quadratics and cubics, which are in $\mathcal{H}^1$, $q = 2$ for Hermite cubics, and $q = 3$ for splines. For $s > q$ the estimate can still be proved element by element, but δ-functions appear at the nodes.

These approximation results lead to the expected rates of convergence for the finite element method, provided that the derivative $u^{(k)}$ has finite energy: the slopes are in error by $O(h^{k-1})$, the strain energy by $O(h^{2(k-1)})$, and the displacements $u - u^h$ by $O(h^k)$. Since these rates are confirmed numerically, there is good reason to compute with refined finite elements.

The final step in this section is to admit differential equations of fourth order,

(45) $$Lu = (ru'')'' - (pu')' + qu = f.$$

The operator L is still formally self-adjoint, since u''' and u' are not allowed to occur by themselves, and it is positive definite if $r \geq r_{\min} > 0$, $p \geq 0$, $q \geq 0$.

The associated energy inner product is

$$a(u, v) = \int (ru''v'' + pu'v' + quv) \, dx,$$

and the Euler equation for the minimization of $I(v) = a(v, v) - 2(f, v)$ is $Lu = f$. For the Ritz method to apply, the trial functions v^h must have finite energy, which means that v^h must lie in $\mathcal{H}^2$. The Hermite and spline cubics are therefore applicable, but not the polynomials which are only continuous. They are "nonconforming," and to use them by ignoring the δ-functions in their second derivatives at the nodes would in one dimension guarantee disaster.

Equation (45) represents the bending of a beam. If it is clamped at $x = 0$, the boundary conditions at that end are

$$u(0) = u'(0) = 0,$$

and these are *essential:* Every trial function must have a double zero at the origin. To see which are the natural conditions at $x = \pi$, we integrate by parts in the equation $a(u, v) = (f, v)$ for the vanishing of the first variation. The result is that for every v in the admissible space $\mathcal{K}_E^2$,

$$(46) \quad \int [(ru'')'' - (pu')' + qu - f]v + ru''v'\,|_\pi + \left(pu' - (ru'')'\right)v\,|_\pi = 0.$$

Thus if no conditions are imposed on v at $x = \pi$, the natural boundary conditions on u are those which physically correspond to a free (or cantilevered) end:

$$u''(\pi) = 0, \qquad (pu' - ru''')(\pi) = 0.$$

There is an additional case of great importance, in which the beam is simply supported: $u(\pi) = 0$. This becomes an essential condition, which the trial functions v must satisfy. Therefore, the last term in the first variation (46) is automatically zero, with no condition on u at π. However, the other integrated term is zero for all v only if $u''(\pi) = 0$, and this natural condition must remain. Thus a simply supported end combines one essential and one natural boundary condition:

$$u(\pi) = u''(\pi) = 0.$$

Other combinations of boundary conditions could be imagined, such as $u'(\pi) = u''(\pi) = 0$. But these are weird, both physically and variationally.

1.8. BOUNDARY-VALUE PROBLEMS IN TWO DIMENSIONS

This section introduces several problems in the plane, or rather in a plane domain Ω bounded by a smooth curve Γ. Our purpose is first of all to match the differential forms of these problems, in which the Laplace operator Δ and the biharmonic operator Δ^2 appear, with the equivalent variational forms. This means that in the variational statement we have to identify the admissible spaces in which the solution is sought. Naturally these spaces depend on the boundary conditions, and as in the two-point boundary-value problem, there will be a distinction between Dirichlet conditions (essential conditions) and Neumann conditions (natural conditions). The examples are extremely conventional, but they are the simplest models for plane stress and plate bending, and it seems worthwhile to illustrate once more the basic ideas:

1. The equivalence of differential and variational problems, the admissible space for the latter being constructed by completion in the energy norm.
2. The vanishing of the first variation, giving the weak form $a(u, v) = (f, v)$ of the equation, and leading to Galerkin's method.
3. The Ritz process of minimization over a subspace.

Then the next section introduces the finite element method in earnest, including many of the most important choices of piecewise-polynomial "elements."

The requirement that the boundary curve Γ be smooth introduces one difficulty while preventing another. On the one hand, it is certain that the interior Ω cannot be subdivided into polygons, say triangles, without leaving a "skin" around the boundary. The approximation theory in Chapter 3 will reflect this difficulty. On the other hand, the smoothness of the boundary makes it reasonable to suppose that the solution is itself smooth. This property follows rigorously from the theory of elliptic boundary-value problems, if the coefficients in the equation and the data are also smooth.

In contrast, consider the problem $u_{xx} + u_{yy} = 1$ in a polygon, with $u = 0$ on the boundary. For the unit square, the solution behaves radially like $r^2 \log r$ near a corner, and the second derivatives break down. (Some breakdown is obvious, since at the corner u_{xx} and u_{yy} both vanish, while their sum is one just inside.) In this problem u does have second derivatives in the mean-square sense; u lies in $\mathcal{3C}^2$ but not in $\mathcal{3C}^3$. This can be verified directly by constructing u as a Fourier series. Therefore, the error estimates for piecewise linear approximation will apply to this problem, but the accuracy associated with higher-order elements cannot be attained without either refining the mesh at the corners, or introducing special trial functions which possess the same kind of singularity as the solution u. For a nonconvex polygon, for example an L-shaped region, u fails even to have second derivatives in a mean-square sense. The solution must lie in $\mathcal{3C}^1$, which is the admissible space (or rather $\mathcal{3C}^1$ *contains* the admissible space, depending on the boundary conditions), but since u behaves like $r^{2/3}$ at the interior angle of the L, it is excluded from $\mathcal{3C}^2$. Such singularities, arising when Γ is not smooth, are studied in Chapter 8.

We begin this section by studying the Dirichlet problem, governed inside the domain by Poisson's equation

$$-\Delta u = f \qquad \text{in } \Omega,$$

and on the boundary by the Dirichlet condition

$$u = 0 \qquad \text{on } \Gamma.$$

The sign in the differential equation is chosen so that the operator

$$L = -\Delta = -\frac{\partial^2}{\partial x^2} - \frac{\partial^2}{\partial y^2}$$

will be associated with a positive, rather than a negative, quadratic form (Lu, u).

As in the one-dimensional example, we need to prescribe some norm for the inhomogeneous term, in other words to choose some set of data f for which the Dirichlet problem is to be solved. Our choice of norm will be the same as before:

$$\| f \|_0 = \left(\iint_\Omega | f(x, y) |^2 \, dx dy \right)^{1/2}$$

If this norm is finite, then f belongs to the data space $\mathcal{IC}^0(\Omega)$. As in one-dimensional problems, this space includes any piecewise continuous f and excludes δ-functions.

As possible solutions to the differential equation we admit all functions u which vanish on the boundary Γ and have derivatives up to second order in Ω. The natural norm for this solution space is

$$\| u \|_2 = \left[\iint_\Omega (u^2 + u_x^2 + u_y^2 + u_{xx}^2 + u_{xy}^2 + u_{yy}^2) \, dx dy \right]^{1/2}.$$

The space of all functions for which *this* norm is finite—functions whose second derivatives have finite energy—is $\mathcal{IC}^2(\Omega)$. The solution space $\mathcal{IC}_B^2$ for the Dirichlet problem is then a subspace of $\mathcal{IC}^2$, determined by the boundary condition $u = 0$ on Γ.

It should be clear from the definitions of the norms that the Laplace operator $L = -\Delta$ is a bounded operator from $\mathcal{IC}_B^2$ to $\mathcal{IC}^0$:

$$\| Lu \|_0 \leq K \| u \|_2.$$

The crucial point of the Dirichlet theory goes in the opposite direction: that the inverse of L—which is given by the Green's function of the problem—yields a solution u which depends continuously on the data f. This means that *to every f there corresponds one and only one u, and for some constant p,*

(47) $$\| u \|_2 \leq p \| f \|_0.$$

The construction of this solution could be attacked directly by finite difference methods, most commonly by the *five-point* difference equation

(48) $$\frac{-U_{i+1, j} + 2U_{i, j} - U_{i-1, j}}{(\Delta x)^2} + \frac{-U_{i, j+1} + 2U_{i, j} - U_{i, j-1}}{(\Delta y)^2} = f_{i, j}.$$

Near the boundary this equation must be modified, but it is possible to preserve both the second-order accuracy of the scheme and the discrete maximum principle which is obvious from (48): If $f = 0$, then $U_{i,j}$ cannot exceed
all four of the values $U_{i\pm1,j\pm1}$. (We do not know whether there is a theoretical
limit to the order of accuracy of schemes satisfying a maximum principle.
It is certain that as the accuracy increases, the boundary conditions on the
difference equation will become extremely complicated.)

Without going into details, we note that the equation implicitly requires
f to be continuous, in order for $f_{i,j} = f(i\,\Delta x, j\,\Delta y)$ to be well defined. When
f is less smooth, some averaging process (which will be built into the variational methods!) may be applied. The best estimate for the five-point scheme
in a square seems to be

$$\max_{i,j} |U_{i,j} - u_{i,j}| \leq Ch^2 |\ln h| \max |f|.$$

Here $h = \Delta x = \Delta y$. The extra factor $|\ln h|$, which was not needed in one
dimension, is required by the $r^2 \log r$ behavior at a corner.

For us the important formulation of Dirichlet's problem is the variational
one: *the admissible functions v vanish on the boundary Γ, and the solution u
is the one which minimizes the quadratic functional*

$$I(v) = \iint_\Omega (v_x^2 + v_y^2 - 2fv)\,dxdy.$$

The first step is to check that a solution u to the differential problem does
minimize I. Perturbing u in any direction v,

$$I(u + \epsilon v) = I(u) + 2\epsilon \iint (u_x v_x + u_y v_y - fv) + \epsilon^2 \iint (v_x^2 + v_y^2).$$

We have only to show that the coefficient of ϵ is zero. Then since the coefficient of ϵ^2 is positive unless v is constant, which would imply by the boundary
condition that $v = 0$, u will be the unique function which minimizes I. The
vanishing of the coefficient of ϵ, that is, of the first variation, means that

(49) $$\iint u_x v_x + u_y v_y = \iint fv \qquad \text{for all admissible } v.$$

To prove that the solution to the Dirichlet problem satisfies this identity,
we apply Green's theorem; or in other words, we integrate by parts over Ω:

(50) $$-\iint_\Omega (u_{xx} + u_{yy} + f)v\,dxdy + \int_\Gamma u_n v\,ds = 0,$$

where u_n is the derivative in the direction of the outward normal. Since $v = 0$

on Γ, and u satisfies Poisson's equation in Ω, the first variation vanishes. Therefore, u is minimizing.

Of course, this process works equally well in the opposite direction. The Poisson equation $-\Delta u = f$ is put into its *weak form* by multiplying by any v which vanishes on the boundary, integrating over Ω, and transforming the left side by Green's theorem. The weak form is exactly the equation (49). It is to this form that *Galerkin's method* applies; this equation is imposed on a subspace S^h and the solution u^h is also in S^h. As we see, when the form $a(v, v)$ is self-adjoint and positive definite, *the Ritz minimization and Galerkin's vanishing of the first variation are equivalent*. The weak equation $a(u, v) = (f, v)$ retains its meaning, however, even without self-adjointness.

This argument depended on the theory of elliptic equations to produce u. Suppose, instead, that we begin with the quadratic $I(v)$ and try to find its minimum directly. Then the first task is to decide exactly which functions v are admissible in the minimization.

The reasoning runs parallel to the one-dimensional example of Section 1.5. Any smooth v which vanishes on Γ must be made admissible, since if f should happen to be $-\Delta v$, this function v would be the minimizing solution we want. On the other hand, the requirement that v be smooth, say that v lie in $\mathcal{3C}_B^2$, is more than is necessary for $I(v)$ to be defined. Therefore, we complete the admissible space: if $a(v - v_N, v - v_N) \to 0$ for some sequence v_N in $\mathcal{3C}_B^2$, then v will also be admissible. This completion process leads to the class which is intuitively natural: v must vanish on Γ (the Dirichlet condition is an essential one), but it has only to possess *first* derivatives in the mean-square sense. In other words, the quadratic term $a(v, v)$ in the functional $I(v)$ should make sense,

$$(51) \qquad a(v, v) = \iint (v_x^2 + v_y^2)\, dx\, dy < \infty.$$

The functions satisfying (51) lie in $\mathcal{3C}^1(\Omega)$; the subspace which satisfies also the Dirichlet condition $v = 0$ on Γ is denoted by $\mathcal{3C}_0^1(\Omega)$. (Our own notation would be $\mathcal{3C}_E^1$, but in the special case of the Dirichlet problem we accept $\mathcal{3C}_0^1$; the notation $\overset{0}{\mathcal{3C}}{}^1$ also appears in the literature.) *This is the space of admissible functions for the Dirichlet problem.*

It is important to emphasize that the boundary condition $v = 0$ was retained for the whole admissible space, not by fiat, but only because it was stubborn enough to remain valid in the limit. In other words, the Dirichlet condition $v = 0$ is *stable* in the $\mathcal{3C}^1$ norm: If the sequence v_N vanishes on Γ and converges in the natural strain energy norm to v, then the limit v also vanishes on Γ.

This stability of the boundary condition is not present for the equation

$$(52) \qquad -\Delta u + qu = f \qquad \text{in } \Omega$$

with a natural boundary condition

$$u_n = 0 \qquad \text{on } \Gamma.$$

The functional for this *Neumann problem* is

$$I(v) = a(v, v) - 2(f, v) = \iint (v_x^2 + v_y^2 + qv^2 - 2fv)\, dxdy.$$

The differential equation looks for a solution u in $\mathcal{K}_B^2$, that is, possessing two derivatives and satisfying the Neumann condition $u_n = 0$. However, every function v in $\mathcal{K}^1$ is the limit of a sequence taken from $\mathcal{K}_B^2$; $\mathcal{K}_B^2$ is *dense* in $\mathcal{K}^1$. After completion, therefore, *the admissible space for the Neumann variational problem is the whole space* $\mathcal{K}^1(\Omega)$. As a result, no boundary conditions are imposed on the trial functions v^h in the Ritz method; any subspace $S^h \subset \mathcal{K}^1$ is acceptable. In practical applications of the finite element method, this means that the values of v^h at boundary points are not constrained, yielding a modest simplification in comparison with the Dirichlet problem. (In fact, the increased numerical instability due to natural boundary conditions leads us to doubt that the advantages are entirely on the side of Neumann.)

Of course the minimizing function u (not the approximate minima u^h) must somehow automatically satisfy the Neumann condition when enough smoothness is present. This is confirmed by equation (50) for the first variation, which now vanishes for all v in the admissible space $\mathcal{K}^1(\Omega)$. In the first place, $w = u_{xx} + u_{yy} + f$ must vanish throughout Ω, since v may be chosen equal to w in any small circle inside Γ and zero elsewhere, giving $\iint w^2 = 0$. Once $w = 0$ is established, it follows easily that $u_n = 0$ on the boundary. Therefore, the Neumann condition holds for u, even though it is not imposed on all admissible v.

If $q = 0$ in (52), then obviously no solution u is unique; $u + c$ will be a solution for any constant c. This single degree of freedom suggests, if Fredholm's theorem of the alternative is to hold, that there must be a single constraint on the data f. Integrating both sides of the differential equation $-\Delta u = f$ over Ω, the left side vanishes by Green's formula (50) with $v = 1$. Therefore, the constraint on f is: No solution to the Neumann problem with $q = 0$ can exist unless $\iint f\, dx\, dy = 0$. This nonexistence can be verified explicitly for the one-dimensional Neumann problem

$$u'' = 2, \qquad u'(0) = u'(1) = 0.$$

The parabola $u = x^2 + Ax + B$, which gives the general solution to $u'' = 2$, cannot be made to satisfy the boundary conditions. Correspondingly,

$$\int u'' = u'(1) - u'(0) = 0 \neq \int 2.$$

In contrast, the solution to the Dirichlet problem is unique.

This difference between the Dirichlet and Neumann conditions must appear also in the theory, at the point of verifying that the energy norm $\sqrt{a(v, v)}$ is equivalent to the standard norm $\|v\|_1$ on the admissible space. Thus the crucial question for existence and uniqueness is whether or not the problem is *elliptic*; is there a constant $\sigma > 0$ such that

(53) $$a(v, v) \geq \sigma \|v\|_1^2 \qquad \text{for all admissible } v?$$

For the equation $-\Delta u + qu = f$, this is the same as

$$\iint v_x^2 + v_y^2 + qv^2 \geq \sigma \iint v_x^2 + v_y^2 + v^2.$$

With $q > 0$ this ellipticity is obvious, and there is a unique minimizing u (or u^h, in the Ritz method). With $q = 0$ ellipticity fails for the Neumann problem; if $v = 1$, the left side is zero and the right side is not. In the Dirichlet problem, the rigid body motion $v = 1$ is not admissible, and an inequality of Poincaré type does hold:

$$\iint v_x^2 + v_y^2 \geq \sigma' \iint v^2 \qquad \text{for } v \text{ in } \mathfrak{IC}_0^1.$$

Thus, ellipticity holds even for $q = 0$ in the Dirichlet problem. In fact, ellipticity simply means that q exceeds $\lambda_{\max}$, the largest eigenvalue of the Laplace operator Δ. In the Neumann case $\lambda_{\max} = 0$, with $v = 1$ as eigenfunction, and $q = 0$ fails to achieve ellipticity. In the Dirichlet case $\lambda_{\max} < 0$, and the problem would even remain elliptic for a range of negative values of q.

It is, of course, possible to impose $u = 0$ on only a part of the boundary, say Γ_1, and to have $u_n = 0$ on $\Gamma_2 = \Gamma - \Gamma_1$. The admissible space for this *mixed problem* contains those v in $\mathfrak{IC}^1$ which vanish on Γ_1, and the solution will have a singularity at the junction between Γ_1 and Γ_2.

A further possibility is that the boundary condition express the vanishing of an *oblique derivative*:

$$u_n + c(x, y)u_s = 0 \qquad \text{on } \Gamma,$$

where u_s is the tangential derivative. This is the *natural* boundary condition associated with the inner product

$$a(u, v) = \iint (u_x v_x + u_y v_y + cu_x v_y - cu_y v_x + c_y u_x v - c_x u_y v).$$

This inner product yields Poisson's equation just as surely as the integral of $u_x v_x + u_y v_y$. In fact, Green's identity transforms $a(u, v) = (f, v)$ into

$$-\iint (\Delta u + f)v + \int (u_n + cu_s)v \, ds = 0.$$

This illustrates that *there is no unique way of integrating* (Lv, v) *by parts to obtain an energy norm* $a(v, v)$. Different manipulations of (Lv, v) lead to different forms for $a(v, v)$ and to correspondingly different natural boundary conditions. This is illustrated again below, where Poisson's ratio enters the boundary condition and the energy form $a(v, v)$ but not the operator $L = \Delta^2$.

Inhomogeneous boundary conditions for the equation $-\Delta u = f$ are of two kinds, either a prescribed displacement

$$u = g(x, y) \qquad \text{on } \Gamma$$

or a prescribed traction, which can be included in the general case of Newton's boundary condition

$$u_n + d(x, y)u = b(x, y) \qquad \text{on } \Gamma.$$

Variationally these two are completely different. The first is an inhomogeneous Dirichlet condition and must be imposed on the trial functions; the solution u minimizes $\iint v_x^2 + v_y^2 - 2fv$ over the class of functions v in $\mathcal{K}^1$ satisfying $v = g$ on the boundary. Notice that this admissible class $\mathcal{K}_E^1$ is no longer a "space" of functions; the sum of two admissible functions would equal $2g$ on the boundary and would be inadmissible. However, the *difference* of any two admissible functions vanishes on the boundary and lies exactly in the space $\mathcal{K}_0^1$. The easiest description of $\mathcal{K}_E^1$ is therefore the following: Choose any one member, say $G(x, y)$, which agrees with g on the boundary, and then every admissible v equals $G + v_0$, where v_0 lies in $\mathcal{K}_0^1$. In short,

$$\mathcal{K}_E^1 = G + \mathcal{K}_0^1.$$

In the Ritz method *it is not required that the trial functions agree exactly with* g *on the boundary*. It is enough if the trial functions are of the form $v^h = G^h(x, y) + \sum q_j \varphi_j^h(x, y)$, where the φ_j^h lie in $\mathcal{K}_0^1$ and G^h *nearly agrees with* g. This means that the trial class is

$$S^h = G^h(x, y) + S_0^h$$

and that S_0^h is a subspace of $\mathcal{K}_0^1$. Section 4.4 verifies that the basic Ritz theorem 1.1 applies to this case (and ultimately we remove even the hypothesis that $S_0^h \subset \mathcal{K}_0^1$, but this leads outside of the Ritz framework).

Finally we consider the boundary condition $u_n + du = b$. This will be a natural boundary condition, provided the basic functional $\iint v_x^2 + v_y^2 - 2fv$ is altered to take account of d and b. Both alterations appear as boundary terms:

$$I(v) = \iint_\Omega v_x^2 + v_y^2 - 2fv + \int_\Gamma (dv^2 - 2bv) \, ds.$$

We emphasize that the first new term enters the energy

$$a(v, v) = \iint (v_x^2 + v_y^2) + \int dv^2,$$

and thus contributes to the stiffness matrix K in the Ritz method. The new linear term $-2bv$ adds a boundary contribution to the load vector F. The admissible space is the whole of $\mathcal{K}^1$, and therefore any piecewise polynomial which is continuous between elements can be used as a trial function.

We want to propose also three problems of fourth order, all governed by the biharmonic equation

$$(54) \qquad \Delta^2 u = u_{xxxx} + 2u_{xxyy} + u_{yyyy} = f \qquad \text{in } \Omega.$$

This equation is a model for the transverse displacement u of a thin plate bent by the body force $f(x, y)$, with bending stiffness normalized to $D = 1$. As usual, the number m of boundary conditions will be half the order of the equation, so that $m = 2$.

A first possibility is to impose Dirichlet conditions, leading physically to the problem of a *clamped plate:*

$$u = 0 \quad \text{and} \quad u_n = 0 \qquad \text{on } \Gamma.$$

An alternative of great interest is the problem of a *simply supported plate*, in which one boundary condition is essential (or forced) and the other is natural. As in the oblique derivative problem described above for Poisson's equation, *the form of the natural condition will depend on the form of the variational integral $I(v)$*. In elasticity this natural condition involves Poisson's ratio v, which measures the change in width when the material is stretched lengthwise; $v = 0.3$ is a common choice. The boundary conditions determined by physical reasoning are

$$(55) \qquad u = 0 \quad \text{and} \quad v\Delta u + (1 - v)u_{nn} = 0 \qquad \text{on } \Gamma.$$

The cases $v = 0$ and $v = 1$, which may arise in other applications, will of course be included. Note that on a straight boundary the tangential derivatives are all zero, $\Delta u = u_{nn}$, and v *disappears from the boundary condition*; the remarkable and apparently paradoxical consequences of this disappearance, for polygonal approximation of a circle, are discussed in Section 4.4.

Finally, there is the pure Neumann problem, corresponding to a *free boundary*. The second of the boundary conditions (55) remains; it is often written as

$$\frac{-M_{nn}}{D} = u_{nn} + v(\varphi_s u_n + u_{ss}) = 0,$$

where φ is the angle between the normal and the x-axis. The condition $u = 0$, which fixes the edge, no longer applies. To fill its place one has to compute the coefficient of δu, when the energy functional in (56) below is varied. This coefficient is given by Landau and Lifshitz explicitly in terms of u and φ; it is normally written in the contracted Kirchhoff form $Q_n + \partial M_{ns}/\partial s = 0$. In practical problems it is unusual for the whole boundary to be free.

All these conditions can be coped with somehow by directly replacing derivatives by finite differences. The construction of accurate equations near the boundary becomes fantastically complicated, however, and a much better starting point is furnished by the variational form.

Variationally, the problem is to minimize

$$
\begin{aligned}
(56) \quad I(v) &= a(v, v) - 2(f, v) \\
&= \iint_\Omega \left(v_{xx}^2 + v_{yy}^2 + 2v v_{xx} v_{yy} + 2(1 - v) v_{xy}^2 - 2fv \right) dx dy
\end{aligned}
$$

subject to the appropriate boundary conditions. In the Neumann problem there are no constraints at the boundary, and the space of admissible v is exactly $\mathcal{3C}^2(\Omega)$. For the clamped plate, the Dirichlet conditions $v = 0$ and $v_n = 0$ are imposed; the subspace satisfying these constraints is $\mathcal{3C}_0^2(\Omega)$. In the intermediate case of the simply supported plate, only the condition $v = 0$ is essential; we denote the corresponding subspace by $\mathcal{3C}_{ss}^2$, noting that $\mathcal{3C}_0^2 \subset \mathcal{3C}_{ss}^2 \subset \mathcal{3C}^2$. Of course, it is Green's theorem which yields the equivalence of these variational problems with their differential counterparts.

We emphasize that the fourth derivatives which appear in the term $\Delta^2 u$ in Green's theorem are not required for the variational formulation—*the theorem of minimum potential energy*—to be valid. In fact, the opposite is true. Any limitation to solutions with four continuous derivatives would be awkward in the extreme, and the whole idea of completion is to arrive at the admissible space under minimum restrictions, requiring only that the essential boundary conditions hold and that the energy $a(v, v)$ be finite. The solution u then satisfies the equation in its weak form $a(u, v) = (f, v)$.

For the theory of the Ritz method, the key point is that the energy norm $\sqrt{a(v, v)}$ should be equivalent to the standard norm $||v||_2$. This is again the condition of ellipticity: For some $\sigma > 0$ and all admissible v,

$$
a(v, v) \geq \sigma ||v||_2^2
$$

or

$$
\iint v_{xx}^2 + v_{yy}^2 + 2v v_{xx} v_{yy} + 2(1 - v) v_{xy}^2
$$

$$
\geq \sigma \iint (v^2 + v_x^2 + v_y^2 + v_{xx}^2 + v_{xy}^2 + v_{yy}^2).
$$

With all edges free ellipticity must fail, since the solution is unique only up to a linear function $a + bx + cy$; with the differential problem altered to $\Delta^2 u + qu = f$, corresponding to a plate under steady rotation with $q = \rho\omega^2 > 0$, this case is also elliptic.

Ellipticity can become difficult to prove when—as in problems of linear elasticity—there are two or three unknowns u_i, and the strain energy involves only certain combinations $\varepsilon_{ij} = \frac{1}{2}(u_{i,j} + u_{j,i})$ of their derivatives. It is *Korn's inequality* which establishes that this strain energy dominates the $\mathfrak{IC}^1$ norm, $\sum \int \varepsilon_{ij}\varepsilon_{ij} \geq \sigma \sum \int u_{i,j}u_{i,j}$.

Before proceeding to the construction of finite elements for the solution of these problems, we want to comment briefly on the function spaces $\mathfrak{IC}^s(\Omega)$. In one dimension they were deceptively simple to describe; v belongs to $\mathfrak{IC}^s[0, 1]$ if it is the s-fold integral of a function f with $\int f^2 \, dx < \infty$. This guarantees that v and its first $s - 1$ derivatives are continuous; only the sth derivative, that is, the original function f, might have jumps or worse.

In the plane it is possible for v to be *discontinuous and at the same time differentiable*. One such function in $\mathfrak{IC}^1$ is $v = \log \log 1/r$, on the circle $r \leq \frac{1}{2}$:

$$\int\int v_x^2 + v_y^2 = \int\int \left[v_r^2 + \left(\frac{v_\theta}{r}\right)^2 \right] r \, dr d\theta$$

$$= \int\int (\log r)^{-2} \, d(\log r) \, d\theta = \frac{2\pi}{\log 2}.$$

Thus the derivatives have finite energy, and so does v itself, but the function is not continuous at the origin. The general rule in n dimensions is the following: *If v is in $\mathfrak{IC}^s$ and $s > n/2$, then v is continuous and*

$$(57) \qquad\qquad \max|v(x_1, \ldots, x_n)| \leq C\|v\|_s.$$

This is the essence of the celebrated *Sobolev inequality*, relating the two properties of continuity and finite energy of the derivatives. If v lies in $\mathfrak{IC}^s$ with $s \leq n/2$, it cannot be guaranteed that v is continuous.

By duality, Sobolev's rule will also decide when the n-dimensional δ-function lies in $\mathfrak{IC}^{-s}$. Of course, this is impossible unless $-s < 0$; it is only after enough integrations that the δ-function might have finite energy. The norm on $\mathfrak{IC}^{-s}$ is defined as in (12):

$$(58) \qquad \|w\|_{-s} = \max_v \frac{|(v, w)|}{\|v\|_s} = \max_v \frac{\left|\int vw \, dx_1 \cdots dx_n\right|}{\|v\|_s}.$$

If w is a δ-function, say at the origin, this gives

$$\|\delta\|_{-s} = \max_v \frac{|v(0)|}{\|v\|_s}.$$

According to Sobolev's inequality (57), this is finite and the δ-function is in $\mathcal{3C}^{-s}$ if and only if $s > n/2$.

In particular, the δ-function in the plane is not in $\mathcal{3C}^{-1}$, and correspondingly the fundamental solution of Laplace's equation, $u = \log r$, is not in $\mathcal{3C}^1$:

$$\int\int u_x^2 + u_y^2 = \int\int \left[u_r^2 + \left(\frac{u_\theta}{r}\right)^2 \right] r\, dr d\theta = \int\int r^{-1}\, dr d\theta = \infty.$$

Therefore *a point-loaded membrane is, strictly speaking, inadmissible in the variational problem.*

For a plate the situation is different, since the differential equation is of order 4. The solution space is $\mathcal{3C}^2$ (with boundary conditions) and the data space is four derivatives lower, that is, $\mathcal{3C}^{-2}$. In the plane the δ-function does lie in $\mathcal{3C}^{-2}$, and a point-loaded plate is acceptable.

There is one additional question about these function spaces $\mathcal{3C}^s$ which is fundamental for finite elements: *What is the condition for an element to be conforming?* In other words, given a differential equation of order $2m$ in n independent variables, *which piecewise polynomials lie in the admissible space $\mathcal{3C}_E^m$?* It is easy to check on the essential boundary conditions; the only question is how smooth the element must be to lie in $\mathcal{3C}^m$.

The standard conforming condition is well known: The trial function and its first $m - 1$ derivatives should be continuous across the element boundaries. This condition is clearly sufficient for admissibility, since the mth derivatives have at worst a jump between elements, and their energy is finite. On the other hand, the example $\log \log 1/r$ may seem to put in doubt the necessity of this condition for conformity; there exist functions which do not have $m - 1$ continuous derivatives, but which nevertheless lie in $\mathcal{3C}^m$ and are admissible. Fortunately, however, such troublesome functions cannot be piecewise polynomials. If a function v is a polynomial (or a ratio of polynomials) on each side of an element boundary, then *v lies in $\mathcal{3C}^m$ if and only if every derivative of order less than m is continuous across the boundary.* The success of the finite element method lies in the construction of such elements, retaining both a convenient basis and a high degree of approximation.

1.9. TRIANGULAR AND RECTANGULAR ELEMENTS

This section describes some of the most important finite elements in the plane. Their construction has been going on for 30 years, if we include Courant's early paper on piecewise linear elements, and it seems to have been one of the most enjoyable occupations in applied mathematics. It involves not much more than high-school algebra, and at the same time the results are of great importance; that is a rare and happy combination. The goal is to choose piecewise polynomials which are determined by a small and convenient set of nodal values, and yet have the desired degree of continuity and of approximation.

There are a great many competing elements, and it is not clear at this writing whether it is more efficient to subdivide the region into triangles or into quadrilaterals. Triangles are obviously better at approximating a curved boundary, but there are advantages to quadrilaterals (and especially to rectangles) in the interior: there are fewer of them, and they permit very simple elements of high degree. These remarks already suggest that the best choice may be a mixture of the two possibilities, provided they can be united by nodes which ensure the required continuity across the junction.

We begin by subdividing the basic region Ω into triangles (Fig. 1.10).

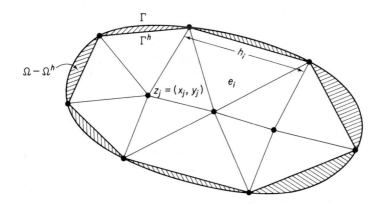

Fig. 1.10   Subdivision of the polygon Ω^h into irregular triangles.

The union of these triangles will be a polygon Ω^h, and in general—if Γ is a curved boundary—there will be a nonempty "skin" $\Omega - \Omega^h$. The longest edge of the jth triangle will be denoted by h_j, and $h = \max h_j$. We assume for convenience that Ω^h is a subset of Ω and that no vertex of one triangle lies part way along the edge of another. In practice—since the computer eventually has to know the location of every vertex $z_j = (x_j, y_j)$—the triangulation should be carried out as far as possible by the computer itself. A fully automatic subroutine can begin by covering Ω with a regular triangular grid and then make the necessary adjustments at the boundary. If the grid is to be very coarse in one part of Ω and more refined in another, or if the domain has corners and other points of singularity, it may be necessary to establish the triangulation by hand. In any case, George has argued [G2] that the *refinement* of a coarse mesh, which can be achieved simply by subdividing each triangle into four similar triangles, is a mechanical procedure for which the computer is ideally suited. This allows the user to begin by describing a coarse mesh to the computer, at a minimum cost in time and effort, and then to work eventually with a fine mesh. Furthermore, he can more easily carry out the one check on his program which is the most convincing and highly recommended of all: to test whether a reduction in h leads to an acceptable or an unacceptable change in the numerical solution.

Given a triangulation, we describe now the simplest and most basic of all trial functions. It is *linear inside each triangle* ($v^h = a_1 + a_2 x + a_3 y$) *and continuous across each edge*. Thus the graph of $v^h(x, y)$ is a surface made up of flat triangular pieces, joined along the edges. This is an obvious generalization of broken-line functions in one dimension. The subspace S^h composed of these piecewise linear functions was proposed by Courant [C11] for the solution of variational problems; it is a subspace of $\mathcal{H}^1$, since the first derivatives are piecewise constant. Its independent development by Turner and others was the beginning of finite elements, and the trial functions are sometimes referred to as *Turner triangles*.

The effect of continuity is to avoid δ-functions in the first derivatives at interelement boundaries; without this constraint the functions are not admissible, and the (infinite) energy over the domain Ω cannot be found by adding the separate contributions from within each element.

The simplicity of Courant's space lies in the fact that within each triangle, the three coefficients of $v^h = a_1 + a_2 x + a_3 y$ are uniquely determined by the values of v^h at the three vertices. This means that *the function can be conveniently described by giving its nodal values;* equivalently, S^h will have a convenient basis. Furthermore, along any edge, v^h reduces to a linear function of one variable, and this function is obviously determined by its values at the two endpoints of the edge. The value of v^h at the third vertex has no effect on the function along this edge, no matter whether this third vertex belongs to the triangle on one side or the other. *Therefore, the continuity of v^h across the edge is assured by continuity at the vertices.*

In case of an essential boundary condition, say $u = 0$ on Γ, the simplest subspace $S^h \subset \mathcal{H}_0^1$ is formed by requiring the functions to vanish on the polygonal boundary Γ^h. Extended to be zero on the skin $\Omega - \Omega^h$, such a function v^h is continuous over the whole domain Ω; it lies in $\mathcal{H}_0^1$, and it is admissible in the Dirichlet problem.

The dimension of the space S^h—the number N of free parameters in the functions v^h—*coincides with the number of unconstrained nodes*. (A boundary node at which v^h is required to vanish, or to equal some other prescribed displacement, is constrained and does not add to the dimension of the subspace.) For proof, let $\varphi_j(x, y)$ be the trial function which equals 1 at the jth node and zero at all other nodes. Then *these pyramid functions φ_j form a basis for the trial space S^h*. An arbitrary v^h in S^h can be expressed in one and only one way as a linear combination

$$v^h(x, y) = \sum_{j=1}^{N} q_j\varphi_j(x, y).$$

In this context the coordinate q_j has direct physical significance, as the displacement v^h at the jth node $z_j = (x_j, y_j)$. This is an important feature of the elements which have been developed for engineering computations: that each q_j can be associated with the value of v^h or one of its derivatives at a specific nodal point in the domain.

The optimal coordinates Q_j are now determined by minimizing $I(v^h) = I(\sum q_j\varphi_j)$, which is quadratic in $q_1, \ldots, q_N$. We emphasize that the minimizing function $u^h = \sum Q_j\varphi_j$ is *independent of the particular basis*; the effect of choosing a basis is simply to put the problem of minimization into the computable (or operational) form $KQ = F$. On the other hand, the actual numerical calculations are strongly dependent on this choice. The stiffness matrix K' from a different choice of basis would be congruent to the first: $K' = SKS^T$ for some S, with F replaced by $F' = SF$. Therefore, the object is to choose a basis in which K is sparse and well conditioned, while keeping the entries of K and F as easy to compute as possible.

It is here that the finite element choice φ_j, based on interpolating nodal values, is an effective compromise. The matrix K is reasonably well conditioned and reasonably sparse, since two nodes are coupled only if they belong to the same element. Furthermore, the inner products $K_{ij} = a(\varphi_i, \varphi_j)$ and $F_j = (f, \varphi_j)$ can be found by a very fast and systematic algorithm. The trick is to compute, not one inner product after another, but the contributions to all the inner products from one triangle after another. This means that the integrals are computed over each triangle, yielding a set of element stiffness matrices k_i. Each k_i involves only the nodes of the ith triangle of the partition; its other entries are all zero. Then the global stiffness matrix K, containing all inner products, is assembled from these pieces. This process was carried out on an interval in Section 1.5 and will be extended to a triangular mesh in the next section.

Courant's triangles yield an interesting stiffness matrix K. For Laplace's equation, the standard five-point difference scheme is produced if the triangles are formed in a regular way, starting with a square mesh and drawing all the diagonals in the northeast direction. (The more accurate nine-point scheme can be constructed in a similar way from bilinear elements [F6], but this is largely a curiosity.) The appearance of such a simple and systematic stiffness matrix means that the *fast Fourier transform* can be used to solve $KQ = F$; this is extremely successful on a rectangle, and non-rectangular applications are being rapidly developed by Dorr, Golub, and others. Mathematically, a key property of the 5-point scheme is the *maximum principle*:

all off-diagonal terms K_{ij} in the stiffness matrix are negative, they are dominated by the diagonal entries K_{ii}, and as a consequence the *inverse matrix
K^{-1} is non-negative*. A simple calculation shows that this property holds for
linear elements on any triangulation, if no angle exceeds $\pi/2$. (The precise
condition is that the two angles subtended by any given edge add to no more
than π.) A similar result holds for linear elements in n dimensions. Because
K^{-1} and all $\varphi_j(x)$ are non-negative, the finite element approximation u^h obeys
the same physical law as the true displacement u: *if the load f is everywhere
positive, then so is the resulting displacement*. Fried has asked whether this
holds also for elements of higher degree; it will not be true that all off-diagonal K_{ij} are negative, but this is not necessary for a positive K^{-1}.

In the Neumann problem there is no constraint on v^h at the boundary
nodes, and the dimension of S^h becomes the total number of interior and
boundary nodes. The basis functions φ_j again equal one at a single node and
zero at the others. In this case, however, defining the trial functions to be
zero on the skin would not leave them all continuous. Instead, we may simply
continue into each piece of $\Omega - \Omega^h$ the linear function defined in the adjacent
triangle.

There are other possibilities, of which we mention one—to ignore the
skin entirely and study the Neumann problem only within the approximate
domain Ω^h. Of course, the approximate solution may be significantly altered,
producing some loss in accuracy in comparison with minimizing the integral
over Ω. Such errors, introduced by modifying the boundary-value problem,
are estimated in Chapter 4.

We turn now to a more accurate and more refined element. This was a
decisive step in the finite element technique, to generalize the basic idea
behind Courant's simple trial functions. Rather than assuming a linear function within each triangle, v^h will be permitted to be a quadratic:

$$(59) \qquad v^h = a_1 + a_2 x + a_3 y + a_4 x^2 + a_5 xy + a_6 y^2.$$

In order to lie in $\mathcal{3C}^1$, v^h must again be continuous across the edge between
adjacent triangles.

To compute with this subspace, a different description is required; we
need a basis for the space. Therefore, we must find a set of continuous piecewise quadratic functions φ_j such that every member of S has a unique expansion as

$$v^h(x, y) = \sum q_j \varphi_j(x, y).$$

There is a beautiful construction of such a basis, if we add to the vertices
a further set of nodes, placed at the midpoints of the edges (as in the first triangle of Fig. 1.11). With each node, whether it is a vertex or the midpoint of
an edge, we associate the function φ_j which equals 1 at that node and zero

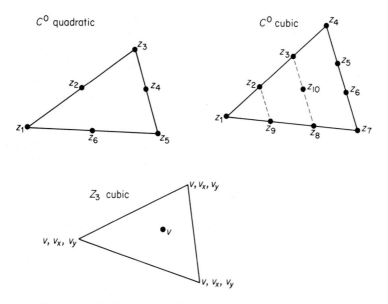

Fig. 1.11   Nodal placement for quadratic and cubic elements.

at all others. This rule specifies φ at six points in each triangle, the three vertices and three midpoints, thereby determining the six coefficients a_i in (59).

The crucial point to decide is whether or not the piecewise quadratic, determined by this rule in each separate triangle, is continuous across the edges between triangles. The proof is deceptively simple. Along such an edge v^h is a quadratic in one variable, and there are three nodes on the edge (one at the midpoint and two at the ends). Therefore, the quadratic is determined by these three nodal values, shared by the two triangles which meet at that edge. The nodal values elsewhere in the two triangles have no effect on v^h along the edge, and continuity holds. For an edge lying on the outer boundary Γ^h in the Dirichlet problem, the three nodal values (and therefore the whole quadratic) are zero.

Any v^h in S^h has a unique expansion into $\sum q_j \varphi_j$, where the coordinate q_j is the value of v^h at the jth interior node (vertex or midpoint). Therefore, these φ_j form a basis for S^h, and *the dimension N equals the number of unconstrained nodes.*

For continuous piecewise cubics, a basis can be constructed in the same way. A cubic in the two variables x, y is determined by 10 coefficients, and in particular by its values at the 10-node triangle of the previous figure. Again the 4 nodes on each edge determine the cubic uniquely on that edge, and continuity is assured. This is the triangular analogue of the one-dimensional cubic element constructed in Section 1.6 (that is, the one which was only continuous, and had two nodes inside each interval).

The same construction extends to polynomials of any degree $k - 1$ in any number of variables $x_1, \ldots, x_n$, provided the basic regions are always simplexes—intervals for $n = 1$, triangles for $n = 2$, and tetrahedra for $n = 3$. In fact, we can obtain discrete analogues of arbitrarily high accuracy in n-space. Unfortunately, there is for practical applications a fatal flaw; the dimension of S^h, which equals the total number of interior nodes, grows enormously fast with k and n. This is an essential problem with finite elements, to impose further constraints on the trial functions (thereby reducing the dimension of S^h) without destroying either the approximation properties or the simplicity of a local basis.

The key is to increase properly the level of continuity required. In the cubic case, we may add the condition that *the first derivatives of v^h should be continuous at every vertex*. This is clearly a subspace of the previous cubic case, in which only continuity of v^h was required. If m triangles meet at a given interior vertex, the continuity of v_x^h and v_y^h imposes new constraints on the trial functions, and the dimension N is correspondingly reduced. To construct a basis we remove the midedge nodes and concentrate a triple node at each vertex. In other words, the 10 coefficients in the cubic are determined by v, v_x and v_y at each vertex, together with v at the centroid; this tenth node has nowhere else to go. A unique cubic is determined by these 10 nodal values, and along each edge it is determined by a subset of 4—the values of v and its edge derivative at each end, a combination which again assures matching along the edge. The result is an extremely useful cubic trial space, which we denote by Z_3.

Piecewise polynomial elements like Z_3 are easy to describe in the interior of the domain; at the boundary we have to be more careful, and there are various alternatives. In case of an essential condition $u = 0$, there is first of all the constraint $v^h = 0$. If v^h is to vanish along all of Γ^h, then also the derivatives along both chords must be set to zero, leaving no parameters free at that vertex. A more satisfactory approximation is obtained by the isoparametric method of Section 3.3, or—if it is preferred to stay with the x–y variables—to impose only the condition that the derivative of v vanish in the direction tangent to Γ. In the latter case we must give up the Dirichlet condition that v^h vanish on Γ^h and by extension on the true boundary Γ. Instead, *the functions v^h will actually violate the essential condition for admissibility*, and in this version of the cubic Z_3, the trial space will not be a subspace of the Dirichlet space $\mathcal{H}_0^1$. Nevertheless, it is possible to give a rigorous estimate of the error induced by such inadmissible elements; this will be a key result of Chapter 4. We may expect v^h to be nearly zero on Γ, and if we go so far as to compute over the curved triangles of Ω rather than the real triangles of Ω^h, the numerical results will be good.

There is an additional point of practical importance about the cubic elements: The unknowns Q_c corresponding to the centroid nodes can be eliminated immediately from the finite element system $KQ = F$, leaving a system with effectively three unknowns per mesh point. This early elimination of unknowns is known as *static condensation*. It depends on the fact that the corresponding basis functions φ_c are nonzero *only inside a single triangle;* each Q_c is coupled only to the other nine parameters of its own triangle. Thus the cth equation can be solved for Q_c in terms of these nine nearby Q_j, and Q_c can be eliminated from the system without increasing its bandwidth. The number of arithmetical operations is directly proportional to the number of centroids which are removed, which is exceptional; in general two-dimensional problems, it is impossible to eliminate n nodes with only αn operations. Physically, the optimal displacement Q_c at the centroid is determined exclusively by the nine parameters which establish the displacement along the boundary of the element.

Mathematically, this process may be regarded as the *orthogonalization* of the basis functions φ_j with respect to each centroid basis function φ_c. Of course, the Gram–Schmidt process could always be used to orthogonalize the whole basis, leaving the trivial stiffness matrix $K = I$, but this would be insane; it is quicker just to solve $KQ = F$ directly. The special orthogonalization against φ_c is possible since it changes only the nine adjacent φ_j, and only within the given triangle. It appears that in general the elimination of unknowns in other than the normal Gaussian order will increase the width of a band matrix, except in special circumstances—such as the *odd–even reduction*, which systematically eliminates every other node in the five-point Laplace difference equation.

There is an important variant of the cubic space Z_3, in which this static condensation is avoided: the centroid is not a node at all, and there are only the nine nodal parameters given by v, v_x, and v_y at each vertex. Correspondingly, one degree of freedom must be removed from the cubic polynomials, and a familiar device is to require equal coefficients in the $x^2 y$ and xy^2 terms:

$$v^h = a_1 + a_2 x + a_3 y + a_4 x^2 + a_5 xy + a_6 y^2$$
$$+ a_7 x^3 + a_8(x^2 y + xy^2) + a_9 y^3.$$

This restriction damages the accuracy of the element; we shall connect the rate of convergence to the degree of polynomial which can be completely reproduced in the trial space, and this degree has dropped from 3 to 2. Nevertheless, this restricted cubic is among the most important elements, and is preferred by many engineers to the full Z_3. Its nodal parameters are excellent. (However, we observe in Section 4.2 that the remaining nine degrees of freedom are not invariant with respect to rotation in the x–y plane, and for some orientations they are not even uniquely determined by the nine nodal parameters. Anderheggen has proposed $\iint v^h = 0$ as an alternative constraint, and Zienkiewicz [22, p. 187] analyzes a third, and very attractive, possibility.)

None of the spaces described so far can be applied to the biharmonic equation, since they are not subspaces of $\mathcal{3C}^2$. (Or rather, their use would be illegal; the restricted Z_3 is frequently used for shells. Elements which are nonconforming along interior edges are discussed in 4.2.) Therefore, denoting by $\mathcal{C}^k$ the class of functions with continuous derivatives through order k, we want now to construct elements which belong to $\mathcal{C}^1$. The essential new condition is that the normal slope be continuous across interelement boundaries. A function in $\mathcal{C}^1$ is automatically in $\mathcal{3C}^2$ and therefore admissible for fourth-order problems; the integrals over Ω can be computed an element at a time, with no δ-functions at the boundaries.

Several possibilities exist. One is to modify the cubic polynomials in Z_3 by *rational functions*, chosen to cancel the edge discontinuities in the normal slopes without affecting the nodal quantities v, v_x, and v_y. This makes v^h piecewise rational instead of piecewise polynomial, and again the accuracy is reduced; the space does not contain an arbitrary cubic. We are opposed to rational functions, however, on other and more important grounds; they are hard to integrate, and even numerical integration has very serious difficulties (Section 4.3).

If we prefer to work exclusively with polynomials, we are led to one of the most interesting and ingenious of all finite elements—the quintics (Fig. 1.12). A polynomial of fifth degree in x and y has 21 coefficients to be determined, of which 18 will come from the values of v, v_x, v_y, v_{xx}, v_{xy}, and v_{yy} at the vertices. These second derivatives represent "bending moments" of physical interest, which will now be continuous at the vertices and available as output from the finite element process; they are given directly by the appropriate weights Q_j. Furthermore, with all second derivatives present there is no difficulty in allowing an irregular triangulation. The quintic along the edge between two triangles is the same from both sides, since three conditions are shared at each end of the edge—v and its edge derivatives v_s and v_{ss}, which are computable from the set of six parameters at the vertex.

It remains to determine three more constraints in such a way that also the normal derivative v_n is continuous between triangles. One technique is to add the value of v_n at the midpoint of each edge to the list of nodal parameters. Then since v_n is a quartic in s along the edge, it will be uniquely fixed by this parameter together with the values of v_n and v_{ns} at each end. (The normal n at the middle of an edge points into one triangle and out of the adjacent one, adding somewhat to the bookkeeping.) This construction produced the complete $\mathcal{C}^1$ quintic, which was devised independently in at least four papers. Its accuracy in the displacement will be $O(h^6)$, if the boundary is successfully approximated; comparable accuracy with finite differences has apparently never been achieved for fourth-order problems in the plane.

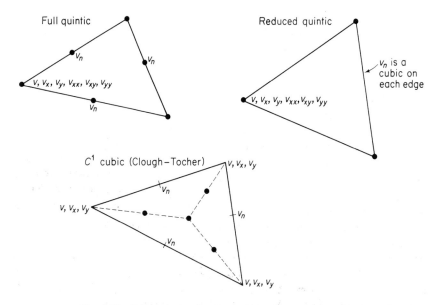

Fig. 1.12   Two quintic triangles and a cubic macro-triangle.

An equally effective way of disposing of the three remaining constraints is to require that v_n reduce to a cubic along each edge, in other words, that the leading coefficient in what would otherwise be a quartic polynomial be made to vanish. With this constraint the nodal values of v_n and v_{ns} determine the cubic v_n along the edge, and the element is in $\mathcal{C}^1$. The complete quintic is lost, reducing the displacement error in $u - u^h$ from h^6 to h^5. At the same time, however, the dimension of S^h is significantly reduced (by a factor of 9:6; see Table 1.1) and this more than compensates for the loss of three quintic terms. In fact, a series of careful numerical experiments [Cl4] has given first prize to this remarkable element. Admittedly, it is not always simple to work with—the continuity of second derivatives must be relaxed at corners in the domain, where the true solution u is less continuous—but it is highly accurate.†

Note that to achieve a $\mathcal{C}^1$ triangular element, it was necessary to prescribe

---

†There are important problems in which the physics introduces side conditions. For incompressible fluids, for example, it is useful to impose div $v^h = 0$ on all the trial functions. The French school (Crouzeix, Fortin, Glowinski, Raviart, Témam) has found that although the Courant triangles are completely inadequate, convergence can be established with the use of

1. standard quadratics in the plane and cubics in 3-space;

2. these $\mathcal{C}^1$ quintics as stream function in two dimensions, with a divergence-free velocity field of quartic polynomials constructed by differentiating the quintics;

3. a non-conforming element, linear on triangles but continuous only at the *midpoints* of the edges (cf. Section 4.2).

even the second derivatives at the vertices. Zenisek has found a remarkable theorem [Z1] in this direction: *To achieve piecewise polynomials of class $\mathcal{C}^q$ on an arbitrary triangulation, the nodal parameters must include all derivatives at the vertices of order less than or equal to $2q$.* He has constructed such elements in n dimensions by using polynomials of degree $2^n q + 1$, conjectured to be the minimum possible.

Fortunately there is a way to avoid this severe limitation and still achieve a conforming $\mathcal{C}^1$ element. It consists of building up a "macroelement" out of several standard elements. One of the best known is the *Clough–Tocher triangle*, formed by combining different cubic polynomials in three subtriangles (Fig. 1.12). The final nodal parameters for the large triangle are v, v_x, and v_y at the vertices and v_n at the midpoints of the edges—12 in all. These guarantee that even the normal slope will match that of the adjoining macrotriangle, since the slope is a quadratic and is therefore determined along an edge exclusively by the parameters on that edge—v_n is given directly at the midpoint, and indirectly at the two vertices as a combination of v_x and v_y. Since each of the three cubics has 10 degrees of freedom, and the final macrotriangle has 12 nodal parameters, 18 constraints may be imposed. It turns out that this is just right to achieve $\mathcal{C}^1$ continuity internally in the triangle. Requiring a common value of v, v_x, and v_y furnishes 3 constraints at each external vertex and 6 at the internal one, and agreement of v_n at the midpoints of the three internal edges makes 18.

Assuming that a given triangulation allows the triangles to be combined three at a time into macrotriangles, we have just found a basis for the space S^h of all piecewise cubics of class $\mathcal{C}^1$. This is one case of an important and apparently very difficult problem—to determine a basis for the space of piecewise polynomials $v(x_1, \ldots, x_n)$ of degree $k - 1$ which have continuity $\mathcal{C}^q$ between simplexes. We have no idea (for $n > 1$) even of the dimension of this space: According to Table 1.1, the $\mathcal{C}^1$ cubics have $M = 6$ parameters for each pair of macrotriangles, and therefore an average of one unknown for each of the original triangles!†

To summarize the polynomials described in this section, we shall tabulate their essential properties. The column headed d gives the number of parameters required to determine the polynomial within each region, that is, the number of degrees of freedom if adjacent elements imposed no constraints. The integer $k - 1$ is the degree of the highest "complete polynomial" which is present; this means that some polynomial of degree k cannot be found in the trial space, and consequently (as we shall prove) that the error in $u - u^h$ is of order h^k. Finally, N is the dimension of the trial space S^h, assuming that Ω is a square which has been partitioned into n^2 small squares and then into

---

†Added in proof: We now have a conjecture about the dimension, but no inkling of a basis, for the space of polynomials of degree $k - 1$ and continuity $\mathcal{C}^q$.

Table 1.1 TRIANGULAR ELEMENTS

Element Type	Continuity	d	k	$N = Mn^2$
Linear	$\mathcal{C}^0$	3	2	n^2
Quadratic	$\mathcal{C}^0$	6	3	$4n^2$
Cubic	$\mathcal{C}^0$	10	4	$9n^2$
Cubic Z_3	$\mathcal{C}^0$, v_x, v_y continous at vertices	10	4	$5n^2$
Restricted Z_3	As above, plus equal coefficients of x^2y and xy^2	9	3	$3n^2$
Quintic	$\mathcal{C}^1$, v_{xx}, v_{xy}, v_{yy} continuous at vertices	21	6	$9n^2$
Reduced quintic	$\mathcal{C}^1$, v_{xx}, v_{xy}, v_{yy} continuous at vertices	18	5	$6n^2$
Cubic macrotriangle	$\mathcal{C}^1$	12	4	$M = 6$

$2n^2$ triangles, each of the small squares being cut in half by the diagonal which has slope $+1$. Only the leading term $N = Mn^2$ is given; there will be correction terms depending on the constraints imposed at the boundary, but the important constant is M. The coefficient M is $p + 3q + 2r$ for an element in which each vertex is a p-tuple node, each edge has q nodes, and the interior of each triangle contains r nodes. The reason is that in any triangulation there are twice as many triangles as vertices; each triangle accounts for 180° in angles and each vertex for 360°. Furthermore, the ratio of edges to triangles is 3:2, since each edge is shared by two triangles. This explains the 1-2-3 weighting of p, r, and q. (*Second proof:* If one new vertex is introduced inside an existing triangle, it brings three new edges and a net increase of two triangles.) This 1-2-3 rule holds on any set of triangles, and it means that on average *there are $M/2$ unknowns per triangle*. It follows that multiple nodes are desirable; an increase in p is economical in its contribution both to dimension and to bandwidth. Internal nodes are next best, especially since they can be removed by static condensation. Nodes along edges are the worst in respect of computation time.

Let us remark that in any theoretical comparison of two finite elements, the one of higher degree $k - 1$ will always be the more efficient, asymptotically as $h \rightarrow 0$. The error decreases like Ch^k, with a constant C which depends both on the particular element and on the kth derivatives of u. Therefore, the only theoretical limitation on the rate of convergence is the smoothness of u, and even this can be avoided (Section 3.2) by grading the mesh.

The balance is altered, however, if one fixes the accuracy required and asks for the element which achieves that accuracy at least expense. The mesh width h will be finite, not infinitesimal. Furthermore, there is the question whether the calculation is to be done only a few times, so that convenience in programming is a necessity, or whether the programming and preparation costs can be amortized over a long program lifetime. It is our judgment that elements of limited order, like those in Table 1.1—together with others constructed to have similar characteristics, such as quadrilateral elements, three-dimensional elements, and shell elements—will provide the basic range of selection for the practical application of the finite element method.

We want now to discuss *rectangular elements*, which are rapidly increasing in popularity. They are of special value in three dimensions, where each cube occupies the same volume as six very elusive tetrahedra. (An irregular decomposition of 3-space into tetrahedra is hard even on the computer.) Furthermore, a number of important problems in the plane are defined on a rectangle or on a union of rectangles. The boundary of a more general region cannot be described sufficiently well without using triangles, but it will often be possible to combine rectangular elements in the interior with triangular elements near the boundary.

The simplest construction, in analogy with the linear element for triangles, is based on functions which are piecewise *bilinear:* $v^h = a_1 + a_2 x + a_3 y + a_4 xy$ in each rectangle. These four coefficients are determined by the values of v^h at the vertices. The overall trial function is continuous and may be applied to differential equations of second order. There is an obvious interpolating basis: φ_j equals one at the jth node and zero at the others. Its graph looks like a pagoda, or at least like our conception of one, and φ_j will be called a *pagoda function*. It is a product $\psi(x)\psi(y)$ of the basic piecewise linear roof functions in one variable, so that the space S^h is a tensor product of two simpler spaces. This is a valuable element.

It is important to notice that for arbitrary quadrilaterals, *these piecewise bilinear functions would not be continuous from one element to the next.* Suppose that two quadrilaterals are joined by a line $y = mx + b$. Then along that edge, the bilinear function reduces to a *quadratic;* it is linear only if the edge is horizontal or vertical. A quadratic cannot be determined from the two nodal values at the endpoints of the edge, and in fact the other nodes do affect the value of v^h. Therefore, bilinear elements may be used only on rectangles. Given a more general quadrilateral, however, it is still possible to change coordinates in such a way that it becomes a rectangle, and then bilinear functions in the new coordinates are admissible. In fact, this coordinate change can itself be described by a bilinear function, so that *the same polynomials are used for the transformation of coordinates as for the shape functions within each element.* This is the simplest of the *isoparametric elements*, which are discussed in detail in Section 3.3.

The bilinear element on rectangles can be merged with Courant's linear elements on triangles, since both are completely determined by the value of v^h at the nodes. Other combinations are also possible: A bilinear function on a triangle, with a midpoint node along one edge, is sometimes merged with a full quadratic on a neighboring element. In general, bilinear trial functions seem slightly superior to linear ones, since they exactly reproduce the twist term xy. Nevertheless, a square "macroelement" can be formed from two triangles and have the same element stiffness matrix. For the Laplacian $L = -\Delta$, the assembled bilinear stiffness matrix K is proportional to 8 along the main diagonal and to -1 at points corresponding to its eight neighbors in the plane. The extension to a *trilinear* function $a_1 + a_2 x + a_3 y + a_4 z + a_5 xy + a_6 xz + a_7 yz + a_8 xyz$ in 3-space is clear; again there is one unknown at each corner.

Just as the bilinear element on rectangles matches the linear element on triangles, there are biquadratics and bicubics which correspond to the $\mathcal{C}^0$ quadratics and cubics on triangles. (Ultimately there are polynomials of degree $k - 1$ in each of the variables $x_1, \ldots, x_n$, possessing k^n degrees of freedom within each "box" element.) The biquadratics are often used and are easy to describe. Figure 1.13 shows the arrangement of nodes. The nine

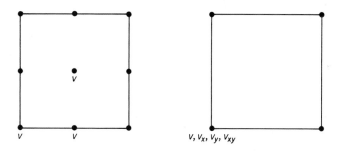

Fig. 1.13 Nodes for biquadratic and Hermite bicubic.

coefficients of a biquadratic are determined by its nodal values, and the three nodes on each edge assure continuity between rectangles. Clearly the biquadratic can be merged with the quadratic element on triangles, since the same vertices and midpoints are nodes. Similar remarks apply to the bicubics.

A simple modification of the biquadratic is to remove the interior node, reducing the number of nodes in each element to eight. To compensate, the biquadratic is restricted by the removal of the $x^2 y^2$ term, whose contribution to the approximation of u was very marginal anyway. This simple and pleasant element belongs to the "serendipity class" described by Zienkiewicz. It is particularly useful for quadrilaterals with curved sides, after a coordinate change transforms the region into a square (Section 3.3); in this context it becomes, along with the bilinear element, the most valuable isoparametric element in the plane.

The corresponding three-dimensional element is the 20-point brick, which again is especially valuable for isoparametrics because it has no internal or multiple nodes. Its nodes lie at the eight vertices and at the midpoints of the twelve edges of the brick. This total matches the number of terms $x^\alpha y^\beta z^\gamma$

which are of second degree in at most one of the variables—x^2yz is allowed, but not x^2y^2z or x^3.

The higher-order serendipity elements in the plane have $4p$ nodes equally spaced along the perimeter of the rectangle, including four at the corners. The shape functions include all terms $x^\alpha y^\beta$ in which neither exponent exceeds p, and the smaller exponent is 0 or 1. The first missing term is therefore x^2y^2, and $k = 4$. These are again especially useful in coordinate changes, where the boundaries of the rectangle can be transformed into arbitrary polynomial curves of degree p, and there are no annoying internal nodes.

There are several fascinating elements (due to Clough, Felippa, and others) in which a quadrilateral is produced as the union of two or more triangular pieces, so that the polynomial varies from piece to piece within the quadrilateral macroelement. The nodal arrangement and continuity between pieces can become rather subtle. However, we shall describe only one other element, the *Hermite bicubic*. The trial functions are again cubic in each variable separately, $v^h = \sum a_{ij}x^iy^j$ for $0 \leq i, j \leq 3$, yielding 16 degrees of freedom in each rectangle. The parameters are determined by the values of v, v_x, v_y, and v_{xy} at the four vertices.† Thus the dimension of S^h is greatly reduced in comparison with the ordinary bicubics described above, which are based on 16 distinct nodes.

This element should be understood as a natural extension of the fundamental one-dimensional element, the Hermite cubic, which was discussed in Section 1.7. In that case $v = a_1 + a_2x + a_3x^2 + a_4x^3$ was determined over each subinterval by the values of v and v_x at the endpoints. This construction left v_x continuous at the nodes, and therefore everywhere, so that the element was in $\mathcal{C}^1$. The standard basis in one dimension contained two kinds of functions, $\psi(x)$ and $\omega(x)$, which interpolated function values and derivatives, respectively:

(60)
$$\psi_j = \begin{cases} 1 & \text{at node } x = x_j, \\ 0 & \text{at other nodes,} \end{cases} \qquad \omega_l = 0 \quad \text{at all nodes,}$$

$$\frac{d\psi_j}{dx} = 0 \quad \text{at all nodes,} \qquad \frac{d\omega_l}{dx} = \begin{cases} 1 & \text{at node } x = x_l, \\ 0 & \text{at other nodes.} \end{cases}$$

These functions span all piecewise cubics of class $\mathcal{C}^1$.

The Hermite bicubic space is a product of two such cubic spaces, and the four parameters at a typical node $z = (x_j, y_l)$ lead to four corresponding basis functions:

(61)
$$\Phi_1 = \psi_j(x)\psi_l(y), \qquad \Phi_2 = \psi_j(x)\omega_l(y),$$
$$\Phi_3 = \omega_j(x)\psi_l(y), \qquad \omega_4 = \omega_j(x)\omega_l(y).$$

†This element was introduced in the engineering literature by Bogner, Fox, and Schmit.

We emphasize that *these nodes must lie on a rectangular grid;* the lines $x = x_j$ in one direction and $y = y_i$ in the other can be spaced arbitrarily, but their intersections absolutely determine the two-dimensional array of nodes. Therefore, the bicubic element is restricted to rectangles (or, after a simple linear transformation of the plane, to parallelograms).

On a sufficiently rectangular region the Hermite bicubic is one of the best elements. Its degree of continuity follows immediately from the basis given in (61); since ψ and ω are both in $\mathcal{C}^1$, all their products inherit this property. *Therefore, this bicubic may be applied to fourth-order equations;* the trial functions lie in $\mathcal{3C}^2$. Furthermore, even the cross derivatives $\partial^2 v/\partial x\,\partial y$ are all continuous. (This suggests a basis-free characterization of the Hermite space S^h; it consists of all piecewise bicubics v such that v, v_x, v_y, and v_{xy} are continuous; we say that v is in $\mathcal{C}^{1,1}$.) The remarkable thing is that this extra degree of continuity does not seem to follow from the usual arguments. The function v_{xy} is quadratic along each edge and is the same for the two rectangles which share that edge, yet only the values of v_{xy} at the two endpoints are automatically held in common—and two values cannot determine a quadratic!

The Hermite idea can be extended to elements of higher degree $2q - 1$. In one dimension there would be q different functions, corresponding to ψ and ω of the cubic case; all their derivatives of order less than q will vanish at the nodes, except that the pth function $\omega_p(x)$ has $(\partial/\partial x)^{p-1}\omega_p = 1$ at the origin. This yields the most natural basis for the space of polynomials of degree $2q - 1$ having $q - 1$ continuous derivatives. In two dimensions we need all products $\omega_p(x)\omega_{p'}(y)$, which means that there are q^2 unknowns at every node; many of them are cross derivatives of high order, and the construction becomes too inefficient after bicubics or biquintics. As always, the number of unknowns can be reduced by imposing additional continuity. The ultimate in this direction is achieved by the *splines*, which are as continuous as possible without becoming a single polynomial over the whole domain. There is only one unknown for each node, with basis function given by the B-spline $\varphi(x)$ on the line and by $\varphi(x)\varphi(y)$ in the plane. The difficulty, of course, is that nodes from different elements become coupled in the construction; the boundary conditions become more complicated, and isoparametric transformations (to get away from rectangles) are impossible.

Because all these spaces are spanned by products of one-dimensional basis functions, the stiffness matrices K may also admit a decomposition into one-dimensional operators. Roughly speaking, this occurs when separation of variables can be applied to the differential operator L. A case of practical importance occurs in parabolic problems, where Galerkin's method leads to implicit difference schemes, with a two-dimensional matrix M—the mass matrix, or Gram matrix, formed from the inner products of the φ_j—to be inverted at each time step. For difference equations it was this kind of difficulty which gave rise to the *alternating direction methods*, in which the inversion is approximated by two one-dimensional computations. For any space formed from the products of one-dimensional elements, the same technique can be applied, with the usual qualification that if the domain is not precisely rectangular, then the success of the alternating direction method is observed but not proved.

In Table 1.2, $M = p + 2q + r$ for an element with p parameters at each

Table 1.2   RECTANGULAR ELEMENTS

Element type	Continuity	d	k	$N = Mn^2$
Bilinear	$\mathcal{C}^0$	4	2	n^2
Biquadratic	$\mathcal{C}^0$	9	3	$4n^2$
Restricted biquadratic	$\mathcal{C}^0$	8	3	$3n^2$
Ordinary bicubic	$\mathcal{C}^0$	16	4	$9n^2$
Hermite bicubic	$\mathcal{C}^{1,1}$	16	4	$4n^2$
Splines of degree $k - 1$	$\mathcal{C}^{k-2,k-2}$	k^2	k	n^2
Hermite, degree $k - 1 = 2q - 1$	$\mathcal{C}^{q-1,q-1}$	k^2	k	q^2n^2
Serendipity, $p > 2$	$\mathcal{C}^0$	$4p$	4	$(2p - 1)n^2$

vertex, q along each edge, and r inside each rectangle. The weights 1, 2, 1 come from the combinatorics of an arbitrary quadrilateral subdivision: The number of vertices equals the number of quadrilaterals equals half the number of edges. There are n subintervals in each coordinate direction.

1.10.   ELEMENT MATRICES IN TWO-DIMENSIONAL PROBLEMS

In this section we shall outline the sequence of operations performed by the computer in assembling the stiffness matrix K, that is, in setting up the discrete finite element system $KQ = F$. The calculation of the *consistent load vector F* will also be briefly described. Our intention is not to give the intimate details which a programmer would require, but rather to clarify the crucial point in the success of the finite element method: polynomial elements like those of the previous section are astonishingly convenient in the practical application of the Ritz method. Perhaps uniquely so.

We recall first the source of the matrix K. The functional $I(v)$ to be minimized has as leading term a quadratic expression $a(v, v)$, representing in many cases the strain energy (or strictly speaking, twice the strain energy). The Ritz method restricts v to a subspace S by introducing as trial functions a finite-dimensional family $v = \sum q_j\varphi_j$. (The superscript h will be dropped in this section, while the Ritz method is applied to a fixed choice of elements.) The substitution of the trial functions v into the strain energy yields a quadratic in the coordinates q_j, describing the energy on the subspace S:

$$(62) \qquad\qquad a(v, v) = q^T K q.$$

The entries of K are the energy inner products $K_{jk} = a(\varphi_k, \varphi_j)$.

In practice these inner products are calculated only indirectly. *The basic computation is the evaluation of the energy integral $a(v, v)$ over each element e,* in other words, over the separate subdomains into which Ω is partitioned. Each such fraction of the overall energy takes the form

$$(63) \qquad\qquad a_e(v, v) = q_e^T k_e q_e,$$

in analogy with (62). The vector q_e now contains only those parameters q_j which actually contribute to the energy in the particular region e. (Of course, the full set of coordinates q_j could be included, at the cost of inserting a host of zeros into the element stiffness matrix k_e. It is much neater to suppress those functions φ_j which vanish in the particular region e, and to maintain the order of k_e equal to the number of degrees of freedom d of the polynomial function within e.) For linear functions $v = a_1 + a_2 x + a_3 y$ on triangles, for example, the matrix k_e will be of order $d = 3$.† With the usual basis, the three components of q_e will be the values of v at the vertices of e. For quadratics $d = 6$, and for the quintic element with 18 degrees of freedom,

$$q_e^{18} = (v^1, v_x^1, v_y^1, v_{xx}^1, v_{xy}^1, v_{yy}^1, v^2, v_x^2, \dots, v^3, v_x^3, \dots).$$

Here, for example, v_x^2 is the x-derivative at the second vertex of the triangle; it equals the weight q_j attached to the basis function φ_j whose x-derivative at this vertex equals one and whose other nodal parameters equal zero. In the following we shall use these quintic functions on triangles as the basic example, since they illustrate most of the difficulties that can arise.

There are now two problems, first to compute the element matrices k_e, and second to assemble them into the overall strain energy

$$a(v, v) = q^T K q = \sum_e q_e^T k_e q_e.$$

The latter is a question of efficient bookkeeping and depends in part on the relative size of the computer and the problem. For problems too large to be handled in core, one possible mode of organization is the *frontal method*, in which the elements (subregions) are ordered rather than the unknowns. The matrices for each element in turn are assembled, and as soon as every element containing some particular unknown Q_n has been accounted for,

---

†The nine entries of the element stiffness matrix for Laplace's equation come directly from the dot products of the sides of the triangle: $k_{ij} = s_i \cdot s_j / 2A$, $A =$ area of the element. Every off-diagonal $k_{ij} \le 0$ unless the triangle is obtuse.

elimination is carried out in the corresponding row of K and the results are stored. In this way, the only unknowns in core at a given moment are those belonging to several elements, some of which have already been assembled and some of which have not.

In this section we shall concentrate on the computation of the element matrices, extending to two dimensions the technique used in Section 1.7 for the Hermite cubic calculations. At the end we comment on the use of numerical integration.

The essential point is that with a polynomial basis φ_j and polygonal (not curved) elements, the energy is a weighted sum of integrals of the form

$$P_{rs} = \iint_e x^r y^s \, dx dy.$$

These integrals will depend on the location of the element e and on constants which can be tabulated. Therefore, the problem is to find a convenient coordinate system in which to describe the geometry of e, and to relate the nodal parameters q_e to the coefficients in the polynomial v. There is general agreement that the "global" x–y coordinates are inappropriate, but there seems to be much less agreement on which local system is the best. We shall therefore describe two or three of the possibilities.

The first is to translate the coordinates so that the origin lies at the centroid (x_0, y_0) of the triangle e. Bell [8] refers to this as the *local–global* system. Since the transformation is linear, a quintic polynomial in x, y remains a quintic in the new variables $X = x - x_0$, $Y = y - y_0$:

$$v = a_1 + a_2 X + a_3 Y + a_4 X^2 + \cdots + a_{21} Y^5.$$

It is easy to find the nodal parameters of q_e^{18} in terms of these coefficients a_i: If the new coordinates of the vertices are (X_i, Y_i), then

$$
\begin{aligned}
v^1 &= a_1 + a_2 X_1 + a_3 Y_1 + a_4 X_1^2 + \cdots + a_{21} Y_1^5, \\
v_x^1 &= v_X^1 = a_2 + 2a_4 X_1 + \cdots, \\
v_y^1 &= v_Y^1 = a_3 + \cdots + 5a_{21} Y_1^4,
\end{aligned}
$$

(64)

and so on. For the 21-degree element the three midedge slopes have also to be computed; the nodal parameters for this element are $q_e^{21} = (q_e^{18}, v_n^1, v_n^2, v_n^3)$. Their relationship with the coefficient vector $A = (a_1, \ldots, a_{21})$ can be written in matrix form as

(65) $q_e^{21} = GA,$

where the first 18 rows of G are given by (64); the last three rows involve not only the coordinates X_i, Y_i of the vertices but also the orientation of the triangle. Inverting (65), we have $A = G^{-1} q_e^{21} = H q_e^{21}$.

For the 18-degree element, the last three rows of G are changed to homogeneous constraints on the coefficients a_i, in order to ensure that the normal slope v_n is a cubic along each edge. This alters the relationship to

$$\begin{pmatrix} q_e^{18} \\ 0 \\ 0 \\ 0 \end{pmatrix} = G_0 A.$$

Inverting, the matrix H which connects q_e to A this time has 21 rows and 18 columns:

$$A = H q_e^{18}.$$

For any element, the first step is to compute this connection matrix H between the nodal parameters q and the polynomial coefficients a.

Now we calculate the energy integral in terms of the coefficients a_i. For the plate-bending example in Section 1.8, this integral is

$$a_e(v, v) = \iint_e \left(v_{XX}^2 + v_{YY}^2 + 2 v v_{XX} v_{YY} + 2(1 - v) v_{XY}^2 \right) dX dY.$$

Substituting the polynomial v, this is easily put in the form

(66) $$a_e(v, v) = A^T N A,$$

where the matrix N requires the evaluation of the integrals P_{rs}. Once these are computed, the calculation of the stiffness matrix is complete. From (66), with $A = H q_e$,

$$a_e(v, v) = q_e^T H^T N H q_e,$$

so that finally

(67) $$k_e = H^T N H.$$

Thus the element stiffness matrix involves two separate computations, the connection matrix H between the nodal parameters and the polynomial coefficients, and a matrix N giving the integrals of polynomials.

A different set of local coordinates was recommended by Cowper, Kosko, Lindberg, and Olson [C14]. In their system the geometrical quantities a, b, c, and θ in Fig. 1.14 become paramount; they are easily computed from the coordinates (x_i, y_i) of the vertices. The trial functions will still be quintic polynomials in the rotated coordinates ξ and η. Therefore, the computation of a new matrix N, giving the integrals of polynomials over the triangle,

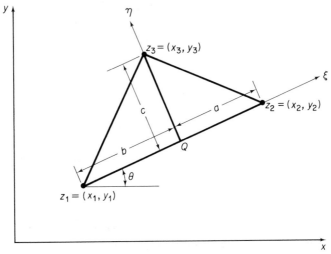

Fig. 1.14   Another choice of local coordinates.

depends on the existence of a convenient formula:

$$\int\!\!\int_{e} \xi^r \eta^s d\xi d\eta = c^{s+1}\big(a^{r+1} - (-b)^{r+1}\big)\frac{r!s!}{(r+s+2)!}.$$

We also need a new connection matrix H, and this is computed in two steps. First, entirely in the ξ–η variables, H' connects the nodal parameters to the polynomial coefficients. Then there is a rotation matrix R to link the local and the global coordinates, and $H = H'R$. The details are clearly explained in [C14] and leave the impression that a cubic variation of the normal slope is simplest to impose in this system.

The computation of $P_{rs} = \int\!\int X^r Y^s\, dX\, dY$ can be carried out without any rotations by introducing *area coordinates*. These are the most natural parameters and are also known to engineers as *triangular coordinates* and to mathematicians as *barycentric coordinates*. In this system each point has actually three coordinates, whose sum is always $\zeta_1 + \zeta_2 + \zeta_3 = 1$. The vertices of e become the points $(1, 0, 0)$, $(0, 1, 0)$, and $(0, 0, 1)$. If the same vertices have rectangular coordinates (X_1, Y_1), (X_2, Y_2), and (X_3, Y_3), then by linearity the coordinates (X, Y) and $(\zeta_1, \zeta_2, \zeta_3)$ of an arbitrary point P are related by

(68)
$$\begin{pmatrix} X \\ Y \\ 1 \end{pmatrix} = \begin{pmatrix} X_1 & X_2 & X_3 \\ Y_1 & Y_2 & Y_3 \\ 1 & 1 & 1 \end{pmatrix}\begin{pmatrix} \zeta_1 \\ \zeta_2 \\ \zeta_3 \end{pmatrix} = B\begin{pmatrix} \zeta_1 \\ \zeta_2 \\ \zeta_3 \end{pmatrix}.$$

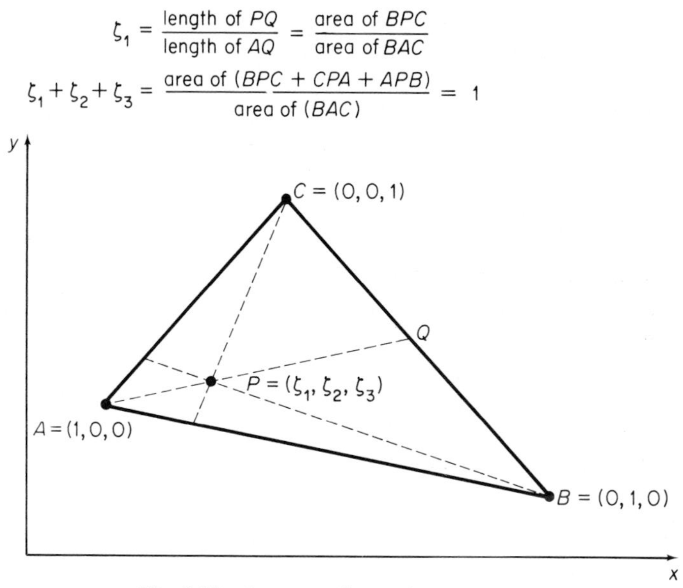

$\zeta_1 = \frac{\text{length of } PQ}{\text{length of } AQ} = \frac{\text{area of } BPC}{\text{area of } BAC}$
$\zeta_1 + \zeta_2 + \zeta_3 = \frac{\text{area of } (BPC + CPA + APB)}{\text{area of } (BAC)} = 1$
y
C = (0, 0, 1)
Q
P = (ζ1, ζ2, ζ3)
A = (1, 0, 0)
B = (0, 1, 0)
x

Fig. 1.15 Area coordinates for a triangle.

Geometrically, ζ_1 is the fraction A_1/A of the area in Fig. 1.15 and also the fraction z_1/h_1 of the distance to the opposite side. In area coordinates, according to (68), the integrals P_{rs} become

$$P_{rs} = \iint (X_1\zeta_1 + X_2\zeta_2 + X_3\zeta_3)^r(Y_1\zeta_1 + Y_2\zeta_2 + Y_3\zeta_3)^s,$$

and we need only the formulas for integration in area coordinates (Holand and Bell [8], p. 84):

$$\iint \zeta_1^m\zeta_2^n\zeta_3^p = \frac{m!n!p!}{(m+n+p+2)!}\det B.$$

This takes into account the Jacobian of the coordinate change; with $m = n = p = 0$, the coefficient 2! appears in the denominator, and the determinant of the matrix B in (68) is recognized as twice the area A of the triangular element.

Holand and Bell [8] have given explicit formulas for the integrals P_{rs}, keeping the origin of the X–Y system at the centroid of the triangle. It is remarkable that for $r + s < 6$, these integrals are all of the simple form

$$P_{rs} = c_{r+s}A(X_1^rY_1^s + X_2^rY_2^s + X_3^rY_3^s).$$

The constants are $c_1 = 0, c_2 = 1/12, c_3 = c_4 = 1/30, c_5 = 2/105$. From these formulas Bell derives the stiffness matrix k_e for the complete quintic applied to the plate-bending problem.

We mention that another local coordinate system could be based entirely on the natural coordinates ζ_i. In this case the parameters q_e, which involve nodal derivatives like v^1_{xy}, must be converted into derivatives with respect to the ζ_i. It is our impression that algebraic manipulations in this system take a considerable amount of time (and practice!), although Argyris and his coworkers did succeed with this system in forming $H = G^{-1}$ analytically. This natural coordinate system really comes into its own with numerical quadrature.

Next we turn to the computation of the *mass matrix M*. This is the matrix which arises from an undifferentiated term $\iint v^2 \, dx \, dy$ in the energy. In other words, writing $v = \sum q_j \varphi_j$, M is the *Gram matrix*† formed from the inner products of the basis functions φ_j:

$$M_{jk} = (\varphi_k, \varphi_j) = \iint_\Omega \varphi_k \varphi_j \, dxdy.$$

This matrix plays a central part in eigenvalue calculations.

Just as with K, the mass matrix is assembled from the separate elements:

$$(69) \qquad \iint_\Omega v^2 = q^T M q = \sum q_e^T m_e q_e = \sum_e \iint_e v^2.$$

The element mass matrix, by our previous computations, can be written as

$$m_e = H^T Z H.$$

Here H is the same matrix connecting the parameters q_e to the coefficients of the polynomial

$$v = \sum_{i=1}^{d} a_i X^{m_i} Y^{n_i}.$$

Z comes directly from the inner products of these monomials:

$$Z_{jk} = \iint_e X^{m_j} Y^{n_j} X^{m_k} Y^{n_k} = P_{m_j + m_k, \, n_j + n_k}.$$

These numbers are tabulated, and any of the local coordinate systems should be acceptable. Thus Z plays the role for mass matrices which N played for stiffness matrices. In both cases we have tacitly assumed constant coefficients

---

†We have tried to think of a pun on the coincidence of mass matrix and Gram matrix, but the editor says it will have to wait.

over the elements; if there are variable material properties which must be taken into account, then numerical integration is recommended.

Finally, there is the calculation of the load vector F, which is the inhomogeneous term in the finite element equations $KQ = F$. This, too, is done an element at a time:

$$(70) \qquad \iint_\Omega fv = q^T F = \sum q_e^T F_e = \sum \iint_e fv.$$

Again the calculation starts with $A = Hq_e$, connecting the nodal parameters of the element to the coefficients a_i in the polynomial. In terms of these coefficients,

$$(71) \qquad \iint_e fv = \sum a_i \iint_e f X^{m_i} Y^{n_i} = A^T \sigma = q_e^T H^T \sigma,$$

where the components of the d-dimensional vector σ are just

$$\sigma_i = \iint_e f X^{m_i} Y^{n_i}.$$

Matching (71) with (70), the required element load vector is

$$F_e = H^T \sigma.$$

Two special cases in the calculation of the σ_i are pointed out in [C14]. The first is that of a uniform load $f(x, y) = f_0$, in which case the tabulated integrals appear again:

$$\sigma_i = \iint f_0 X^{m_i} Y^{n_i} = f_0 P_{m_i, n_i}.$$

The second is that of a point load $f(x, y) = f_0 \delta$, concentrated at the point (X_0, Y_0). If this point is outside e, then of course $F_e = 0$; if it is inside, then

$$\sigma_i = f_0 X_0^{m_i} Y_0^{n_i}.$$

Again these calculations can be done in any local system.

For a more general loading f, the integrals σ_i may be computed either by a quadrature formula over each element, or else by interpolating f by an element which lies in the subspace S^h. In the latter case we compute the nodal parameters f_j—the values of f and its derivatives at the nodes—and construct the *interpolating element*

$$(72) \qquad f_I = \sum f_j \varphi_j.$$

The change from f to f_I gives a perturbed load vector $\tilde{F}$ with components

$$\tilde{F}_k = \iint f_I \varphi_k = \iint \sum f_j \varphi_j \varphi_k = M f'.$$

M is the mass matrix described above and f' is the vector formed from the nodal parameters f_j. This is a comparatively easy computation.

So far we have considered only analytical computation of the element matrices, based on exact integration formulas for polynomials over polygons. *In more and more cases, the element matrix computations are being carried out approximately, by some form of Gaussian quadrature.* For curved elements, arising from shell problems or from curved boundaries in planar problems, this numerical quadrature is virtually a necessity. In this case there will be a perturbation not only from F to $\tilde{F}$ but also from K to $\tilde{K}$. This means that the corresponding $\tilde{Q}$ is the finite element solution for a perturbed problem. In a later chapter we shall estimate the error $Q - \tilde{Q}$; it must depend on the accuracy of the quadrature formula.

We regret to report that these inexact numerical integrations have even been shown in some cases to *improve* the quality of the solution. This is one instance (nonconforming elements are another) in which computational experiments yield results which are frustrating to the mathematical analyst but nevertheless numerically valid and important. The improvement for finite h is due partly to the following effect: The strict Ritz procedure always corresponds to an approximation which is *too stiff*, and the quadrature error reduces this excess stiffness.

Stiffness is an intrinsic property of the Ritz method. By limiting the displacements v to a finite number of modes $\varphi_1, \ldots, \varphi_N$, instead of allowing all admissible functions, the numerical structure is more constrained than the real one. In eigenvalue problems, the result of this constraint is that λ_j^h always lies above the true λ_j. In static problems, the potential energy $I(u^h)$ exceeds $I(u)$, since u^h is obtained by minimization over the subspace spanned by $\varphi_1, \ldots, \varphi_N$. This overestimate of I corresponds to an *underestimate* of the strain energy a, as proved in the corollary to the fundamental theorem 1.1:

$$(73) \qquad\qquad a(u^h, u^h) \leq a(u, u).$$

In the special case of a point load $f = \delta(x_0)$, where the displacement $u(x_0)$ is proportional to the strain energy, this displacement is also underestimated by the Ritz method; the overstiff numerical structure has less "give" under a point load than the true structure. For distributed loads, the tendency is the same: The finite element displacement $u^h(x)$ is generally below the true displacement $u(x)$. This is not a rigorous theorem, because the Ritz method minimizes energy, and its link to displacement is not strictly monotone.

In other words, u^h may exceed u in some part of the structure and still have smaller derivatives in the mean-square sense. Nevertheless, one-sided estimates of both displacement and slope are common with finite elements. This can be changed either by a fundamental alteration of the Ritz process (the stress, mixed, and hybrid methods are described in the next chapter) or by intentionally committing numerical errors in the right direction.

The latter effect is the one produced by Gaussian quadrature. Algebraically, this effect appears as a decrease in the positive definiteness of K; the approximate $\tilde{K}$ satisfies

(74) $$0 \leq q^T \tilde{K} q \leq q^T K q \qquad \text{for all } q.$$

It may seem paradoxical that this leads to an *increase* in the strain energy is the Ritz solution, but the paradox is easily explained. The new solution is $\tilde{Q} = \tilde{K}^{-1} F$. Its strain energy is $\tilde{Q}^T \tilde{K} \tilde{Q} = F^T \tilde{K}^{-1} F$, whereas the old strain energy was $Q^T K Q = F^T K^{-1} F$. Since the inequality (74) on the stiffness matrix is equivalent (after some argument) to the opposite inequality on their inverses,

$$F^T \tilde{K}^{-1} F \geq F^T K^{-1} F \geq 0 \qquad \text{for all } F,$$

and it follows that the strain energy is increased.

In case of a point load β at the jth node, this leads as in the continuous case to a similar conclusion about the displacement. The only nonzero component of the load vector F is $F_j = \beta$, and the displacement at the jth node is $Q_j = (K^{-1} F)_j = \beta (K^{-1})_{jj}$. If K is decreased to $\tilde{K}$, this displacement is increased to $\beta (\tilde{K}^{-1})_{jj}$.

It remains to see why Gaussian quadrature tends to have this desirable effect of reducing K. Obviously K will be reduced if every element matrix k_e is reduced. A rigorous proof of the latter effect was found by Irons and Razzaque [16] in the one-dimensional case, by imagining that $(v^h)'$ is expanded in terms of Legendre polynomials. The strain energy over $[-1, 1]$ is

$$q^T k_e q = \int ((v^h)')^2 = \int (\alpha_0 + \alpha_1 P_1(x) + \cdots + \alpha_n P_n(x))^2$$
$$= 2\left(\alpha_0^2 + \frac{\alpha_1^2}{3} + \cdots + \frac{\alpha_n^2}{2n+1}\right).$$

The effect of n-point Gaussian quadrature is to preserve all the terms through α_{n-1}^2, since they come from polynomials of degree less than $2n$, and to annihilate the term in α_n^2, since P_n vanishes at the Gauss points. (This is how they are defined.) Therefore in this special case the integral is clearly reduced, and the same tendency reappears in more general problems.

The *least stiffness matrix* criterion is a valuable basis for the comparison of different elements. Mathematically, it must reflect the approximation qualities of these elements, and in particular the numerical constants which enter approximation of polynomials one degree higher than those reproduced exactly by the element. Computationally, it is a clearly visible effect, and a number of theoretically useful elements have been discarded as too stiff. The criterion raises the possibility of optimizing the stiffness matrix under the constraint of reproducing polynomials of degree $k - 1$.

2 A SUMMARY OF THE THEORY

2.1. BASIS FUNCTIONS FOR THE FINITE ELEMENT SUBSPACES S^h

In this chapter we collect in one place some of the main results of analyzing the finite element method. The goal is to describe a general framework for the analysis, in which each of the various error estimates has its place. Then succeeding chapters will take up these estimates one by one, in more detail.

The first step is to decide which subspaces S^h to study. Therefore, we look at the examples already given in Chapter 1 and try to draw out the properties which are mathematically essential.

One general description which fits these examples is the following, which we call the *nodal finite element method*. It will be the foundation for our whole analysis. Every trial function v^h is determined by its nodal parameters, which are the unknowns q_j of the discrete problem. Each of these nodal parameters is the value, at a given node z_j, of either the function itself or one of its derivatives. Thus the unknowns are

$$q_j = D_j v^h(z_j),$$

where the differential operator D_j is of order zero ($D_j v^h = v^h$) in case the parameter is just the function value, and otherwise it may be $\partial/\partial x, \partial/\partial y, \partial/\partial n$, $\partial^2/\partial x\, \partial y$, and so on.

To each of these nodal parameters q_j we associate a trial function φ_j, determined by the following property: At z_j the value of $D_j \varphi_j$ is 1, whereas all other nodal parameters of φ_j are zero. Note that it is possible for the node z_j to be shared by several parameters, in other words to be a multiple node; it is not z_j but the pair (z_j, D_j) which uniquely defines the parameter q_j. Thus the key property is that each φ_j satisfies $D_j\varphi_j(z_j) = 1$ but gives zero when matched with a different pair (z_i, D_i):

(1) $$D_i\varphi_j(z_i) = \delta_{ij}.$$

These functions φ_j form an *interpolating basis* for the trial space, since every trial function can be expanded as

$$v^h = \sum q_j\varphi_j.$$

By applying D_i to both sides and evaluating at z_i, the parameters q_i are exactly what they should be:

$$q_i = D_iv^h(z_i).$$

The whole object of the Ritz method is to find the optimum values Q_j for these parameters, by minimizing $I(v^h)$. Then the particular trial function which possesses these optimal parameters is the finite element approximation

$$u^h = \sum Q_j\varphi_j.$$

Let us fit two or three examples into this general framework. For the basic piecewise linear functions on triangles (Turner triangles, or Courant triangles) the nodes z_j are the vertices in the triangulation, and all derivatives D_j are of order zero, $D_jv^h = v^h$. The unknowns are $q_j = v^h(z_j)$, and the basis is formed from the pyramid functions determined by $\varphi_j(z_i) = \delta_{ij}$. The same is true for bilinear functions on quadrilaterals and for quadratics on triangles, this time with midedge nodes included in the z_j. For the Hermite cubics in one dimension, the derivatives do enter: Every node z_j appears in both pairs (z_j, I) and $(z_j, d/dx)$. We distinguished the two kinds of basis functions φ_j by ψ_j and ω_j. For Hermite bicubics there are four parameters per node, corresponding to v, v_x, v_y, and v_{xy}, that is, to $D = I, \partial/\partial x, \partial/\partial y$, and $\partial^2/\partial x\, \partial y$. For the cubic space Z_3 on triangles, the vertex nodes are triple and the centroids are only simple.

To complete the description of the nodal finite element method, we say something about the geometry. The domain Ω is partitioned into a union of closed subdomains, or elements, overlapping only at the interelement boundaries. *Each element e contains no more than d of the nodal points z_j, and all the basis functions φ_j, except those which correspond to these d nodes, are zero throughout e.* Thus φ_j is also a *local basis*. This framework seems to be sufficiently general to include most of the finite element spaces in current use. We note, however, that cubic splines are not included, since their basis functions (the B-splines) are nonzero over several elements.

We call attention to one additional property of the φ_j which can be put to use in the analysis. This property emerges only when the elements are formed in a geometrically regular pattern, where the domain can be thought of as covered by a regular mesh of width h, and the nodes fall in the same positions in each mesh square. The property of the basis is then one of *translation invariance*: If a basis function φ is associated with a pair (z, D), and this node z is translated to lie at the new point $z^* = z + lh$, then *the basis function φ^* associated with (z^*, D) is just the translate of φ*:

$$(2) \qquad\qquad \varphi^*(x) = \varphi(x - lh).$$

In n dimensions, $l = (l_1, \ldots, l_n)$ is a vector of integers, that is, a multiinteger. Clearly $D\varphi^*(z + lh) = D\varphi(z) = 1$; the interpolating function φ is just shifted to give a function φ^* interpolating at z^*. Of course, this pattern of translation will have to break down at the boundary of Ω unless the boundary conditions happen to be periodic; in that case we can imagine that the basis has a completely periodic pattern.

Let us take as an example the space of continuous piecewise quadratic functions which was described in Section 1.7. The nodes fall on a regular triangular mesh, as indicated in Fig. 2.1, where the mesh width is normalized

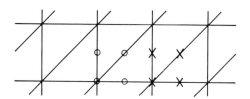

Fig. 2.1 Nodes for C^0 quadratics on a square mesh.

to unity. Note that we can associate with each mesh point a unique set of four nodes; the four associated with the origin are circled and those for the mesh point $(1, 0)$ are crossed. Note also that this number $M = 4$, which gives the number of basis functions φ_j and unknown weights q_j per mesh-square, differs from the number d of degrees of freedom; $d = 6$ in each triangle. *M is the coefficient in the last column of the table in Section 1.9, giving the dimension of the space S^h.* On a unit square, with periodicity, this dimension is M/h^2.

We shall denote by $\Phi_1, \Phi_2, \Phi_3,$ and Φ_4 the basis functions which interpolate at the nodes z_1 to z_4, respectively. They are piecewise quadratic, and vanish at every node except their own; this determines them completely. The basis functions at the four nodes associated with the mesh point (l_1, l_2) are found just by translation as described above; these shifted functions are $\Phi_i(x - l_1, y - l_2)$, $i = 1, 2, 3, 4$. If the meshwidth is reduced to h, this just rescales the independent variables; the four basis functions associated with the mesh point (l_1h, l_2h) are $\Phi_i(x/h - l_1, y/h - l_2)$.

This pattern of translation is so useful and important that we shall make it the foundation for a second general description of the finite element method. This description applies on a regular mesh, and we call it the *abstract finite element method*. In n dimensions it begins with the choice of M functions $\Phi_1(x), \ldots, \Phi_M(x)$. These will eventually lead to M unknowns per mesh cube, and the finite element equations $KQ = F$ will take the form of M *coupled finite difference equations*.

To form the basis functions associated with one particular grid point, the one at the origin, we simply rescale the variable $x = (x_1, \ldots, x_n)$, obtaining $\Phi_1(x/h), \ldots, \Phi_M(x/h)$. To form the basis functions associated with a different grid point $lh = (l_1, \ldots, l_n)h$, we translate the functions just constructed. Thus to denote all the basis functions constructed in this way, by scaling and translation, we need two subscripts i and l:

$$\varphi_{i,l}^h(x) = \Phi_i(x/h - l), \qquad i = 1, \ldots, M.$$

We must add the requirement that *the original functions $\Phi_i(x)$ shall vanish outside some sphere $|x| \leq R$*. Then $\varphi_{i,l}^h$ will vanish outside the local sphere $|x - lh| \leq Rh$, and we shall again have a local (but possibly not a strictly local) basis.

Note that in the abstract finite element method, we do *not* require the Φ_j to interpolate at some node z_j, and we do not require them to be piecewise polynomials. (We prove, however, that the latter are the most efficient.) The interpolating property will be possessed by every basis which falls also within the scope of the nodal finite element method, but our theory in the abstract case will not use this property. As an example we consider the case of cubic splines in one dimension. In this case $M = 1$, and a suitable choice for Φ_1 is the B-spline of Fig. 1.9. The space S^h of all combinations $\sum q_l \varphi_{1,l}^h$ is then exactly the spline space of all twice differentiable piecewise cubic functions with joints at $x = 0, \pm h, \pm 2h, \ldots$. It is essential to see that the B-spline Φ_1 is not interpolating; it is nonzero over *four* intervals instead of the two allowed in the nodal method, and in particular it is nonzero at three nodes instead of one.

The spline subspaces S^h included in the abstract finite element method have been used with some success in one-dimensional applications. The assembly of the equation $KQ = F$ and the treatment of the boundary require modifications in the technique which is standard in the nodal case, but the ground rules are still those common to any form of the Ritz method. The chief advantage in splines is that the extra continuity reduces the dimension of the trial space, without destroying the degree of approximation. In the nodal case, the Hermite cubic polynomials are determined over each subinterval by v and v' at the endpoints; this means $M = 2$ parameters for each mesh point. Their generating functions Φ_1 and Φ_2 are the ψ and ω of Fig. 1.8.

Experiments have been carried out with splines in two dimensions, where the reduction in M is still greater, but the boundary is almost forced to be regular if there is to be any comparison with the simplicity of the nodal method. The bandwidth of the stiffness matrix K is of course not necessarily decreased by a reduction in M; for cubic splines in one variable there will be seven nonzero entries in each row of K, since the B-spline Φ_1 stretches over four intervals. This is virtually the same bandwidth as in the Hermite case, where the usual ordering of unknowns gives a matrix of the form

$$K = \begin{pmatrix} & \cdot & & \cdot & & \cdot & \\ & \begin{pmatrix} x & x \\ x & x \end{pmatrix} & \begin{pmatrix} x & x \\ x & x \end{pmatrix} & \begin{pmatrix} x & x \\ x & x \end{pmatrix} & \\ & & \begin{pmatrix} x & x \\ x & x \end{pmatrix} & \begin{pmatrix} x & x \\ x & x \end{pmatrix} & \begin{pmatrix} x & x \\ x & x \end{pmatrix} & \\ & & \cdot & & \cdot & & \cdot \end{pmatrix}.$$

The order of the matrix K is of course directly affected by the number M, and splines do appear to be more efficient (at least in solution time) when the domain Ω is conveniently rectangular. This is confirmed in the calculations of Chapter 8.

We frankly believe that the nodal method allows such flexibility in the geometry that it will continue to dominate the use of splines. Since the interpolating property is not essential to part of the analysis, however, it will be convenient when on a regular mesh to be able to use the more general description in terms of functions $\Phi_1, \ldots, \Phi_M$. Since everything in the construction of S^h depends on these functions, they must contain in themselves the answers to all our questions about approximation and numerical stability. Therefore, this abstract approach turns a significant part of the finite element analysis into an agreeable problem in function theory.

2.2.   RATES OF CONVERGENCE

Suppose that u is the solution to an n-dimensional elliptic variational problem of order m. This means that u minimizes $I(v)$ over an admissible class $\mathcal{H}_E^m$, restricted by homogeneous or inhomogeneous essential boundary conditions, and (by ellipticity) that the strain energy is positive definite: $a(v, v) \geq \sigma \|v\|_m^2$. Suppose also that u^h is the minimizing function over the trial class S^h and that $\tilde{u}^h$ is the solution when the problem is perturbed by numerical integration; the coordinates of $\tilde{u}^h$ satisfy $\tilde{K}\tilde{Q} = \tilde{F}$. Suppose, finally, that $\bar{\bar{u}}^h$ is the solution which is actually computed, differing from $\tilde{u}^h$ because of roundoff error in the numerical solution. Obviously the three approximations u^h, $\tilde{u}^h$, $\bar{\bar{u}}^h$ include progressively more sources of error. We want to summarize, for problems with smooth solutions and for typical finite elements, what are the orders of magnitude of these errors.

The starting point is always the same: If S^h is a subspace of $\mathcal{H}_E^m$, then by the fundamental Theorem 1.1

$$(3) \qquad a(u - u^h, u - u^h) = \min_{v^h \text{ in } S^h} a(u - v^h, u - v^h).$$

The theorem still holds with inhomogeneous essential conditions, where the difference of any two functions in $\mathcal{H}_E^m$ lies in V_0 and the difference of any two functions in S^h lies in S_0^h, as long as S_0^h is a subspace of V_0. (For proof see Section 4.4.) Therefore, the strain energy in $u - u^h$ is a question in pure approximation theory, to estimate the distance between u and S^h. The basic hypotheses will be, first, that S^h *is of degree* $k - 1$—it contains in each element the complete polynomial of this degree, restricted only at the boundary by essential conditions—and second, that *its basis is uniform* as $h \to 0$. The latter is effectively a geometrical condition on the element regions: If the diameter of e_i is h_i, then e_i should contain a sphere of radius at least τh_i, where τ is bounded away from zero. This forbids arbitrarily small angles in triangular elements. Angles near π must also be forbidden in quadrilaterals. Under these conditions, the distance between u and S^h is

$$a(u - u^h, u - u^h) \leq C^2 h^{2(k-m)} |u|_k^2.$$

Therefore, $h^{2(k-m)}$ is the rate of convergence in strain energy.

The form of this error bound is typical of the results of numerical analysis. There are three factors. The power of h is the simplest to find, since it depends only on the degree of the polynomials, and it indicates the rate of convergence as the mesh is refined; this effect should be clearly observable numerically. The constant C depends on the construction of the element and its nodal parameters. For regular geometries a good asymptotic value of C can be computed as the error in approximating polynomials of the next higher degree k (Section 3.2). The third factor, $|u|_k$, reflects the properties of the problem itself—the degree to which the solution is smooth, and therefore easy to approximate accurately. This norm is the mean-square value of the kth derivatives of u and therefore—according to partial-differential-equation theory—directly related to the derivatives of order $k - 2m$ of the data f.

Notice that *convergence occurs if and only if* $k > m$; this is the constant strain condition, that *the elements should reproduce exactly any solution which is a polynomial of degree* m. This requirement for convergence appeared very gradually in the engineering literature, partly developing from intuition and partly fron the numerical failures which were observed when it was violated. (Most notably in the biharmonic case $m = 2$ of plate bending, when the twist term xy was left out of S^h.) We shall describe a rigorous proof—as far as we know, the first proof—that this condition is necessary for convergence in the case of a regular mesh. Such a theorem fits naturally into the abstract finite element theory, which admits the most general trial functions on a regular subdivision.

We have stated the constant strain condition as if it were necessary also for irregular meshes, but it is not. At least it is not quite necessary. Ω could be mapped into another domain Ω' by a fixed smooth invertible transformation T which does not preserve polynomials, and then elements which satisfy the constant strain condition on Ω will not do so in the new variables. However, convergence on Ω implies convergence of the corresponding problem on Ω', since we can go freely back and forth. The effect of the Jacobian on errors for isoparametric elements is considered in Section 3.3.

Convergence in strain energy is essentially convergence of the mth derivatives of u^h to those of u. This derivative is therefore special, because the Ritz method is minimizing in energy. For the sth derivative, where s may be smaller or larger than m, convergence cannot be faster than $O(h^{k-s})$; this is the order of the best approximation to u from a space S^h of degree $k - 1$. This rate of convergence will normally be attained by the Ritz solution u^h. Using the "Nitsche trick," which gave the $O(h^2)$ rate in displacement for one-dimensional linear elements in Section 1.6, we shall show that

$$(4) \qquad\qquad \|u - u^h\|_s = O(h^{k-s} + h^{2(k-m)}).$$

In almost all cases the first exponent is smaller, and approximation theory governs the convergence rate. [For $s = -1$, the left side is effectively the mean error in displacement over an element, and we see that it may well be one order better (h^{k+1}) than the displacement itself.] There are, however, cases in which the $h^{2(k-m)}$ term would enter: If we imagined cubic splines applied to a sixth-order problem, or more realistically a plate-bending (fourth-order equation) element complete only through quadratics, then the rate would be limited to $2(k - m) = 2$ even in displacement.

The convergence rate at individual points can be expected to be the same, provided u has k derivatives at every point. (With singularities, the order of differentiability and therefore the rate of convergence are quite different in the mean-square and pointwise senses. We shall not give a detailed proof of optimum pointwise error bounds.) *At special points the error may actually converge faster than it does in the mean.* With linear elements for $-u'' = f$, for example, the nodes themselves were special: $u^h = u_I$, and the Ritz solution is exact at the nodes. This is true in general if the elements are solutions to the homogeneous differential equation [H6, T4]. For the heat equation, Thomée noticed a special convergence rate at the nodes of splines; Douglas and Dupont are extending this principle to their collocation methods.

The rate of convergence is maintained with inhomogeneous boundary conditions (Section 4.4), provided the boundary data are interpolated (or approximated) by polynomials of at least the same degree $k - 1$.

For a domain Ω of irregular shape, a new type of approximation error may enter. It will generally be necessary to approximate the boundary Γ by a piecewise polynomial boundary Γ^h. In the simplest case Γ^h is piecewise linear; Ω is replaced by a polygon Ω^h. Such a polygon can be carved into triangles, and the finite element method may proceed by ignoring the skin $\Omega - \Omega^h$ between the true boundary Γ and the polygon. Therefore, it is as if the original differential problem were moved to Ω^h. In Section 4.4 we investigate the effect of this *change of domain*. Briefly, the mth derivative is in error by $O(h)$ at the boundary, but this error decays rapidly in the interior. There is a boundary-layer effect, and the mean error is $O(h^{3/2})$. Since the strain energy depends on the square of this derivative, the error in energy due to computing on Ω^h (with natural or essential boundary conditions) is $O(h^3)$. This estimate applies to Ω as well as Ω^h if the finite element solution is extended in the natural way, by extending each polynomial out to Γ. Otherwise all the energy in the skin will be lost, and this amounts to $O(h^2)$—proportional to the volume of the skin.

Note that the h^3 error due to change of domain will dominate when the finite elements used in the interior are of degree higher than m. If the polynomials are of the minimum degree m which is required for convergence, then the change of domain effect will be submerged (at least in the interior of Ω^h) by the $h^{2(k-m)} = h^2$ error which arises from ordinary approximation theory.

If the boundary Γ is approximated by piecewise polynomials of higher degree l, there is a corresponding reduction in the change of domain error. The error in the mth derivative (the strain) at the boundary is $O(h^l)$, and in the overall strain energy in Ω^h it is $O(h^{2l+1})$. This assumes, in case of an essential condition $u = 0$, that the condition is exactly satisfied by the polynomial trial functions on the approximate boundary. Mitchell has found a neat construction of cubic elements which vanish on a boundary made up of piecewise hyperbolas, with h^5 error in strain energy and h^3 in displacement.

For an essential condition on a curved boundary, say $u = 0$, an alternative is to use any of the standard elements and to require that they interpolate the conditions at boundary nodes. In this case the trial functions will not satisfy the essential condition along the whole boundary; each trial function may vanish along some curve close to Γ, but this zero curve will vary from one function to the next. As a result, the Ritz theory will not apply; the trial functions are not in $\mathcal{H}_E^m$, either on the exact Ω or on an approximate Ω^h. It no longer follows that u^h, which is still chosen to minimize $I(v)$, is the closest trial function to u. Nevertheless, it is possible to estimate the error, taking into account that each v^h is nonvanishing but small on the boundary. The best error bound on the strain energy—which surprised us by being rather low—is normally of order h^3. Roughly speaking, if we work with cubics on triangles and interpolate the essential boundary conditions, the worst of these cubics will still violate the conditions by $O(h^{3/2})$ between nodes; in Section 4.4, we deduce from this the $O(h^3)$ error in energy.

There is still another alternative, which is undoubtedly the most popular. Curved elements can be straightened out by a change of coordinates. Such a transformation may even be necessary to achieve continuity between quadrilaterals whose sides are *already* straight, unless they happen to be rectangles. These coordinate changes are a central technique in the finite element method.

In theory we can straighten almost any boundary curve, but of course in practice that is absurd. Piecewise polynomials are the best element boundaries for the same reasons that they are the best approximations to the displacements: they can be handled efficiently by the computer. In fact, we may describe the coordinate changes by the same class of polynomials that are used as trial functions; this is the method of *isoparametric transformations*. It is a brilliant idea. Coordinate changes lead to the same difficulties as for trial functions: The mapping must be continuous across element boundaries, so that elements which are adjacent in the original x–y plane will remain adjacent in the ξ–η plane. If the transformations $x(\xi, \eta)$, $y(\xi, \eta)$ are constructed from nodal parameters in the standard way, and if we are assured of continuity in ξ and η (as we are for the standard rectangular or triangular elements), then the isoparametric mappings will succeed even for elements whose boundaries are polynomials of degree $k - 1$ in x and y. This technique raises new questions of approximation theory, since polynomials in ξ and η are no longer polynomials in x and y. Nevertheless, isoparametric transformations need not decrease the order of accuracy; the full order h^{k-s} in the sth derivative is achieved, provided these transformations are uniformly smooth (Section 3.3). In this sense *the isoparametric technique is the best one* for second-order equations and curved boundaries. An essential boundary condition $u = g$ can be handled naturally and efficiently, with no loss in the fundamental order of accuracy.

The errors enumerated so far have all contributed to $u - u^h$; we have assumed the Ritz approximation to be calculated exactly. In practice, however, there will also be errors in numerical integration (yielding $\tilde{u}^h$ instead of u^h) and in the solution of the final linear system (yielding $\bar{u}^h$ instead of $\tilde{u}^h$). It is essential to know the scale of the integration errors, so that a quadrature rule can be chosen which is neither wasteful of time nor destructive of accuracy. We mentioned in the example of Chapter 1 the two leading possibilities: (1) the inhomogeneous data and any material coefficients which vary with x can be replaced by interpolating polynomials and the integrals for the resulting problem computed exactly; or (2) integrations over all elements can be carried out from the beginning by a standard quadrature rule, say Gaussian quadrature. For a simple problem like $-(pu')' + qu = f$, the user has some freedom of choice, and we mention one result of the analysis: Both interpolation by polynomials of degree $k - 1$, and Gauss quadrature with $k - 1$ points, yield errors in the strains of order h^k. For more complicated problems, method (2) is strongly indicated, and in fact numerical integration has become one of the major components of the finite element method. It produces a solution $\tilde{u}^h$ whose convergence to u requires a certain accuracy in the numerical quadrature: *The mth derivatives of all trial functions must be integrated exactly.* For each additional degree of exactness, the error $\tilde{u}^h - u^h$ due to numerical quadrature improves by a power of h. The proof depends on an identity given in Section 4.3, and we also determine the extra accuracy required for isoparametric elements.

Finally we come to the roundoff error. This has a character completely different from the others; it is proportional to a *negative* power of h. As h decreases there is a region of crossover, prior to which roundoff is negligible and uninteresting, and after which it is all-important. The roundoff does not depend too strongly either on the degree of the polynomials used or on the number of dimensions; the key factor h^{-2m} is set by the mesh width and the order of the equation itself. For second-order equations the crossover usually occurs below the mesh widths which are typical in practical problems, but for equations of order 4 it is not so delayed; accurate calculations may well require double precision. We discuss in Chapter 5 both a priori estimates of the condition number and a posteriori estimates of the roundoff actually committed in a given problem.

These are the principal errors to be analyzed for linear static problems $Lu = f$. In every case, the analysis is based on the fundamental variational equation, the vanishing of $a(u, v) - (f, v)$. In comparison with finite differences, there can be no dispute that this produces a more coherent and satisfactory mathematical theory. In part, the same techniques can be extended to *nonlinear equations;* this must surely be at present the outstanding theoretical problem, to isolate classes of nonlinear problems which are both physically important and mathematically amenable, and to study the dependence of the approximation on the trial space, domain, and coefficients. Lions has made tremendous progress in this direction (see [9]).

Two classes have already been isolated, as yielding the simplest generalizations of linear elliptic equations. We cannot say how comprehensive are their applications in engineering and physics, but they appear to be very natural. One is the class of *strictly monotone operators*, satisfying

$$\int_\Omega [M(u) - M(v)](u - v)\, dx \geq \sigma \, \|u - v\|_m^2.$$

The other (closely related) class contains the *potential operators*, those for which M is the gradient of some nonquadratic convex functional $I(v)$. We refer to the book of Vainberg [18] for an introduction to the theory.

Galerkin's method applies directly to these operators; we seek a function $u^h = \sum Q_j \varphi_j$ in the subspace S^h such that

$$(M(u^h), \varphi_k) = 0 \qquad \text{for all } \varphi_k.$$

The number of (nonlinear) equations equals the number of unknown coefficients; that is, it equals the dimension of the trial space S^h. For the two classes described above, it is possible to prove the existence of such a solution u^h and its convergence to u—provided a suitable continuity condition is imposed on the operator. In fact, one possible existence proof for the solution u follows these lines; the existence of u^h is demonstrated in finite dimensions, as well as an a priori bound which establishes that all u^h lie in some compact set. Then the u^h must have a limit point as $h \longrightarrow 0$, and this is u. Ciarlet, Schultz, and Varga [C4] have shown that the error estimates differ very little between linear and monotone nonlinear problems.

The finite element method is in constant use for nonlinear problems, for example in elastic-plastic materials or in thermo-viscoelasticity. In addition to the text by Oden [15], there is a rapidly growing engineering literature; we mention among many others the early survey articles by Martin [M2] and Marcal [M1]. It seems fruitless for us to repeat here the possible finite element formulations, distinguishing between the techniques for geometric nonlinearities due to large deflections and those for material nonlinearities, without doing something substantial about the mathematics. It is absolutely clear that the convergence problems are well posed, extremely interesting, and ripe for solution. We hope that ultimately this mathematical theory will become sufficiently complete to justify a book (by someone else).

We do want to include one warning about nonlinear equations. It applies to problems such as nonlinear elasticity, and we take as a simple model the minimization of a functional like

$$I(v) = \int [p(v, v_x)v_x^2 - 2fv] \, dx.$$

Since $I(v)$ is not quadratic in v, the vanishing of the first variation will not be a linear equation for u. Therefore some iteration will be considered, the simplest being the method of *successive substitution*: evaluate the nonlinear coefficient at the nth approximation u_n, and determine u_{n+1} as the solution of a linear problem.

Our warning is this: if such an iteration is applied to the variational problem, so that u_{n+1} minimizes

(d)
$$\int \left[p\left(u_n, \frac{du_n}{dx}\right)v_x^2 - 2fv \right] dx,$$

then *the iteration will converge to the wrong answer*. The reader can easily verify that the limit u^* of such an iteration satisfies

$$-\frac{d}{dx}\left(p\left(u^*, u_x^*\right)\frac{du^*}{dx}\right) = f,$$

which is not the equation of virtual work for $I(v)$. (Take $p = v_x$.) The mistake was to perform successive substitutions before taking the first variation; if we first establish the nonlinear equation for the minimizing function u, and *then* solve that equation by iteration, the limit will be the right one.

There is one particular nonlinear problem, describing the deformation of an elastic-plastic material, which illustrates both the possibilities and the difficulties in nonlinear convergence proofs. It might be useful to present some of the details. For simplicity of notation we consider a one-dimensional model, with strain du/dx; the same arguments apply to the system of strains ϵ_{ij} occurring in two- or three-dimensional elasticity. The deformation is a nonlinear function of the external forces, and cannot be determined from a knowledge only of the final loading. Instead it depends on the history of the problem, that is on the chronological order in which the forces are applied over the domain. This introduces an artificial "time" parameter, and at a given instant t the rate of change $\dot{u}$ of the deformation is the function which minimizes

$$I(\dot{v}) = \int [p(\dot{v}_x)^2 - 2\dot{f}\dot{v}] \, dx.$$

The crucial quantity is the elastic modulus $p(x)$. If this coefficient is independent of the stresses within the material, then there is no need for the artificial time; the final deformation $u(T)$ can be determined by a single minimization, as in the rest of this book, using the final load $f(T)$. In the case of a nonlinear stress-strain law, the coefficient p at time t depends on the stresses, and not only on their values at the given instant: *it depends on the whole stress history.* This "path-dependence" enters physically in several ways. For example, once the elastic limit has been exceeded, a loading followed by an equal unloading leaves a net change in the state of stress in the material. This phenomenon actually creates a nonlinear problem for $\dot{v}$ at each instant of time, since the elastic modulus then depends not only on the past history but also on the current rate of change; p is influenced by the *sign* of $v_x\dot{v}_x$, and has one value for loading and another for unloading. For simplicity we shall avoid this extra difficulty, and assume there is no unloading. We do not require, however, that the stress be a single-valued function of the strain—only that it can be computed from a knowledge of the strains $u_x(\tau)$ at all times $\tau \le t$.

In the Ritz approximation, $\dot{u}^h$ is the function in the trial subspace S^h which minimizes

$$I(\dot{v}^h) = \int [p^h(\dot{v}_x^h)^2 - 2\dot{f}\dot{v}^h] \, dx.$$

At a given t, the coefficient $p^h(x)$ will not have the same value as the true modulus p. Instead it will depend on the history of the Ritz approximations $u_x^h(\tau)$, $\tau \leq t$. To prove convergence we must assume that if these approximate strains have remained close to the true ones, then the coefficient p^h at the given instant is close to the true p:

$$\max_x |p(x) - p^h(x)| \leq C \int_0^t \|\dot{u}_x(\tau) - \dot{u}_x^h(\tau)\| \, d\tau.$$

To prove convergence, we must estimate how the function $\dot{u}^h$ which minimizes $I(\dot{v}^h)$, and its x-derivative the rate of change of strain, depend on the coefficient p in the variational principle. Our plan is to split the error at time t into two parts,

$$\dot{u}_x(t) - \dot{u}_x^h(t) = (\dot{u}_x(t) - \dot{w}_x^h(t)) + (\dot{w}_x^h(t) - \dot{u}_x^h(t));$$

$\dot{w}^h$ is the function which minimizes $I(\dot{v}^h)$ over all trial functions in S^h, using the *true* elastic coefficient p at time t. In other words, the first error is due to approximation, and the second to a change in the elastic coefficient. For the first we can appeal to the standard approximation theory; it is a linear problem at each instant of time, and the error in the first derivatives for a space of degree $k - 1$ is

$$\|\dot{u}_x(t) - \dot{w}_x^h(t)\| \leq C' \, h^{k-1}.$$

For the second part we must turn to Section 4.3. The corollary in that section asserts that the effect of a change in coefficient is bounded by

$$\|\dot{w}_x^h(t) - \dot{u}_x^h(t)\| \leq C'' \max |p(x) - p^h(x)|.$$

Substituting these last two bounds, and writing $\dot{e} = \dot{u}_x - \dot{u}_x^h$,

$$\|\dot{e}(t)\| \leq C'h^{k-1} + CC'' \int_0^t \|\dot{e}(\tau)\| \, d\tau.$$

This is exactly the situation in which to apply the argument which leads to *Gronwall's lemma:* dividing through by the right side of the inequality, multiplying by CC'', and integrating with respect to t,

$$\log\left(C'h^{k-1} + CC'' \int_0^t \|\dot{e}(\tau)\| \, d\tau\right) - \log C'h^{k-1} \leq CC''t.$$

Taking the exponential, and substituting into the previous inequality,

$$\|\dot{e}(t)\| \leq C'h^{k-1} \exp(CC''t).$$

Finally, an integration with respect to t yields

$$\|e(t)\| \leq \int \|\dot{e}(t)\| \, dt \leq C'''h^{k-1}.$$

Therefore the rate of convergence in h is the same for this nonlinear plastic problem as it is for linear elasticity.

Notice that we have assumed a Ritz approximation computed *continuously in time;* the only discretization so far has been the change from the full admissible space to its subspace S^h. This is in keeping with the presentation of initial value problems in Chapter 7, where the Ritz error is separated from the error due to a finite difference method (or other procedure) in the time direction. For the nonlinear problem there has been considerable discussion of the best "incremental method," but we believe all of the leading possibilities to be convergent. They should simply contribute a new error—proportional to a power of Δt, in case of a difference equation.

There is one technical difficulty in the proof, however, which an over-zealous conscience will not allow us to ignore. It is a question of the choice of norm: if the double bars represent a mean-square norm, then the pointwise bound given for $p - p^h$ is not strictly valid. On the other hand, to use maximum norms throughout requires a fresh examination of the h^{k-1} estimate for $\dot{u}_x - \dot{w}_x^h$. This bound followed from mean-square approximation theory, and perhaps the simplest remedy is to establish instead a pointwise estimate of the Ritz error in static linear problems. Another possibility is to use an idea proposed by the first author (for subsequent applications see [W3]) which permits switching back and forth between these two norms when the solution is smooth; such a technique is frequently required in nonlinear problems when the error estimates are global but instabilities can arise locally. Or a third possibility is to improve the corollary in Section 4.3, to depend instead on the mean-square norm of the perturbation $p - p^h$. We are confident that the basic proof is correct, and that a combination of experiment and analysis will soon lead to a much fuller understanding of nonlinear errors.

This summary of the theory must include also *eigenvalue problems* and *initial-value problems.* The finite element method applies directly and successfully to both. For self-adjoint eigenvalue problems, the computation of upper bounds by minimizing the Rayleigh quotient over a subspace is a classical technique; it leads to a discrete eigenvalue problem $KQ = \lambda MQ$, where K and M are precisely the same stiffness and mass matrices already encountered. We devote Chapter 6 to deriving this discrete form and to estimating the errors in eigenvalue and eigenfunction which depend on approximation theory—they are due to replacing the true admissible space $\mathcal{3C}_E^m$ by S^h. The results are simple to descibe: $\lambda - \lambda^h$ is of order $h^{2(k-m)}$, and for $k \geq 2m$ the eigenfunction errors in the sth derivative are of the maximum order h^{k-s} permitted by approximation theory.

For initial-value problems $u_t + Lu = f$, the position is equally favorable. The finite element solution has the form $u^h(t, x) = \sum Q_j(t)\varphi_j(x)$; the time variable is left continuous, while the dependence on x is discretized in terms of the standard piecewise polynomial basis functions φ_j. The coefficients $Q_j(t)$ are determined by a system of N ordinary differential equations, expressing

Galerkin's rule: The residual $u_t^h + Lu^h - f$ will not be identically zero unless the true solution u happens to lie in the trial space S^h, but its component in S^h should vanish. Thus the original equation holds "on the subspace."

In practice, time must also be made discrete. This is assumed to be done by a finite difference method. The Crank–Nicolson scheme, for example, is centered at the time $t_{n+1/2}$ when $u^h(t_{n+1})$ is computed from $u^h(t_n)$ and is therefore accurate to order Δt^2. Thus the final computed approximation includes this error as well as the Galerkin error due to discretization in x. The latter is the one which we analyze in detail, in order to show that for $k \geq 2m$, it, too, is of the optimal order h^{k-s} in the sth derivative. This result applies to equations of *parabolic* type, for example the heat equation; L is the same kind of elliptic operator that occurs in static problems. In the case of *hyperbolic* equations, which include no dissipative term, the power of the finite element method is somewhat diminished; the price to be paid in comparison with explicit difference methods may be too high. Nevertheless, even in this case the comparatively automatic treatment of boundaries is a tremendous advantage, and we include in Chapter 7 a sketch of the hyperbolic finite element theory.

For want of space, the analysis of change of domain, numerical integration, and roundoff will be limited to the static case $Lu = f$. The results for eigenvalue and initial-value problems are certain to be very similar, and in the case of quadrature errors these extensions of the theory have been carried out by Fix (Baltimore Finite Element Symposium).

In the final chapter, we present the results of a rather extensive series of numerical experiments. One of the most interesting concerns a problem with strong singularities, produced by a crack which runs into the material. It is a classical problem in fracture mechanics to compute the stress intensity factor at the head of the crack; around that point the stress varies like $r^{-1/2}$. Therefore, a number of questions arise:

1. Do our error estimates yield the observed rate of convergence, both pointwise and in mean-square?
2. The singularity reduces the smoothness of u and therefore the rate of convergence; does a reduced rate apply even where the solution is smooth, so that the singularity pollutes the whole computation?
3. Is it possible, by grading the mesh or by introducing special trial functions at the singularity, to recover the normal rate of convergence of the method?

The answer in each case is "yes," and the numerical results are very convincing. Both for static and for eigenvalue problems, an excellent remedy for singularities which are introduced by sharp angles in the domain is the introduction of trial functions which mirror the singularity correctly.

Straight interfaces between materials create a slightly different problem.

There is a jump in the derivatives of u across the interface, and we strongly recommend the following simple solution: relax any continuity imposed on the derivatives of the trial functions, so that u^h is free to copy the singularity in u. We do not believe that in normal circumstances the jump condition—or any other natural boundary condition—should be imposed.

Finally, we have chosen to give the theoretical background for one comparatively new computational technique—the Peters–Wilkinson algorithm for the matrix eigenvalue problem $Kx = \lambda Mx$. The solution of the linear system $KQ = F$ is of course a still more fundamental problem, and it is subject to considerable refinement in the ordering of the unknowns or in the choice of a gradient procedure; but it is comparatively well understood. The eigenvalue problem is more subtle, and without an efficient algorithm the number of unknowns will be artificially limited—below the number required to represent the physics of the problem. Therefore, we have described the Peters–Wilkinson idea (as well as some more established algorithms) in Chapter 6, and applied it to the numerical experiments of Chapter 8.

2.3. GALERKIN'S METHOD, COLLOCATION, AND THE MIXED METHOD

As we have described it, the Ritz technique applies only to problems of the classical variational type, in which a convex functional is to be minimized. The corresponding Euler differential equation is self-adjoint and elliptic. It is well known, however, that equations of quite general type can also be written in a weak form, and that this form suggests a generalization from the Ritz to the Galerkin method. Applied to initial-value problems, this is the subject of Chapter 7. Here we discuss two types of static problems, first those in which derivatives of odd order spoil the self-adjointness of an elliptic equation, and then those in which the associated functional is not positive definite—the problem is to find *a stationary point rather than a minimum* of $I(v)$. This arises in the Hellinger–Reissner principle in elasticity, and in the corresponding *mixed method* for finite elements. It leads to some difficult mathematical questions in the proof of convergence.

First some comments on the weak forms of a differential equation $Lu = f$. There are several, but they all share the following basic principle: The equation is multiplied by test functions $v(x)$ and integrated over the domain Ω, to yield

$$(Lu, v) = (f, v).$$

This is to hold for each function v in some test space V, and everything hinges on the choice of V. If V includes all δ-functions, then the equation $Lu = f$ will have to hold in the most classical (some would say old-fashioned) sense, at every point. The discrete form of this test space leads to the *collocation method*, discussed below. At the other extreme, V may contain only infinitely smooth functions which vanish in a boundary strip. A formal integration by parts shifts all the derivatives off u and onto v, leading to the equation

$$(u, L^*v) = (f, v), \qquad L^* = \text{formal adjoint of } L.$$

In this weakest form, u satisfies the equation only "in the sense of distributions." No doubt a discrete form has been studied.

Between these extremes there lie a great many possibilities. If $V = \mathcal{K}^0(\Omega)$, the equation is said to hold in the strong sense, or sometimes in the L^2 sense. More generally the choice $V = \mathcal{K}^s(\Omega)$ permits s of the derivatives to be shifted from u to v; if L is of order $2m$, then presumably the solution u will be sought in $\mathcal{K}^{2m-s}$. The case $s = m$ is of prime importance, and it has already been given the notation

$$a(u, v) = (f, v).$$

In all these manipulations, the boundary conditions must play their part. With the strong form ($s = 0$), the full set of boundary conditions is imposed on u; the solution must lie in $\mathcal{K}_B^{2m}$. As s increases and u needs only $2m - s$ derivatives to qualify for the solution space, the only boundary conditions which make sense are those of order less than $2m - s$; for $s = m$, these are the essential boundary conditions,† and u lies in $\mathcal{K}_E^m$. At the same time, the number of conditions which are imposed on v is increasing. These are governed by the derivatives of order less than s which appear in Green's formula; for $s = m$, v too lies in $\mathcal{K}_E^m$. Thus the Ritz case is symmetric between the trial space and the test space.

Galerkin's method is the obvious discretization of the weak form. In general it involves two families of functions—a subspace S^h of the solution space and a subspace V^h of the test space V. Then the Galerkin solution u^h is the element of S^h which satisfies

$$(5) \qquad (Lu^h, v^h) = (f, v^h) \qquad \text{for all } v^h \text{ in } V^h.$$

---

†In mathematical terms, an essential condition $Bv = g$ is one specified by a bounded linear operator B on the space $\mathcal{K}^m$ of all functions with finite strain energy. The functions which satisfy $Bv = 0$ form a closed subspace. Natural conditions are those which are not, and cannot be, imposed on every v; but because of the special form given to $I(v)$, they are satisfied by the minimizing u. The usual test for essential conditions is that B must involve only derivatives of order less than m, but this condition is neither necessary nor sufficient. In a two-dimensional problem, for example with $m = 1$, we cannot impose that $v = 0$ at a given single point P. The value at P is not a bounded functional (it may be arbitrarily large, as with log log r, while the function has unit strain energy) and the functions which satisfy $v(P) = 0$ do not form a closed subspace. In fact, they come arbitrarily close in strain energy to trial functions which do *not* vanish at P, and to minimize over them would give the same result as to minimize over all of $\mathcal{K}^1$.

The left side will need s integrations by parts, as in the continuous problem. When S^h and V^h are of equal dimension N, this Galerkin equation goes into operational form in the usual way: If $\varphi_1, \ldots, \varphi_N$ is a basis for S^h and $\psi_1, \ldots, \psi_N$ is a basis for V^h, the solution $u^h = \sum Q_j \varphi_j$ satisfies

$$(\sum Q_j L\varphi_j, \psi_k) = (f, \psi_k), \qquad k = 1, \ldots, N.$$

In matrix form this is $GQ = F$, with

$$G_{kj} = (L\varphi_j, \psi_k), \qquad F_k = (f, \psi_k).$$

The Ritz case is of course the one in which $S^h = V^h \subset \mathcal{3C}_E^m$, $\varphi_j = \psi_j$, and $(L\varphi_j, \varphi_k) = a(\varphi_j, \varphi_k)$; G becomes the stiffness matrix K.

Suppose that S^h is a finite element space of degree $k - 1$, and V^h a finite element space of degree $l - 1$. Then the expected rate of convergence of the sth derivatives in the Galerkin method might be

(6) $$\|u - u^h\|_s = O(h^{k-s} + h^{k+l-2m}).$$

As in (4), the first exponent of h reflects the best order of approximation which is possible from S^h, and the second exponent is influenced also by approximation in the test subspace and the order $2m$ of the differential equation. In theory it appears possible to make some economies by choosing $l < k$, say $k = 4$ (cubic splines) and $l = 2$ (linear test functions) in a second-order problem ([B21]). The convergence rate is as good as with $l = 4$ and the bandwidth is reduced; however, G is no longer symmetric, even in self-adjoint problems, and we are dubious.

A related possibility is to "lump" some terms in the discretization, thereby departing from the Ritz equations $KQ = F$ and $KQ = \lambda MQ$, by using subspaces of lower degree on the lower-order terms in the equation. In the past this was useful in eigenvalue problems, in order to replace the *consistent mass matrix* M by a diagonal *lumped mass matrix*. In the newest algorithms for the eigenvalue problem (Chapter 6) it is no longer so important that the mass matrix be diagonal, and we shall refer to [T8] for an error analysis of the lumping process.

In the collocation method, the test subspace V^h has a basis of δ-functions: $\psi_j(x) = \delta(x - x_j)$. The Galerkin equation (5) then demands that the differential equation hold at each node: $Lu(x_j) = f(x_j)$. In the error bound (6), the method will normally act as if $l = 0$ and converge at the rate h^{k-2m}. There are special collocation points, however, which increase this order of convergence and make the method very interesting [D7, B24]. G has a smaller bandwidth than K, and there are no inner products or element matrices to compute; for complicated nonlinear problems, these advantages may well compensate for the heavier approximation and smoothness demands on S^h.

Problems which are not positive definite symmetric.

We propose to analyze in more detail two cases in which the trial space S^h coincides with the test space V^h—there are m integrations by parts, so the weak form is essentially the $a(u, v) = (f, v)$ of the Ritz method—but the problem departs from the classical one of minimizing a positive-definite functional. The first departure arises when the equation is not self-adjoint, as in the constant-coefficient example

$$Lu = -pu'' + \tau u' + qu = f.$$

The adjoint operator L^* will have the opposite sign multiplying the derivative of odd order:

$$\int (-pu'' + \tau u' + qu)v = \int u(-pv'' - \tau v' + qv).$$

This suggests an energy inner product

$$a(u, v) = \int pu'v' + \tau u'v + quv,$$

which itself is not symmetric: $a(u, v) \neq a(v, u)$. In fact, the new term is *skew-adjoint*; it corresponds to the *imaginary part* of the operator L, whose real part $-pu'' + qu$ is as positive as ever. This appears most clearly if the inner product is extended to complex functions,

$$a(u, v) = \int pu'\bar{v}' + \tau u'\bar{v} + qu\bar{v}.$$

The real part of $a(u, u)$ is $\int p|u'|^2 + q|u|^2$, and the new term is purely imaginary.

We want to show that the rate of convergence in slope can be rigorously established as h^{k-1}, just as before, but that for large τ the finite element method may be very poor. The argument is quite general. The convergence proof when the real part (self-adjoint part) is elliptic includes the Ritz method as a special case, but the Galerkin method may be unsatisfactory when the odd-derivative term (the imaginary or skew-adjoint part) is of significant size.

THEOREM 2.1

Suppose that $|a(u, v)| \leq K\|u\|_m\|v\|_m$ and that the real part of the problem is elliptic: $\mathrm{Re}\, a(u, u) \geq \sigma\|u\|_m^2$ for u in $\mathcal{H}_E^m$. Let u^h be the function in $S^h \subset \mathcal{H}_E^m$ which satisfies the Ritz–Galerkin equation $a(u^h, v^h) = (f, v^h)$ for all v^h in S^h. Then the order of convergence in energy (equivalently, the order of convergence *of the mth derivatives) equals the best possible order of approximation which can be achieved by S^h:*

(7) $$\| u - u^h \|_m \leq \frac{K}{\sigma} \min_{S^h} \| u - v^h \|_m.$$

For a finite element space of degree $k - 1$, this order is h^{k-m}.

Proof. Since $a(u, v) = (f, v)$ for all v in $\mathcal{H}_E^m$, subtraction yields $a(u - u^h, v^h) = 0$ for all v^h. Therefore,

$$\sigma \| u - u^h \|_m^2 \leq \operatorname{Re} a(u - u^h, u - u^h)$$
$$= \operatorname{Re} a(u - u^h, u) = \operatorname{Re} a(u - u^h, u - v^h) \leq K \| u - u^h \|_m \| u - v^h \|_m.$$

Dividing by the common factor $\| u - u^h \|_m$, the proof is complete. Nitsche's method could now be used to establish the usual rate of convergence (4) for the displacement.

In spite of this convergence, the Galerkin method may in practice be a bad choice. Suppose that S^h is the standard piecewise linear subspace. Then in our example the Galerkin equations for $Q_j = v^h(x_j)$ will be just the difference equations

$$p \frac{-Q_{j+1} + 2Q_j - Q_{j-1}}{h^2} + \tau \frac{Q_{j+1} - Q_{j-1}}{2h} + q \frac{Q_{j+1} + 4Q_j + Q_{j-1}}{6}$$
$$= h^{-1} \int f\varphi_j \, dx.$$

Notice that the first derivative is replaced by a centered difference, regardless of the sign of τ. The true solution, however, may depend very strongly on this sign; as $|\tau| \to \infty$, the advection term $\tau u'$ dominates the second derivative, and the solution u is of boundary-layer type. Over most of the interval u is essentially the solution to an *initial-value* problem, for which the centered difference is completely inappropriate; at the far end there is a rapid variation in order to satisfy the other boundary condition, and an extremely fine mesh is required. The need for *one-sided* (upstream) differences is well known in chemical engineering. Mathematically the dominance of τ is reflected in a large value for K/σ in the error estimate (7).

A second departure from the classical Ritz formulation arises when the potential-energy functional $I(v)$ is not convex—$a(u, u)$ is not positive definite—and the task is to find a stationary point rather than a minimum. This occurs naturally in the *mixed method*, when both the displacement and its derivatives are taken as *independent unknowns*. The potential energy involves products which are just as likely to be negative as positive; it is like going from $x^2 + y^2$ to xy, which has a saddle point instead of a minimum at the origin.

A good example is given by the equation $w^{(iv)} = f(x)$, for the bending of a beam under loading. Suppose that the moment $M = w''$ is introduced as a new unknown. Then the original equation becomes $M'' = f$, and we have a system of two second-order equations:

$$(8) \qquad \begin{pmatrix} 0 & 1 \\ 1 & 0 \end{pmatrix} \begin{pmatrix} M \\ w \end{pmatrix}'' - \begin{pmatrix} 1 & 0 \\ 0 & 0 \end{pmatrix} \begin{pmatrix} M \\ w \end{pmatrix} = \begin{pmatrix} 0 \\ f \end{pmatrix}.$$

This reduction of order brings several advantages. The trial functions in the variational form need only be continuous between elements, whereas the fourth-order equation was associated with the potential energy $I(v) = \int (v'')^2 - 2fv$, and this is finite only if the slope v' is also continuous. Furthermore, the condition number of the stiffness matrix is completely altered by a reduction in the order of the differential equations; the fourth differences are replaced by second differences, and the condition number goes from $O(h^{-4})$ to $O(h^{-2})$. This may seem miraculous, that such an improvement can be achieved by a purely formal introduction of the derivative M as a new unknown, but the improvement is quite genuine.

We examine this roundoff error in two ways. Consider a simply supported beam, with u and M vanishing at both ends. Then the equations $M'' = f$ and $w'' = M$ can be solved separately, first for M and then for w, using either finite differences or finite elements. Suppose that the approximate solution of $M'' = f$ includes a roundoff error ϵ_1, which will normally be of order $h^{-2}2^{-t}$ if the computer word length is t. Then the approximate solution of $w'' = -f$ will include first of all its own roundoff error ϵ_2, of this same order, and also an inherited error ϵ_3. This error satisfies $\epsilon_3'' = \epsilon_1$, or rather it satisfies exactly the discretization of this equation which is being used, and therefore ϵ_3 is also of order h^{-2}. *The roundoff error is not compounded to h^{-4}.*

As a second check, we shall compute the condition number of the discrete system. In the finite difference case, the matrix form is

$$(9) \qquad \begin{pmatrix} -1 & \delta^2 \\ \delta^2 & 0 \end{pmatrix} \begin{pmatrix} M \\ w \end{pmatrix} = \begin{pmatrix} 0 \\ f \end{pmatrix},$$

and the eigenvalues μ of this block matrix are related to the eigenvalues λ of δ^2 by

$$\mu^2 + \mu = \lambda^2.$$

We know that the eigenvalues λ of the second difference operator δ^2 range from $O(1)$ to $O(h^{-2})$. Solving the quadratic, the eigenvalues μ fall in the same range, and the condition number μ_{max}/μ_{min} of the coupled system is indeed of order h^{-2}.

We wish to offer an explanation for this miracle. It rests on the observation that the usual measure of condition number for these matrices is unnatural. We are regarding them for numerical purposes as transforming Euclidean space (discrete $\mathfrak{IC}^0$) into itself, and therefore we use the same norm for both the residual in the equation and the resulting error in the solution. This is completely contrary to what is done in the differential problem, or for that matter in estimating discretization error; in such a case f is measured in $\mathfrak{IC}^0$, M and its error in $\mathfrak{IC}^2$, and w and its error in $\mathfrak{IC}^4$. (In the variational problem, these may be $\mathfrak{IC}^{-2}$, $\mathfrak{IC}^0$, and $\mathfrak{IC}^2$, respectively.) In fact, the operator $L = d^2/dx^2$, with any of the usual boundary conditions, is perfectly conditioned as a map from $\mathfrak{IC}^2$ to $\mathfrak{IC}^0$. This was the essential point of Section 1.2, that both L and L^{-1} are bounded. We can show that the same is true of the difference operator δ^2, and of any reasonable finite element analogue, provided these natural norms are retained. Therefore, there ought to exist an algorithm for solving $KQ = F$ which reflects this property, and in this case the miracle would disappear; the errors in M and w would be appropriate to their station. As long as a standard elimination is used, however, there will be a difference in roundoff error between fourth-order and second-order problems.

Before going further, the coupled differential equations of the mixed method should be put into variational form. Multiplying the first equation in (8) by M and the second by w, and integrating by parts,

$$\int (w''M + M''w - M^2 - 2fw)\, dx$$
$$= -\int (2w'M' + M^2 + 2fw)\, dx + w'M + M'w \,|_0^\pi.$$

In the simply supported case w and M vanish at each end—the admissible space satisfies full Dirichlet conditions—and the integrated term disappears. For the clamped beam a remarkable thing occurs: The condition $w = 0$ is imposed at each end, but *the vanishing of w' yields a natural boundary condition for M*. The stationary point of

$$I(\bar{v}) = -\int (2w'M' + M^2 + 2fw),$$

if the admissible space V contains all pairs (M, w) with M in the Neumann space $\mathfrak{IC}^1$ and w in the Dirichlet space $\mathfrak{IC}_0^1$, solves exactly the problem of the clamped beam. Roughly speaking, if M is unconstrained at the endpoints and the first variation of I is to vanish, then the factor w' which multiplies M must be zero. We emphasize that as in the Reissner principle of elasticity, the functional $I(\bar{v})$ has no definite sign; the problem is one of a stationary point, and not a minimum.

Let us note the computational consequences of this variational form. With piecewise linear elements and $Nh = \pi$, M^h and w^h are expressed by

$$M^h = \sum_0^{N+1} z_j \varphi_j(x), \qquad w^h = \sum_1^N q_j \varphi_j(x).$$

(Because M is unconstrained, it requires two extra basis functions.) There are now $2N + 2$ unknowns, but *the system can no longer be solved in series*, first for M^h and then for w^h. With $N = 2$ the coefficient matrix is

$$\begin{pmatrix} -G & D \\ D^* & 0 \end{pmatrix}, \quad \text{where } G = \frac{h}{6} \begin{pmatrix} 2 & 1 & 0 & 0 \\ 1 & 4 & 1 & 0 \\ 0 & 1 & 4 & 1 \\ 0 & 0 & 1 & 2 \end{pmatrix} \quad \text{and} \quad D = \frac{1}{h} \begin{pmatrix} -1 & 0 \\ 2 & -1 \\ -1 & 2 \\ 0 & -1 \end{pmatrix}.$$

G is the mass matrix (or Gram matrix) of the φ_j, and D is a rectangular second difference matrix. For large N the unknowns should be ordered by z_0, q_1, z_1, q_2, ... to reduce the band width. The condition number is again $O(h^{-2})$; with G replaced by the identity, the eigenvalues are $\mu = 1$, twice, and the roots of $\mu^2 + \mu = \lambda$, where λ is an eigenvalue of the usual fourth difference matrix $D^* D$. The errors in M^h and w^h will be of order h^2, and their derivatives will be accurate to order h. Of course the second derivative of w^h will not equal M^h.

For two-dimensional problems, Hellan and Herrmann have developed simple elements which are appropriate for the mixed method. Rather than to catalogue these elements, however, we prefer to return instead to what is mathematically the main point: the difficulty of establishing convergence when the functional is indefinite. The problem is easiest to explain in terms of an infinite symmetric system $Lu = b$ of simultaneous linear equations. Suppose that S^h is a finite-dimensional subspace and P^h the symmetric projection operator onto S^h: $P^h v$ is the component of v in S^h. *Then the Galerkin method* (5) *is identical with the problem of finding the approximate solution u^h in S^h which satisfies*

(10) $$P^h L P^h u^h = P^h b.$$

The operator $P^h L P^h$ of the discrete problem, which is just our stiffness matrix K, will be positive definite if L is; this is the Ritz case. In fact, $P^h L P^h$ will be even more positive than L itself; the minimum eigenvalue increases when the problem is constrained to a subspace. This is easy to see: If v is in the subspace S^h, then $(P^h L P^h v, v) = (L P^h v, P^h v) = (L v, v)$. Suppose L is positive definite on the whole space: $(Lv, v) \geq \sigma(v, v)$ for all v. Then this positive definiteness is inherited by P^hLP^h on the subspace S^h, and with no decrease in σ.

If L is symmetric but indefinite, then the same is to be expected of the Galerkin operator P^hLP^h. (This assumes that the test space V^h and trial space S^h are the same. Otherwise, if Q^h is the projection onto V^h, the Galerkin equation is $Q^hLP^hu^h = Q^hf$, and Q^hLP^h is not even symmetric.) It is natural to hope that with increasing S^h, u^h will approach u. This convergence is *not automatic*, however, and in searching for the right hypotheses, we are led back to the fundamental theorem of numerical analysis: *Consistency and stability imply convergence.*

THEOREM 2.2

Suppose that Galerkin's method is
(a) *consistent: for every v, $\|v - P^hv\| \to 0$*
and
(b) *stable: the discrete operators are uniformly invertible, $\|(P^hLP^h)^{-1}\| \le C$.*

Then the method converges: $\|u - u^h\| \to 0$.

Proof. Denote $(P^hLP^h)^{-1}$ by R. Since $Lu = f$, we have

$$P^hLP^hu + P^hL(u - P^hu) = P^hf,$$

or

$$u + RP^hL(u - P^hu) = RP^hf.$$

Subtracting $u^h = RP^hf$,

$$\|u - u^h\| = \|RP^hL(u - P^hu)\| \le C\|L\|\|u - P^hu\| \to 0.$$

Note that the rate of convergence depends both on the stability constant C and on the approximation properties of S^h, exactly as in Theorem 2.1. (That was effectively a special case of the present theorem, with $C = 1/\sigma$ and $\|L\| = K$. A more general result is given by Babuska [B4].) We could extend the theory to show that consistency and stability are also *necessary* for convergence, and that the existence of the solution u need not be assumed; Browder and Petryshyn have shown how to deduce the invertibility of the original operator L.

We point out again the special role of positive definiteness: It makes stability automatic. That is why the Ritz method is so safe. In the indefinite case, suppose that S^h is the subspace formed from the first N coordinate directions; P^hv is given by the first N components of v. Then P^hLP^h is the Nth principal minor of L—the submatrix in the upper left corner of L—and stability means that these $N \times N$ submatrices are uniformly invertible. It sounds as if this might follow from the invertibility of the whole matrix L, but it does not. The following invertible matrix

$$L = \begin{pmatrix} \begin{pmatrix} 0 & 1 \\ 1 & 0 \end{pmatrix} & & & \\ & \begin{pmatrix} 0 & 1 \\ 1 & 0 \end{pmatrix} & & \\ & & \cdot & \\ & & & \cdot \\ & & & & \cdot \end{pmatrix}$$

is a good example; *for every odd N, the leading principal minor is singular.* Its last row is zero, and for $f = (1, 1/2, 1/4, \ldots)$, the functional $(Lv, v) - 2(f, v)$ on the subspace S^h will not have a saddle (stationary) point. (We thank C. McCarthy for his help with these questions.) Even in the 2 by 2 case, the indefinite quadratic $2xy$ breaks down completely on the subspace $x = 0$.

Note that a reordering of the unknowns yields $L = \begin{pmatrix} 0 & I \\ I & 0 \end{pmatrix}$, which is very close to the coupled system of the mixed method example. We claimed convergence in that example, but this is actually justified only for a proper choice of the subspaces S^h. *Apparently the finite element construction does yield such a choice.* C. Johnson has just succeeded in proving convergence for two of the most important mixed elements, in which the moments are respectively constant and linear in each triangle. His proof establishes a special property, after the displacement unknowns have been eliminated and the moment unknowns are seen to be determined as the minimizing functions for a positive definite expression (the complementary energy). This property is exactly the familiar Ritz condition, that the trial moments in the discrete case are contained in the space of admissible moments (those in equilibrium with the prescribed load) in the full continuous problem. Therefore, as in the Ritz method, convergence depends on approximation theory and can be proved. †

Convergence for mixed and hybrid elements is becoming very much clearer, thanks to the work of Brezzi and others; the essential extra hypothesis is that of stability.

To achieve numerical stability in the mixed method, where the coefficient matrix is indefinite, pivoting (row and column exchanges) must be permitted during the Gauss elimination process. Computational results have borne out the predicted decrease in condition number and roundoff error, although Connor and Will have reported some unsatisfactory results with elements of high degree: the mixed method has enjoyed only mixed success. Nevertheless reduction of order is such a valuable property that the development of the idea must continue.

† We believe that for other mixed (and hybrid) elements, the natural proof is a direct verification of consistency (or approximability) and stability. Then Babuska's general theorem [6, B4] yields convergence. Brezzi has established stability for one hybrid element, and his technique needs to be generalized; these methods have recently been discussed by Miyoshi, Kikuchi, and Ando in Japan.

2.4. SYSTEMS OF EQUATIONS; SHELL PROBLEMS; VARIATIONS ON THE FINITE ELEMENT METHOD

It may be objected that our finite element theory deals always with a single unknown, while most applications involve a system of r equations for a vector unknown $\vec{u} = (u_1, \ldots, u_r)$. Fortunately, the distinction is often not essential. The variational principle for a system again minimizes a quadratic $I(\vec{v})$, and the error estimates depend exactly as before on the approximation properties of S^h.

A typical example is furnished by two- or three-dimensional elasticity. The unknowns at each point are the displacements in the coordinate directions, and the finite element solution $\vec{u}^h$ is again the trial function closest to the true solution in the sense of strain energy—which is now a quadratic function involving all the unknowns. The approximation theory in the next chapter will show that, independent of the number of dimensions and unknowns, the rate of convergence depends on the degree of the finite elements and on the approximation to the domain.

Some new and much more difficult problems arise for shells. The theory is normally constructed as a limiting case of three-dimensional elasticity, in which the domain Ω becomes very thin in one direction (normal to the surface of the shell). The result of the limiting process is to introduce second derivatives of the transverse displacement w into the potential-energy functional; the differential equations are of fourth order in w, and of second order in the in-plane displacements. This imbalance is the price to be paid for the reduction from a three-dimensional to a two-dimensional problem.

Let us note immediately the increasing popularity of three-dimensional elements, in which this reduction *is not made*. It is by no means automatic, when a limiting process simplifies the search for exact solutions to problems with special symmetries—as it certainly does in shell theory—that the same process will simplify the numerical solution of more general problems. (The same question arises with the Airy stress functions of plate bending; is it numerically sensible to reduce the number of unknowns and increase the order of the equations? We doubt it.) Obviously a thin shell will never present a typical three-dimensional problem; there will always be difficulties with a nearly degenerate domain. Not only the isoparametric technique but also special choices of nodal unknowns and of reduced integration formulas in the normal direction are being tried experimentally. From a theoretical viewpoint, it will become necessary to estimate the effect of a small thickness parameter t (Fried has done so, in respect of numerical stability and the condition number), but in general the approximation-theoretic approach can proceed in the usual way.

There are also a number of curved shell elements, in two independent variables, to which our analysis can be directly applied—even though they involve derivatives of different orders. These elements are constructed on the basis of a fully compatible shell theory: The surface of the shell is described parametrically by a set of three equations $x_i = x_i(\theta_1, \theta_2)$, and the independent variables in the shell equations are θ_1 and θ_2—which simply vary over a standard plane domain Ω. The curvature of the shell appears only through the derivatives of the x_i, which enter as variable coefficients in the differential equations. Numerical integration will almost certainly be required for the evaluation of the element stiffness matrices, but the problem is in every essential respect like any other problem in the plane. Cowper and Lindberg have constructed a conforming and highly accurate element known as CURSHL [C12], by combining the $\mathcal{C}^1$ reduced quintics and $\mathcal{C}^0$ complete cubics of Section 1.9 for the normal and in-plane displacements, respectively. Since the difference $k - m$ equals 3 for all components, the rate of convergence in strain energy will be h^6.

Such a construction, whose accuracy has been confirmed by numerical experiments, ought in theory to end the search for good shell elements—but it has not. For most applications and most programs, CURSHL appears to be quite complicated; it has not been widely accepted. On the one hand, there are many problems with special symmetries, in which cylindrical or shallow shell elements can be used [O4]. On the other hand, it is possible to approximate a general shell by an assemblage of flat pieces. Each of these pieces becomes effectively a simple plate element, and the bending and stretching deformations are coupled only by the fitting together of these plates. Some variant of this approach—which inevitably involves nonconforming elements from the viewpoint of pure shell theory—appears to be surviving as the simplest and most practical way to deal with complicated shells.

We would very much like to analyze the convergence or nonconvergence of these combinations of flat plate elements. It has been conjectured that under reasonable conditions the deformation of a polyhedral shell (such as a geodesic dome) approaches that of a genuinely curved shell, but we are unable to verify that conjecture. The mathematical problems are novel and extremely interesting.

In the one-dimensional case, when a curved arch is replaced by a polygonal frame, we can be more explicit. The strain energy for the arch is a sum of bending and extensional energies, and can be normalized to

$$a(v, v) = C \int \left(\frac{dv_1}{ds} + \frac{v_2}{r} \right)^2 + \frac{t^2}{12} \left(\frac{1}{r} \frac{dv_1}{ds} - \frac{d^2 v_2}{ds^2} \right)^2$$

The radius of curvature is r, the thickness is t, and the admissibility conditions are that v_1 and v_2 lie in $\mathcal{H}^1$ and $\mathcal{H}^2$, respectively; the first derivative $\partial v_2 / \partial s$

is effectively the rotation of the vector normal to the arch, and this is not permitted to have a jump discontinuity. The pair $v_1 = u$ and $v_2 = w$, which minimize the potential energy $[I(\vec{v}) = a(\vec{v}, \vec{v}) +$ linear terms], are the tangential and normal displacements of the arch. Note that they are determined by differential equations, the Euler equations of $I(\vec{v})$, which are of different orders.

For an approximating frame, made up of straight lines, the radius of curvature becomes $r = \infty$. This uncouples the two components of $\vec{v}$ in the strain energy, and therefore leads to simple and independent differential equations for u and w. The coupling reappears, however, at the joints of the frame, in order to prevent the structure from coming apart in its deformed state. In other words, each trial function $\vec{v} = (v_1, v_2)$, as $s = s_0$ is approached from the left and from the right, must move the joint to the same place and thereby leave the frame continuous; this represents two conditions relating both components of $\vec{v}_-$ and $\vec{v}_+$ to the joint angle θ. (The conditions are essential, not natural.) Furthermore, there is an essential continuity condition on the rotation $\partial v_2 / \partial s$, preserving the angle at each joint. (Note that v_2 itself need not be continuous; it will include δ-functions at the joints, whose derivatives from left and right are both zero.)

With finite elements these conditions constrain the polynomials meeting at the nodes. Typically v_2 is a modified Hermite cubic: the continuity of the rotation is left unchanged, but discontinuities in function value are allowed, and related to those in v_1. Obviously such trial functions are inadmissible in the arch problem, but since the strain energy is also modified by the removal of r, the convergence question remains open. Direct evidence *against convergence* was provided by Walz, Fulton, and Cyrus (Wright-Patterson Conference II) for the case of a circular arch and a regular polygon. The finite element equations are simply difference equations, and they were found to be consistent with the *wrong differential equation*. The leading terms were correct—the radius of curvature reappears through the agency of the angle θ in the frame continuity conditions—but there were, for the particular element chosen, also unwanted terms of zero order in h.† This suggests the possibility of a convergence condition resembling the patch test of Section 4.2.

For a shell the problems are inevitably more complex. Even with a polyhedron built from flat plates, the continuity requirements are difficult to impose on polynomial elements; continuity of the derivatives is always difficult to achieve with anything less than quintics. One useful but mathematically cloudy approximation has been the *discrete Kirchhoff hypothesis:* The rotations (the rate of change of normal displacement in the two coordinate direc-

---

†These unwanted terms are, however, comparable to others which appear in relating the arch equations to a full two-dimensional elasticity theory. Thus the leading terms in all theories, including the frame approximation, may coincide—and the same may apply to shells.

tions on the shell) are taken to be independent unknowns, and their true relationship to the normal displacement is imposed only as a constraint at the nodes, not over the entire surface. Approximations of this kind will continue to be practical and valuable; there are too many kinds of shell problems, and too many alternative elements, to decide that a single approach is best. Even conforming elements such as CURSHL have had their detractors, on the ground that rigid body motions are not reproduced exactly. The argument is that a tall structure leaning in the wind might have such large motions, and as a consequence such large finite element errors, as to obscure the internal bending, which is of prime importance. We are inclined to trust in the accuracy, and to be concerned rather about the convenience, of high-degree elements.

Obviously shells have provided finite elements with a very severe test—of all linear problems perhaps the most severe. The same is true of the mathematical analysis. But these are problems which were virtually inaccessible by earlier techniques, and we believe they are now on their way to being understood.

In addition to systems of equations, there is another major omission in the finite element analysis described so far; we have concentrated only on the *displacement* method, in which an expression $\sum q_j \varphi_j$ is assumed for the displacement, and the optimal coefficients Q_j are determined by the Ritz–Galerkin principle. The numerical effectiveness of this method is well established, and this is largely a period of perfecting the elements and automating the input and output—rapid developments in computer graphics have been required to make the output intelligible. There remain, however, some imperfections which can be removed only by modifying the mathematical method itself. In the remainder of this section we want to describe some of the proposed changes, still remaining in the framework of the *method of weighted residuals*. This means that the approximate solution will still be a combination of trial functions, but the rule for choosing among the possible coefficients q_j may be different, or the unknown itself may be altered.

The most important variant is the *force method*, in which the strains—the derivatives of u, rather than u itself—are taken as unknowns. In many problems these are the quantities of greatest importance, and it is natural to approximate them directly. The result is quite different from the mixed method, in which *both* displacements and strains are unknowns, and the functional is indefinite. Here there is a *complementary energy functional* to be minimized, and apart from a change in the admissibility conditions—the trial functions have new interelement and boundary constraints—the mathematics is the same. The order of approximation achieved by the trial subspace is still decisive.

As an example we take Laplace's equation

$$\Delta u = 0 \quad \text{in } \Omega, \qquad u = g \quad \text{on } \Gamma,$$

in which u is determined by minimizing $I(v) = \iint v_x^2 + v_y^2$. In the complementary energy principle these derivatives $v_x = \epsilon_1$ and $v_y = \epsilon_2$ are taken as the primary dependent variables. The pair (ϵ_1, ϵ_2) is no longer constrained to be the gradient of some v; in other words the cross-derivative identity $(\epsilon_1)_y = (\epsilon_2)_x$, or *compatibility condition*, is no longer imposed. (It follows, of course, that when approximations ϵ_1^h and ϵ_2^h are determined, there is no unique way of integrating to find a corresponding u^h. In the exact application of the complementary principle the optimal ϵ_1 and ϵ_2 *will* be the derivatives of the true displacement u, but in the discrete approximation this link between gradient and displacement is lost.)

In the Laplace example, the trial functions for ϵ_1 and ϵ_2 need only to lie in $\mathcal{3C}^0$ for the quadratic functional to be finite. There is a constraint of *equilibrium*, however, provided by the differential equation itself: $u_{xx} + u_{yy} = 0$ leads to

(11) $$(\epsilon_1)_x + (\epsilon_2)_y = 0.$$

This actually means that the pair $(\epsilon_2, -\epsilon_1)$ will be the gradient of some function w: $w_x = \epsilon_2$ and $w_y = -\epsilon_1$. We may interpret the complementary process in this special example as follows: Instead of looking for the harmonic function u, we are computing the associated stream function w. The function w is conjugate to u, meaning that $u + iw$ is analytic.

This leaves the two principles very nearly parallel in the interior of Ω. On the boundary, however, there is a marked distinction; the Dirichlet condition on u is exchanged for a Neumann condition on w. To see this, recall that along any curve, u and its stream function are related by $w_n = -u_s$. (This follows from the Cauchy–Riemann equation $w_x = -u_y$ when the curve is vertical, from $w_y = u_x$ when it is horizontal, and from a linear combination in the general case.) Therefore, on Γ the essential condition $u = g$ is replaced by $w_n = -g_s$. As in any inhomogeneous Neumann problem, this means that the admissible space is not constrained at the boundary, but the functional I is changed to

$$I' = \iint (w_x^2 + w_y^2) \, dx \, dy + \int g_s w \, ds.$$

Integrating the last term by parts to obtain $-\int g w_s \, ds$, w can be eliminated in favor of its derivatives $w_x = \epsilon_2$ and $w_y = -\epsilon_1$. The complementary principle now asserts that the pair (ϵ_1, ϵ_2), which minimizes I', is the gradient of the displacement u which minimizes I.

In finite element approximation the trial functions for w must lie in $\mathcal{3C}^1$, in order for I' to be finite. What does this imply about the variables ϵ_1 and ϵ_2? First, since the trial functions for w are continuous across inter-element boundaries, so are their derivatives in the direction along the edge. Therefore, the tangential component of the gradient $(\epsilon_2, -\epsilon_1)$, which is the normal component of (ϵ_1, ϵ_2), must be continuous. *It is this constraint which imposes the greatest difficulty in approximating the complementary principle.* The separate functions ϵ_1 and ϵ_2 need not be continuous, but the normal component of the vector (ϵ_1, ϵ_2) must be.

It is important to establish this constraint directly from the *equilibrium condition* (11), without arguing by way of a stream function, since in a more general problem the idea of a stream function is not so relevant. First we have to attach some meaning to equation (11), $(\epsilon_1)_x + (\epsilon_2)_y = 0$, when ϵ_1 and ϵ_2 lie only in $\mathfrak{K}^0$. They may have no derivatives, and therefore the condition must be understood in a weak sense; we multiply by a smooth function z which vanishes on Γ, and integrate by parts. The proper constraint is then given by this weak form of equation (11):

$$\int\int \left(\epsilon_1 \frac{\partial z}{\partial x} + \epsilon_2 \frac{\partial z}{\partial y}\right) dx\, dy = 0 \qquad \text{for all such } z.$$

For a trial space of piecewise polynomials, Green's theorem can be applied to this form of the constraint, taking it one element at a time. Then inside each element the original condition (11) must hold, and on each interelement boundary, there must be a cancellation of the boundary integrals or *tractions*, arising from the two adjacent regions. It is precisely this cancellation which requires the continuity of the normal component of (ϵ_1, ϵ_2).

The literature on the complementary method apparently begins with Trefftz [T9], who established that it yields a lower bound to the energy integral. Friedrichs [F19] then discovered that as in the Laplace example, the underlying idea was exactly the Ritz method applied to a conjugate problem, and that boundary conditions which are essential in one problem become natural in its conjugate. The orthogonality of the admissible spaces for the two problems was developed by Synge [17] into the hypercircle method, after a fundamental paper with Prager [P9]. A combination of the direct and conjugate principles leads to upper as well as lower bounds for the strain energy, and also for the displacement u; an abstract account is given by Aubin and Burchard [A10], and the idea is applied to some model problems by Weinberger [W2]. It has been programmed by Mote and Young.

The finite element method was first applied to the stresses, that is, to the complementary principle, by de Veubeke [F11]. By now there is an enormous literature.†

Always the interelement conditions present practical difficulties, and this has led to the construction of *multiplier* and *hybrid* methods. For example,

---

†The analysis of the force method proceeds from the same principles, and in fact the same approximation theorems, that are applied in this book to the displacement method. There is not space to develop this whole theory in parallel.

Anderheggen [A3], in applying to a fourth-order plate problem the standard Ritz method of minimizing the potential energy functional $I(v)$, proposes to use cubic polynomials for which the normal slope will ordinarily be discontinuous between elements. Imposing the constraint of slope continuity associates with each element edge a Lagrange multiplier and alters the Ritz method into a constrained minimization. The required programming changes are very simple. The stiffness matrix becomes indefinite, however, and (because of the edge unknowns) the computation time for cubics seems to be comparable to the usual stiffness method for the reduced quintics proposed in Section 1.9.

A second interesting modification is the hybrid method pioneered by Pian and Tong [P4, P5]. This copes with the interelement difficulties in an ingenious way, by constructing a family of approximations both to the stress field within each element and to the displacement on the element boundaries. The stress fields satisfy the differential equation within each element (so that the homogeneous case $f = 0$ is by far the simplest) and the displacements—which are given by an independent set of piecewise polynomials—are continuous. For each displacement pattern the complementary energy is first minimized within the elements separately. This leads to a family of displacements v^h, defined now not only on the element boundaries but throughout the region Ω, to which the Ritz method can be applied; the final hybrid approximation is found by minimizing over these v^h. The energy generally lies between the lower bound provided by the pure Ritz method and the upper bound of its conjugate, and in many cases will yield a substantially better approximation than either.

We return once more to the basic Ritz method and reconsider the following question: Can the variational principle be altered in such a way that essential boundary conditions need not be imposed? The complementary principle is one possible answer, but there are others. In fact, there is now a standard device for dealing with unsatisfied constraints: to insert a *penalty function* into the expression to be minimized. (This was the main theme of Courant's remarkable lecture [C11]; the finite element method was an afterthought!) For $-\Delta u = f$, with $u = g$ on Γ, $I(v)$ is altered to

$$I^h(v) = \iint_\Omega (v_x^2 + v_y^2 - 2fv) + C_h \int_\Gamma (v - g)^2 \, ds.$$

The exact minimum over the unconstrained admissible space $\mathcal{K}^1$ is achieved by the function U^h, which satisfies

$$-\Delta U^h = f, \qquad C_h^{-1} U_n^h + U^h = 0 \qquad \text{on } \Gamma.$$

Therefore, if $C_h \rightarrow \infty$, these solutions U^h converge to the solution u, which vanishes on Γ.

The Ritz method now minimizes the functional I^h, including the penalty term, over a space S^h with no boundary constraints. We suppose it to be a piecewise polynomial space, containing the complete polynomial of degree $k - 1$. Then there will be a balance between the error $u - U^h$, due to the unsatisfied boundary condition and the penalty, and the error $U^h - u^h$ involved in minimizing only over a subspace. This balance led Babuska [B5] to determine an optimal dependence of C_h on $h: C_h = ch^{1-k}$.

Babuska has also given rigorous error estimates for the related method of Lagrange multipliers, in which again the trial functions are unrestricted at the boundary. For Poisson's equation, this method searches for a stationary point $(u(x, y), \lambda(s))$ of the indefinite functional

$$F(v, \Lambda) = \int_\Omega (v_x^2 + v_y^2 - 2fv)\, dx\, dy - 2 \int_\Gamma \Lambda(v - g)\, ds.$$

The Lagrange multiplier runs over all admissible functions defined on Γ, and at the true stationary point it is related to the solution by $\lambda = \partial u/\partial n$.† The error in the stationary point (u^h, λ^h) over a finite element subspace can be estimated [B6].

The essential boundary conditions can also be made to fall away by a quite different approach, one with a long history: *the method of least squares.* Instead of minimizing $I(v) = (Lv, v) - 2(f, v)$, this method minimizes the residual $\| Lw - f \|^2$. Ignoring boundary conditions for a moment, the underlying functional is

$$\| Lw - f \|^2 = (Lw, Lw) - 2(f, Lw) + (f, f)$$
$$= (L^*Lw, w) - 2(L^*f, w) + (f, f).$$

The last term is independent of w, and therefore *the least-squares method actually minimizes $I' = (L^*Lw, w) - 2(L^*f, w)$, which is exactly the Ritz functional for the problem $L^*Lu = L^*f$.* This new problem is automatically self-adjoint, but we note that the order of the equation has doubled.

With boundary conditions, the least-squares method was properly analyzed for the first time in the recent work of Bramble and Schatz [B30, B31]. Taking as an example the Dirichlet problem $-\Delta u = f$ in Ω, $u = g$ on Γ, they introduce the functional

(12) $$I''(w) = \iint (\Delta w + f)^2\, dx\, dy + ch^{-3} \int_\Gamma (w - g)^2\, ds.$$

---

†We believe this method to be a promising variant of the standard displacement method, because it attacks directly the problem of computing the solution (or rather its normal derivative) at the boundary. Often it is exactly this information which is desired, and to find it from an approximate solution in the interior is numerically unstable and not entirely satisfactory.

The factor h^{-3} is not related to the degree k of the subspace S^h, as it was in the penalty method. Instead, it achieves a natural balance between the dimensions of the interior and boundary terms. The minimization of I'' becomes now a problem of *simultaneous approximation* in Ω and on the lower-dimensional manifold Γ.

It is reasonable to expect piecewise polynomial spaces to give the same degree of approximation on Γ as in Ω. If Γ were straight, to take the simplest case, then a complete polynomial in $x_1, \ldots, x_n$ reduces on Γ to a complete polynomial in the boundary variables $s_1, \ldots, s_{n-1}$. With a curved boundary we believe the approximation theory to be effectively the same; u differs from its interpolate on Γ by $O(h^{k-s})$ in the sth derivative. However, given only that approximation of degree k can be achieved in Ω, simultaneous approximation with coefficient h^{-3} is a much more subtle problem; its solution by Bramble and Schatz showed that the balance of powers in (12) is exactly right. Provided that $k \geq 4m$, their error estimate for the least-squares solution u^h_{LS} is the optimal

$$(13) \qquad \qquad \| u - u^h_{LS} \|_0 \leq ch^k \| u \|_k.$$

The practical difficulties in the least-squares method are exactly those associated with an increase in the order of the differential equation from $2m$ to $4m$. The trial functions must lie in $\mathcal{H}^{2m}$ in order for the new functional to be finite. This means that the admissibility condition is continuity of all derivatives through order $2m - 1$, which is difficult to achieve. Furthermore, the bandwidth of K is increased and its condition number is essentially squared, going from $O(h^{-2m})$ to $O(h^{-4m})$.[†] Therefore, the numerical solution must inevitably proceed more slowly. The rate of convergence in $\mathcal{H}^s$ will normally be h^p, where p is the smaller of $k - s$ and $2k - 4m$.

Finally, we mention three of the techniques recently invented for problems which have homogeneous differential equations, say $\Delta u = 0$, and inhomogeneous boundary conditions. There are a number of important applications in which u and $\partial u/\partial n$ are wanted only on the boundary, and it seems rather inefficient to compute the solution everywhere in Ω.

One possibility is to find a family of exact solutions $\varphi_1, \ldots, \varphi_N$ to the differential equation, and choose the combination $\sum Q_j \varphi_j$ which satisfies the boundary conditions as closely as possible. This means that some expressions for the boundary error is to be minimized. With least squares this leads to a linear equation for the Q_j; Fox, Henrici, and Moler [F10] had remarkable success by using instead a minimax principle, minimizing the error over a discrete set of boundary points. If the φ_j are eigenfunctions of the differential operator, there may be enormous simplifications at some points;

[†]This objection has been met by a modification due to Bramble and Nitsche.

again the serious difficulties are at the boundary. We refer to Mikhlin's books [11–13] for a clear and detailed discussion of this more classical circle of ideas.

A second way to use exact information about the homogeneous equation is to introduce its Green's function and transform the problem into an integral equation over the boundary Γ. In some initial trials the approximate solution on Γ has been taken as a piecewise polynomial and its coefficients determined by collocation. And a third idea, suggested by Mote, is to use the old-fashioned global trial functions in the interior of Ω, with finite elements at the boundary. This makes sense, because the smoothness of u in the interior permits good approximation by comparatively few global functions, and the hard work is done in a "boundary layer". There appears to exist almost no theoretical analysis of this idea, but its time will come.

3 APPROXIMATION

3.1. POINTWISE APPROXIMATION

This section begins our discussion of the question which lies mathematically at the center of the whole finite element analysis—approximation by the spaces S^h. We start with approximation in a pointwise sense, where the special role of polynomials will be easy to recognize. Then in the remainder of the chapter, the same pattern is extended to the spaces $\mathcal{JC}^s(\Omega)$, in other words, to the approximations in energy norms on which the Ritz–Galerkin–finite element method is based.

Suppose to start with that we are given a smooth function $u = u(x)$, defined at every point $x = (x_1, \ldots, x_n)$ in the n-dimensional domain Ω. Suppose also that S is a nodal finite element space, spanned by $\varphi_1(x), \ldots,$ $\varphi_N(x)$. As in Section 2.1, this means that to each φ_j there is associated a node z_j and a derivative D_j such that

$$(1) \qquad\qquad D_i\varphi_j(z_i) = \delta_{ij}.$$

Suppose, finally, that *the space has degree $k - 1$: Every polynomial in $x_1, \ldots, x_n$ of total degree less than k is a combination of the basis functions φ_j, and therefore lies in S.* (The total degree of $x_1 x_2$, for example, is 2; its presence is required in a space of degree $k - 1 = 2$, but not in a space of first degree even though it is linear in x_1 and x_2 separately.) If $P(x)$ is such a polynomial, and the linear combination which produces it is $P = \sum p_j\varphi_j$, then by the interpolation property (1) the weights are just the nodal values of the polynomial:

$$p_j = D_j P(z_j).$$

The nodal derivatives D_j should be of order $\leq k$.

The degree of S is usually trivial to compute. For linear or bilinear approximation the degree is one (in other words, $k = 2$), for cubics and bicubics $k = 4$, for the reduced quintic k is only 5, and so on.

From these hypotheses we want to deduce the order of approximation achieved by S. Recalling that the domain Ω is partitioned into elements $e_1, e_2, \ldots$, we shall use the distances

$$h_i = \text{diameter of } e_i, \qquad h = \max h_i,$$

in measuring the accuracy of approximation. We have in mind a *sequence* of nodal subspaces S^h, parameterized by h, and we hope to establish that the approximation error decreases like a power of h. This will require a uniformity assumption on the functions φ_j^h, which can be expressed as follows: *The basis functions φ_j^h are uniform to order q provided there exist constants c_s such that for all h, i, and j,*

$$(2) \qquad \max_{\substack{x \text{ in } e_i \\ |\alpha| = s}} |D^\alpha \varphi_j^h(x)| \leq c_s h_i^{|D_j| - s}.$$

This condition is imposed on all derivatives $D^\alpha = \partial^{|\alpha|}/\partial x_1^{\alpha_1} \ldots \partial x_n^{\alpha_n}$ up to order q; that is, it holds for all α such that $|\alpha| = \alpha_1 + \ldots + \alpha_n \leq q$. Recall that $|D_j|$ is the order of the derivative which is interpolated by the basis function φ_j^h at its node z_j^h; $|D_j| = 0$ if φ_j^h is associated with a function value at a vertex, $|D_j| = 1$ if it is associated with v_x or v_y or v_n, and so on.

A good example in one dimension is furnished by the Hermite cubic polynomials ($k = 4$), determined over each interval by their values and their derivatives at the endpoints. On a unit interval the two basis functions ψ and ω are displayed in Fig. 1.8. Over the reduced interval $[0, h]$ these become $\psi^h = \psi(x/h)$ and $\omega^h = h\omega(x/h)$, respectively; one sees the extra factor h appearing in ω^h, in order to keep its slope equal to 1 at the origin. This is the reason for the term $h^{|D_j|}$ in (2), to make the inequality dimensionally correct. These basis functions are uniform up to $q = 2$, but for third derivatives a δ-function enters at the nodes and uniformity breaks down. This is typical of finite elements; the qth derivatives involve step functions, which we accept in (2), but higher derivatives are disallowed. Thus q is the parameter associated with the smoothness of the subspace; S^h is contained in the space $\mathcal{C}^{q-1}$ of functions with $q - 1$ continuous derivatives, and more importantly, it is contained in the space $\mathcal{K}^q$ of functions with q derivatives in the mean-square sense. The conforming condition, in order to apply the Ritz method to a differential equation of order $2m$, is simply $q \geq m$.

The uniformity condition becomes especially significant in two or more dimensions. Normally it can be translated directly into a geometrical constraint on the elemental regions e_i. First one considers a standard element, say the right triangle T with vertices at $(0, 0)$, $(1, 0)$, and $(0, 1)$. With respect to T the basis functions and their derivatives up to order q should be bounded. Then a change of coordinates, taking T into the given triangle e_i and involving the Jacobian, leads to the following result: *The basis is uniform provided that as $h \to 0$, all angles in the triangulation exceed some lower bound θ_0.* In this case it is not difficult to find constants c_s bounded by

$$c_s \leq \frac{\text{constant}}{(\sin \theta_0)^s}.$$

We emphasize that the effect of the element geometries on approximation is entirely contained in this estimate. For quadrilaterals the angles must also be bounded away from $180°$, to avoid degeneration into triangles.

We should note a situation in which this bound on c_s is not valid. It will certainly fail if the polynomial is not uniquely determined within the element by the nodal parameters, that is, if the matrix H in Section 1.10 which links the polynomial coefficients a_i to the nodal parameters q_j becomes unbounded for some element configuration. Such singular cases, once recognized, have been avoided in the literature. The danger is greatest when e_i is originally curved, and is mapped into a polygon in order to use one of the standard finite element constructions. If the Jacobian of this mapping is allowed to vanish (see Section 3.3, for example), the mathematical properties of the construction are lost.

The geometrical condition is easy to understand for linear functions on triangles. The pyramid φ_j^h equals 1 at the jth vertex and zero at the others. In between one always has $|\varphi_j^h(x)| \leq 1$, so that uniformity holds for the zeroth

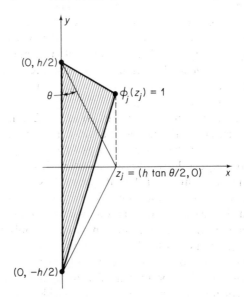

Fig. 3.1   Linear shape function in a thin triangle.

derivative: $c_0 = 1$. Consider, however, the x-derivative in Fig. 3.1. Since φ_j equals 1 at z_j and zero at P, its slope between these points is

$$\frac{\partial}{\partial x}\varphi_j = \left(\frac{h}{2}\tan\theta\right)^{-1}.$$

Comparing this with the uniformity condition (2), c_1 will be at least $2/\tan\theta$. Thus, if the triangles are allowed to degenerate into thin "needles," the basis will not be uniform.

On such a triangle the linear interpolate can lead to significant errors. Consider, for example, the function $u(x, y) = y^2$. The interpolate u_I vanishes at z_j and equals $h^2/4$ at the other two nodes (and therefore at P). Consequently the derivative along the x-axis, which should vanish since $u = y^2$ is independent of x, may be seriously in error for the interpolate:

$$\frac{\partial}{\partial x}(u - u_I) = \frac{h^2/4}{(h/2)\tan\theta} = \frac{h}{2\tan\theta}.$$

This derivative error is $O(h)$ only if the angle θ is bounded below. We note that $u - u_I = O(h^2)$ regardless of θ; the difficulty comes from the dimensionless factor "$h_{\max}/h_{\min}$" introduced by the derivative.

There is also another reason to avoid degenerate triangles of any kind: They may injure the numerical stability of the Ritz method, as reflected in the condition number of K. Therefore, several algorithms have been constructed to triangulate an arbitrary domain Ω, keeping all the angles greater than $\theta_0 \approx \pi/8$. For a more general n-dimensional element, the geometrical requirement can be expressed in terms of inscribed spheres rather than angles: e_i should contain a sphere of radius at least νh_i, ν a fixed constant.

We are now ready for pointwise approximation in the nodal method. The given function u will be approximated by its interpolate in the space S, in other words, by the function u_I which shares the same nodal parameters as u itself:

$$u_I(x) = \sum D_j u(z_j)\varphi_j(x).$$

We suppose that u has k derivatives in the ordinary pointwise sense, and estimate the sth derivatives of $u - u_I$.

THEOREM 3.1

Suppose that S is of degree $k - 1$ and its basis satisfies the uniformity condition (2). Then for $s \leq q$,

(3) $$\max_{\substack{x \text{ in } e_i \\ |\alpha|=s}} |D^\alpha u(x) - D^\alpha u_I(x)| \leq C_s h_i^{k-s} \max_{\substack{x \text{ in } e_i \\ |\beta|=k}} |D^\beta u(x)|.$$

Proof. We choose an arbitrary point x_0 in e_i, and expand u in a Taylor series:

$$u(x) = P(x) + R(x),$$

where P is a polynomial of degree $k - 1$ and R is the remainder term. We shall need the standard estimate for the Taylor remainder R and its derivatives,

(4)
$$\max_{\substack{x \text{ in } e_i \\ |\alpha| = s}} |D^\alpha R(x)| \le Ch_i^{k-s} \max_{\substack{x \text{ in } e_i \\ |\beta| = k}} |D^\beta u(x)|.$$

This can be proved by expressing R as a line integral from x_0 to x.

The interpolate of u decomposes, by linearity, into the sum of two interpolates:

$$u_I(x) = P_I(x) + R_I(x).$$

The crucial point is that P_I is identical with P; any polynomial of degree $k - 1$ is exactly reproduced by its interpolate. This is where polynomials play a special role. In other words, $P - P_I$ coincides in e_i with a trial function for which all the determining nodal parameters are zero; therefore, $P \equiv P_I$ in e_i.

This means that $u - u_I = R - R_I$, and the only terms left to estimate are the derivatives of R_I:

$$D^\alpha R_I(x) = \sum D_j R(z_j) \cdot D^\alpha \varphi_j(x).$$

At most d nonzero terms can appear in this sum, since the other basis functions vanish in the element. Therefore, if (4) is combined with the uniformity condition (2),

$$|D^\alpha R_I(x)| \le dCh_i^{k-|D_j|} \max |D^\beta u| \cdot c_s h_i^{|D_j|-s}$$
$$\le C' h_i^{k-s} \max |D^\beta u|.$$

Thus R_I is bounded exactly as R was, and it follows immediately that

$$|D^\alpha(u - u_I)| = |D^\alpha(R - R_I)| \le (C + C')h_i^{k-s} \max |D^\beta u|.$$

This is the estimate (3), and the proof is complete.†

Note that this is a completely local estimate; u_I imitates the properties of u within each element. This is a property which we cannot expect of the

---

†The uniformity of the basis is *not* a necessary condition for the approximation theorems to hold. For bilinear or bicubic approximation on rectangles, the choice of a very fine mesh in one coordinate direction will spoil the uniformity but not the order of approximation. Even for triangles, Synge's condition in [17] is that the largest angle be bounded below π, i.e., that the *second smallest* angle in each triangle exceed some θ_0.

finite element solution u^h, since it is obtained by minimizing a global function $I(v)$, the potential energy over the whole domain Ω. In fact, a singularity in u is indeed propagated by u^h—sometimes with frustratingly slow decay—throughout the entire domain. This is analyzed in Chapter 8.

There is a similar approximation theorem for the abstract finite element method. We shall take time to prove this second theorem, even though the two overlap for the usual nodal method on a regular mesh. Splines were not covered by the previous theorem, because they lack a local interpolating basis, yet their approximation properties are of considerable importance. In fact, the special regularity of the mathematical structure allows a more elegant result. *Approximation on a regular mesh hinges on the presence of a trial function ψ with the following remarkable property: For any polynomial P of degree less than k,*

$$(5) \qquad\qquad P(x) \equiv \sum_l P(l)\psi(x - l).$$

Such a superfunction ψ can be found in the trial subspace S if and only if S is of degree $k - 1$. ψ will be nonzero over a small patch of elements.

In the simplest one-dimensional example, S^h is made up of continuous piecewise linear functions. There is one unknown per mesh interval ($M = 1$), and the basis is generated by the roof function Φ_1. In other words, the basis functions are $\varphi_i^h(x) = \Phi_1(x/h - l)$, and S^h is of degree 1. In this example Φ_1 itself will serve as the function ψ: For any linear polynomial $P(x) = \alpha x + \beta$,

$$P(x) = \sum P(l)\Phi_1(x - l).$$

The two sides agree at every node and are linear in between, so they must be identical. The same will be true in two dimensions both in the case of linear functions on triangles, where the graph of $\Phi_1 = \psi$ is a pyramid, and of bilinear functions on rectangles, where the graph is a "pagoda." In these cases the approximation u_Q in the following theorem coincides with the interpolate u_I.

The construction of ψ is not so trivial for cubics. The two generating functions for Hermite cubics, as well as the B-spline Φ_1, were sketched in Section 1.7. None of these functions satisfies the identity (5) up to $k = 4$, which is required of ψ. However, there exists a cubic spline which does, and it is nonzero over six intervals (Fig. 3.2). This ψ is a combination of the B-spline and its two neighbors, and also (since every spline is automatically a Hermite cubic) it will serve as the superfunction for the Hermite space, too. We must emphasize here that it is never essential to know what ψ is; what matters is only that somewhere in the trial space lies a function which is responsible, by itself, for the approximation properties of the space.

Given the existence of $\psi(x)$, vanishing for $|x| \geq \rho$ and satisfying the

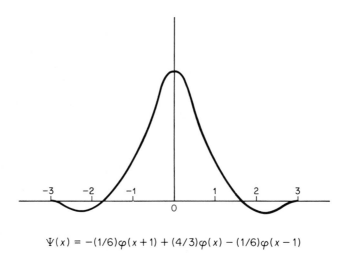

$$\Psi(x) = -(1/6)\varphi(x+1) + (4/3)\varphi(x) - (1/6)\varphi(x-1)$$

Fig. 3.2   The superfunction for cubics.

superfunction identity, we approximate a given function u by the following member of S^h:

$$u_Q(x) = \sum_l u(lh)\psi\left(\frac{x}{h} - l\right).$$

We refer to u_Q as a *quasi-interpolate* based on ψ. It depends as u_I did on the local properties of u, but does not quite interpolate u at the nodes. Its advantage is that it can be written down so easily and explicitly, whereas interpolation by splines requires the solution of simultaneous equations (several B-splines are nonzero at each node). Since approximation only demands that there exists *some* function in S^h which is close to u, we are free to work with a convenient choice.

THEOREM 3.2

Suppose the abstract finite element space S^h is of degree $k-1$ and ψ has bounded derivatives up to order q. Then for any derivative D^α of order $|\alpha| \le q$,

(6) $$\max_{\substack{0 \le x_i < h \\ |\alpha| = s}} |D^\alpha u(x) - D^\alpha u_Q(x)| \le c_s h^{k-s} \max_{\substack{|x| \le ph \\ |\beta| = k}} |D^\beta u(x)|.$$

Proof. The argument is nearly the same as in Theorem 3.1. A Taylor expansion around the origin gives $u(x) = P(x) + R(x)$, where P is of degree $k-1$ and R is the remainder. Splitting u_Q into $P_Q + R_Q$, we note first that P_Q coincides with P:

$$P(x) = P_Q(x) = \sum_l P(lh)\psi\left(\frac{x}{h} - l\right).$$

This is the identity (5) with x scaled by the factor h; that is, (5) is applied to the polynomial $p(x) = P(xh)$, and then $x \longrightarrow x/h$.

The Taylor remainder R was estimated in the previous theorem, and this leaves only R_ϱ:

$$D^\alpha R_\varrho(x) = \sum R(lh)D^\alpha \left[\psi\left(\frac{x}{h} - l\right) \right].$$

For x in the mesh cube $0 \leq x_i < h$,

$$|D^\alpha R_\varrho(x)| \leq \sum C\,|lh|^k \max_{\substack{|x| \leq \rho h \\ |\beta| = k}} |D^\beta u(x)|\, h^{-|\alpha|}C',$$

where C' bounds the derivatives of ψ, the factor $h^{-|\alpha|}$ comes from the differentiation D^α, and the rest is an upper bound for $R(lh)$. As in Theorem 3.1, it is essential that the sum is finite; since ψ is nonzero only for $|x| \leq \rho$, only finitely many values of l actually appear. We now have the required estimate, with $|\alpha| = s$:

$$|D^\alpha(u - u_\varrho)| = |D^\alpha(R - R_\varrho)| \leq c_s h^{k-s} \max |D^\beta u|.$$

The same result follows in any other mesh cube by expanding u around one of the vertices.

3.2. MEAN-SQUARE APPROXIMATION

We want to prove that also in the mean-square norms, the same degree of approximation is possible under the same hypotheses on S^h—that it has degree $k - 1$ and a uniform basis. It is the variational principle which makes it natural and inevitable to work with these $\mathcal{3C}^s$ norms; the strain energy itself is exactly such an expression, an integral of the square of derivatives of u. Since u^h is the element closest to u in this energy norm, the key error estimate in the whole subject is this one.

In general, the $\mathcal{3C}^m$ norm involves all derivatives D^α of order $|\alpha| \leq m$:

$$||v||_{m,\Omega}^2 = \sum_{|\alpha| \leq m} \int_\Omega \left| \frac{\partial^{|\alpha|} v(x_1, \ldots, x_n)}{\partial x_1^{\alpha_1} \ldots \partial x_n^{\alpha_n}} \right|^2 dx.$$

We shall have use also for the *seminorm* $|v|_{m,\Omega}$, for which only the derivatives of order exactly $|\alpha| = m$ are included in the sum. This is a seminorm, because it shares the properties $|cv| = |c|\,|v|$ and $|v + w| \leq |v| + |w|$ of a norm, but it fails to be positive definite: For a true norm, $||v|| = 0$ can happen only if $v = 0$, but for these seminorms, $|P|_m = 0$ whenever P is a polynomial of degree less than m. Obviously $||v||_m^2 = |v|_0^2 + \ldots + |v|_m^2$.

In one sense it is completely trivial to extend the pointwise results to these new norms. Using the interpolate u_I of the previous section, we may simply square the pointwise error and integrate: If $|\alpha| = s$,

$$\int_\Omega |D^\alpha u - D^\alpha u_I|^2 \, dx \leq (\text{vol } \Omega) \max |D^\alpha u - D^\alpha u_I|^2$$

$$\leq C_s^2 (\text{vol } \Omega) h^{2(k-s)} \max_{\substack{x \text{ in } \Omega \\ |\beta| = k}} |D^\beta u(x)|^2.$$

Thus if the kth derivatives of u are bounded, the error is of the same order h^{k-s} as before. This settles the rate of convergence to smooth solutions.

Such an estimate, however, can never be entirely satisfactory. It involves two quite different norms, and we have made a pointwise assumption on the kth derivatives in order to obtain a mean-square error bound on the sth derivatives. One has a right to expect a more symmetric theorem, in which only norms of the same kind appear. Furthermore, such an improvement is made necessary by singularities. For a plate with a fracture, the solution u increases only like the square root of the distance from the end of the crack. The function $r^{1/2}$ is not even differentiable; the pointwise error will behave like $h^{1/2}$ or perhaps $h^{1/2} \log h$. However, the solution does lie in $\mathcal{H}^1$; it must. In fact, u nearly has "one and a half derivatives" in the mean-square sense; the sth derivative behaves like $r^{(1/2)-s}$, and for any $s < \frac{3}{2}$,

$$\int |r^{(1/2)-s}|^2 \, r \, dr \, d\theta < \infty.$$

In short, we would like to prove that *the mean-square error is roughly of order $h^{3/2}$*, and such a result can only follow from a theorem in which hypotheses are made on the *mean-square differentiability* of u.

THEOREM 3.3

Suppose that S^h is of degree $k - 1$, and its basis is uniform to order q. Suppose also that the derivatives D_j associated with the nodal parameters are all of order $|D_j| < k - n/2$. Then for any function $u(x_1, \ldots, x_n)$ which has k derivatives in the mean-square sense, and any derivative D^α of order $s \leq q$,

(7) $$\int_{e_i} |D^\alpha u(x) - D^\alpha u_I(x)|^2 \, dx \leq C_s^2 h_i^{2(k-s)} |u|_{k,e_i}^2.$$

Since the integral over Ω is the sum of the integrals over e_i,

(8) $$|u - u_I|_{s,\Omega} \leq C_s h^{k-s} |u|_{k,\Omega}.$$

Remark 1. The assumption $|D_j| < k - n/2$ is necessary in order that the interpolate u_I be defined. Sobolev's lemma (mentioned in Section 1.8) assures that since $u(x_1, \ldots, x_n)$ has k mean-square derivatives, the derivative $D_j u$ is well defined at any point z_j:

$$(9) \qquad\qquad |D_j u(z_j)| \leq c \|u\|_k.$$

Fortunately, this assumption $|D_j| < k - n/2$ is fulfilled for all practical elements, and the theorem leads directly to the rate of convergence of the finite element method.

To find a more awkward case, we turn to approximation by piecewise constants ($k = 1$) in the plane ($n = 2$). Then, as for the function log log r in Section 1.8, Sobolev does not guarantee that the interpolate of u makes sense; if a node happened to fall at the origin, what would be log log 0? Even in this case, however, S^h *does* contain a function which gives the correct order h of least-square approximation; u can be suitably smoothed before it is interpolated. The same construction succeeds in the general case [S6]: *If S^h has degree $k - 1$ and a uniform basis, then for any dimension n,*

$$|u - \bar{u}_I|_s \leq C_s h^{k-s} |u|_k, \qquad \bar{u} = \text{smoothed } u.$$

Thus each finite element space S^h contains a function, usually u_I but if necessary $\bar{u}_I$, which approximates u to the expected order. The bound for u_I is much simpler to prove and is sufficient for all practical purposes—or it would be, if it were extended also to fractional derivatives, as required by the example of a cracked plate. This extension is actually an immediate corollary of the theorem, given the theory of "interpolation spaces"—which we omit as being too technical.

Remark 2. The proof even of the simpler theorem does still require one technical lemma. The Taylor expansion on which pointwise approximation was based will not work here; u may have enough derivatives for its interpolate to be defined, but not for a Taylor expansion with remainder h^k. Therefore, we must rely on functional analysis to produce a polynomial which is as close to u in the mean-square sense as the leading Taylor-series terms were in the pointwise sense. The key result, simplified from [14] and [B27], is this: *In every element there exists a polynomial P_{k-1} such that the remainder $R = u - P_{k-1}$ satisfies*

$$(10a) \qquad\qquad |R|_{s,e_i} \leq c h_i^{k-s} |u|_{k,e_i}, \qquad s \leq k,$$

and at every node z_j,

$$(10b) \qquad\qquad |D_j R(z_j)| \leq c' h_i^{k-|D_j|-n/2} |u|_{k,e_i}.$$

On a region of diameter $h_i = 1$, (10a) is a standard lemma [14]; (10b) follows from (10a) and Sobolev's inequality (9). Then the given powers of h_i appear when the independent variables are rescaled so as to shrink the region down to e_i.[†] The constants c and c' will depend on the smallest angle of e_i; it is simplest to assume this angle to be bounded below, as is normally required for the basis to be uniform.

Proof of the theorem. The argument will be exactly the same as in the pointwise case; that is the happy effect of the lemma. In each element e_i, we write $u = P_{k-1} + R$; R satisfies the inequalities (10). The interpolate u_I is, by linearity, the sum of the two interpolates:

$$u_I = (P_{k-1})_I + R_I = P_{k-1} + R_I,$$

since the polynomial P_{k-1} is exactly reproduced by interpolation. Therefore $u - u_I = R - R_I$, and we consider

$$R_I(x) = \sum (D_j R)(z_j)\varphi_j(x).$$

Only a finite number d of these terms can be nonzero in e_i. Applying the uniformity of the basis, as well as (10b), any derivative of order $|\alpha| = s$ gives

$$|D^\alpha R_I(x)| \le dc' h_i^{k-|D_j|-n/2} |u|_{k,e_i} \cdot c_s h_i^{|D_j|-s}.$$

Now if we square, integrate over e_i, and take the square root,

$$\left(\int_{e_i} |D^\alpha R_I(x)|^2 \, dx \right)^{1/2} \le c'' h_i^{k-s} |u|_{k,e_i}.$$

This is exactly the estimate given for R by (10a), and exactly the same as the right side of (7). Therefore, the proof is complete. The technique, depending on (10) and on the special role of polynomials, is known as the *Bramble–Hilbert lemma.*

The same theorem and proof hold for abstract finite elements on a regular mesh, in particular for splines; we again use u_Q instead of u_I [S5].

The theory of partial differential *inequalities* leads to questions of *one-sided approximation.* We have established for the standard linear elements that, given $u \ge 0$, the estimates of Theorems 3.1–3.3 continue to hold if the approximate v^h is required to satisfy $0 \le v^h \le u$. (The interpolate $v^h = u_I$ is of course no use, since it need not lie below u.) We note that Duvaut-Lions were able to formulate as variational inequalities several important physical problems, including elastic-plastic phenomena—which in differential form lead to extremely awkward elliptic-hyperbolic systems separated by an unknown free boundary. Mosco and Strang, and independently Falk, have confirmed the usual h^2 error in energy for linear approximation of the *St. Venant torsion problem*, which is typical of the class of variational inequalities known as *obstacle problems.*

An additional question arises if the elements contain a few polynomial

---

†h_i^k arises from rescaling the kth derivatives, and $h_i^{n/2}$ from the square root of the volume of e_i.

terms of the next higher degree k: Are all derivatives of this order needed on the right side of (7)? Bilinear elements, for example, reproduce the twist term xy, and therefore it seems superfluous to include the cross derivative u_{xy} in the error bound $ch^{2-s}|u|_2$. This point was settled by Bramble and Hilbert [B28]: It is indeed sufficient to include only u_{xx} and u_{yy} in the error bound. This will have valuable consequences in the following section for the theory of rectangular isoparametric elements.

It may well happen that (7) holds inside each element for derivatives of all orders $s \leq k$. The limitation to $s \leq q$ in (8) arises when u_I has only $q-1$ derivatives across element boundaries.

It would be useful to know something about the constants C_s in the theorem. They are a direct indication of the properties of the particular element; if they are larger for one element than for another of the same degree, then the first element is comparatively inaccurate or "stiff." For piecewise linear approximation on the line, with equally spaced nodes $x_j = jh$, these optimal constants can be computed. There are two functions of special interest: the first trigonometric polynomial $f(x) = \sin \pi x/h$ that vanishes at every meshpoint, and the first algebraic polynomial $g(x) = x^2$ not reproduced by its linear interpolate. In the case of f, both its interpolate and its best linear approximation are identically zero. Therefore, the error is f itself, and an easy computation gives

$$|f|_0 = \frac{1}{\pi^2} h^2 |f|_2, \qquad |f|_1 = \frac{1}{\pi} h |f|_2.$$

It follows from the proof of Theorem 1.3 in Section 1.6 that these constants are optimal, since for any u the interpolate u_I was shown to satisfy

$$|u - u_I|_0 \leq \frac{1}{\pi^2} h^2 |u|_2, \qquad |u - u_I|_1 \leq \frac{1}{\pi} h |u|_2.$$

Therefore, $C_0 = 1/\pi^2$ and $C_1 = 1/\pi$ for linear approximation.

What role is left for the quadratic $g(x) = x^2$? It yields, for each *fixed* function u, the constant which is *asymptotically correct as* $h \to 0$. In the previous case we fixed h and sought out the worst function $\sin \pi x/h$. In this case we fix u and ask for the limits

(11) $\qquad c_0 = \lim_{h \to 0} \frac{\min |u - v^h|_0}{h^2 |u|_2}$ and $\quad c_1 = \lim_{h \to 0} \frac{\min |u - v^h|_1}{h |u|_2}.$

The minimum is taken over all v^h in S^h, that is, over all piecewise linear functions. For every h these ratios are bounded by $1/\pi^2$ and $1/\pi$, respectively, and therefore the limits cannot exceed these constants; we may expect them to be smaller, since a fixed u cannot resemble all the oscillatory functions $\sin \pi x/h$ at once. These new constants c_0 and c_1 (if they exist) seem to be more natural in assessing the improvement to be expected from mesh refinement in a practical problem, since it is the solution which is fixed and h which changes.

Consider first the particular function $u = g = x^2$. On the interval $[-1, 1]$, the best linear approximations to displacement and slope minimize

$$\int_{-1}^{1} |x^2 - a_1 - a_2 x|^2 \, dx \quad \text{and} \quad \int_{-1}^{1} |2x - a_2|^2 \, dx,$$

respectively. These expressions equal

$$\frac{2}{5} - \frac{4a_1}{3} + 2a_1^2 + \frac{2a_2^2}{3} \quad \text{and} \quad \frac{8}{3} + 2a_2^2.$$

In both cases a_2 should be zero; the best approximation to an even function on a symmetric interval $[-1, 1]$ is even. The optimal value of a_1 is $\frac{1}{3}$, and the ratios which enter c_0 and c_1 (with an interval of length $h = 2$) become

$$\frac{|x^2 - \frac{1}{3}|_0}{2^2 |x^2|_2} = \frac{1}{12\sqrt{5}}, \quad \frac{|x^2 - \frac{1}{3}|_1}{2 |x^2|_2} = \frac{1}{2\sqrt{3}}.$$

The latter happens not to be very different from $1/\pi$; it is smaller, as it should be.

Notice that $x^2 - \frac{1}{3}$ is the second Legendre polynomial; it is the error after least-square approximation by linear functions of the quadratic $g(x) = x^2$. Notice also that this error has the same value $\frac{2}{3}$ at the two ends of the interval. This makes it easy to connect it with the best approximation to x^2 on the *next* interval $1 \leq x \leq 3$. On this interval the error function after optimal approximation is just $(x - 2)^2 - \frac{1}{3}$, the same Legendre polynomial translated two units along the axis. This pattern continues: On every interval $[2n - 1, 2n + 1]$, the optimal error function $x^2 - v^h$ is given by $(x - 2n)^2 - \frac{1}{3}$, and the error function is periodic with period $h = 2$ (Fig. 3.3). The constants $1/12\sqrt{5}$ and $1/2\sqrt{3}$ are the same on any union of these intervals. *Furthermore, these constants are dimensionally correct, and they will not change if the independent variable is rescaled to give an arbitrary initial interval $[-h/2, h/2]$, and then translated to give an arbitrary position of the origin.* The x-axis in the figure is scaled by a factor of $h/2 : 1$ and the error function on the y-axis is scaled by $(h/2)^2 : 1$. It is easy to check that the ratios are invariant; the reason x^2 is so special is that translation to $(x - x_0)^2$ alters it by a linear expression $2xx_0 - x_0^2$, which can be taken up by the trial functions. Note that *the optimal approximation is by no means the linear interpolate*; the interpolate yields the right exponent of h but too large a constant. The Ritz procedure achieves the best constant, because it minimizes over S^h.

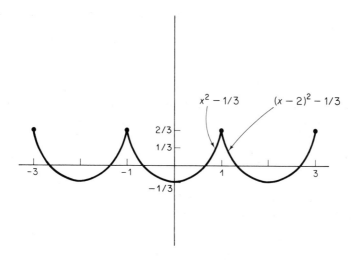

$x^2 - 1/3$

$(x - 2)^2 - 1/3$

Fig. 3.3   Error in linear approximation to x^2.

The remarkable thing is that the asymptotic constants are independent of the function u.

THEOREM 3.4

For an arbitrary function $u(x)$ in $\mathfrak{IC}^2$, the limiting values of the ratios as $h \longrightarrow 0$ are the same as for the special function x^2: $c_0 = 1/12\sqrt{5}$ and $c_1 = 1/2\sqrt{3}$.

This theorem means that in the limit as $h \longrightarrow 0$, every function behaves like a piecewise quadratic; the error function after linear approximation locally resembles Fig. 3.3. In other words, *the more closely you look at a function, the more it looks like a polynomial.* This is the underlying foundation for Taylor-series expansions. The constant c_0 is therefore asymptotically correct for least-square approximation, not just to some extreme choice of u but rather to all choices. Similarly, c_1 is asymptotically correct for the finite element error $u - u^h$ in the second-order example of Chapter 1. In general, c_1 will be decisive *even in displacement error* for $u - u^h$, since u^h is chosen to minimize the strain energy of the error; Nitsche's argument in Theorem 1.5 indicates how c_1^2 rather than c_0 enters the displacement error.

Theorem 3.4 will be proved first for smooth functions u. In each interval of length h, with midpoint x_0,

$$(12) \qquad u(x) = u(x_0) + (x - x_0)u'(x_0) + \frac{(x - x_0)^2}{2}u''(x_0) + O(h^3).$$

The optimal (least-squares) linear approximation to the first three terms is

$$l(x) = u(x_0) + (x - x_0)u'(x_0) + \frac{1}{3}\left(\frac{h}{2}\right)^2 \frac{u''(x_0)}{2}.$$

Using these linear functions l in each subinterval, we are effectively back to the quadratic case. The ratios in each interval, taking account of the $O(h^3)$ error from the remainder in the Taylor expansion, still satisfy

$$|u - l|_0 = \frac{h^2 |u|_2}{12\sqrt{5}}(1 + O(h)), \qquad |u - l|_1 = \frac{h|u|_2}{2\sqrt{3}}(1 + O(h)).$$

Squaring, and summing over all the intervals, the piecewise linear function L formed from all these pieces l satisfies

(13)
$$\frac{|u - L|_0}{h^2 |u|_2} = \frac{1 + O(h)}{12\sqrt{5}}.$$

There remains one difficulty: L *is not continuous at the nodes*. The linear approximations l depend on Taylor expansions in their own interval and cannot be expected to connect with their neighbors. The discrepancy, however, is only of the order

$$\frac{1}{3}\left(\frac{h}{2}\right)^2 \left(\frac{u''(x_0)}{2} - \frac{u''(x_0 + h)}{2}\right) = O(h^3),$$

and therefore an $O(h^3)$ alteration in each linear piece l will produce a continuous $\tilde{L}$. (In the 1-norm the alteration is $O(h^2)$.) After such an alteration, both (13) and its analogue in the 1-norm continue to hold for $\tilde{L}$. Therefore, as $h \rightarrow 0$, the ratios approach the constants $1/12\sqrt{5}$ and $1/2\sqrt{3}$. It is clear that no other choice of v^h in (11) could yield a smaller constant, because these constants were already correct when L was formed from the optimal l on each subinterval. The theorem is therefore proved when u is smooth enough to permit the Taylor expansion (12).

The extension to all u in $\mathfrak{IC}^2$ is a standard technical problem. We define the linear operator P_s^h, from $\mathfrak{IC}^2$ to $\mathfrak{IC}^s$, by

$$h^{s-2} P_s^h u = \text{component of } u \text{ perpendicular to } S^h.$$

Two properties of this operator are already proved, that

1. $|P_s^h u|_s \leq C_s |u|_2$ (Theorem 3.3, with $k = 2$).
2. $|P_s^h u|_s \rightarrow c_s |u|_2$ for smooth u (see previous paragraph).

By a standard density argument, which is uninteresting to an engineer and boring to a mathematician, property 2 holds for all u and Theorem 3.4 is proved.

We point out now (and shall do so again) that the zeros of the function $x^2 - \frac{1}{3}$ are special points. Of course, they are the zeros of the Legendre polynomial, and therefore appear in Gauss quadrature—on an interval $[jh, (j+1)h]$ they move to $(j + 1/2 \pm 1/\sqrt{3})h$. For finite element purposes, however, they are special in another way: The best approximation to a quadratic vanishes at these points, and u^h is likely to be of exceptional accuracy. (In collocation this is known to be the case; see Section 2.3.) We refer to them as *displacement points*. There are also *stress points*, discovered by Barlow, which are of still greater importance. These are the points at which the derivatives of the error function vanish ($x = 0$ in our simple example) and we show in Section 3.4 that *the stress errors at these points are smaller by an extra power of h.*

The previous theorem extends to any finite element on a regular mesh in n dimensions and more generally to any example of the abstract finite element method. In n variables, there are several derivatives D^β of order $|\beta| = k$, possibly associated with different error constants in approximation. In fact, if x^β happened to be present in S^h, the corresponding constant would vanish. Locally, we imagine the function u as expanded in a Taylor series through terms of degree k; those of degree $k - 1$ are reproduced exactly by the trial space, and the approximation depends asymptotically only on the derivatives $D^\beta u$ of order k. This generalization of Theorem 3.4 can be expressed in the following way, with matrices K_s in place of the scalar constants c_s.

THEOREM 3.5

If S^h is of degree $k - 1$ on a regular mesh, there exist nonnegative definite matrices K_s such that for every u in $\mathfrak{IC}^k$,

$$(14) \qquad h^{2(s-k)} \min_{S^h} |u - v^h|_s^2 \longrightarrow \sum_{|\alpha| = |\beta| = k} K_s^{\alpha\beta} \int (D^\alpha u)(D^\beta u)\, dx.$$

The diagonal entries $K_s^{\beta\beta}$ can be determined by approximation of the monomial $u = x^\beta = x_1^{\beta_1} \ldots x_n^{\beta_n}$, for which $D^\beta u$ is a constant and the other derivatives $D^\alpha u$ of order k are zero.

It was convenient to square the expressions which occurred in Theorem 3.4; thus $K_0 = c_0^2 = \frac{1}{720}$ and $K_1 = c_1^2 = \frac{1}{12}$. For linear approximation in in two dimensions, the matrices K_s will be of order 3, corresponding to the three derivatives $\partial^2/\partial x^2$, $\partial^2/\partial x\, \partial y$, and $\partial^2/\partial y^2$ of order $k = 2$. For spline spaces of arbitrary degree the constants $K_s^{\alpha\beta}$ have been computed [S5] in terms of the Bernoulli numbers; we believe they are the same for the Hermite spaces of corresponding degree. Also the minimal constants C_s, for a fixed h and varying u, are known for splines (Babuska). In this case the extreme functions are again sines of wavelength $2h$, their best approximations are identically zero, and the constants are $C_s = \pi^{s-k}$.

It is possible to use Theorem 3.5 for a quantitative comparison of two different polynomial elements, or of the same element with two different geometrical configurations. Consider two regular triangulations of the plane, one involving diagonals in both directions, the other in only one direction (Fig. 3.4). Combinatorially the two are quite different. In **(A)** some

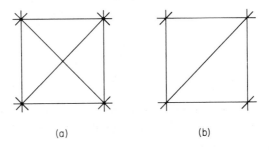

$$\text{(a)} \hspace{5cm} \text{(b)}$$

Fig. 3.4   Two possible triangulations.

nodes are connected to four neighbors, and others to eight; in (B) every node has six neighbors. Consider Courant's space S^h of continuous piecewise linear functions on these triangles. Since (A) has twice as many nodes as (B), the dimension of S^h_A will be twice that of S^h_B. Furthermore, the space S^h_A contains S^h_B and is therefore at least as good in approximation. The question is whether it is *twice as good*, to compensate for having twice the number of parameters.

In two dimensions, the three quadratic monomials are x^2, xy, and y^2. For each of them we propose to find the element u^h of S^h which minimizes $|u - u^h|_1$. This u^h will be exactly the finite element solution for Poisson's equation, when the true solution u is a quadratic.

We begin with $u = x^2$, on a square of side $h = 2$ centered at the origin. From considerations of symmetry, the optimal u^h takes the same value at the four vertices, say α in (A) and β in (B); at the center, it must equal β in case (B) but may have a different value γ in (A). This means that in configuration (B), u^h is a constant on this particular square—the same constant $\beta = \frac{1}{3}$ as in the one-dimensional case computed above. The error over the mesh square is

$$|u - u^h|_1^2 = \iint (x^2 - \beta)_x^2 + (x^2 - \beta)_y^2 = \iint (2x)^2 \, dx \, dy = \tfrac{16}{3}.$$

This equals $h^2 |u|_2^2/12$, exactly as in one dimension: $K_1^{11} = \tfrac{1}{12}$. In case (A) there is a minimization with respect to γ; the constant must be smaller, and turns out to be

(15) $$|u - u^h|_1^2 = \frac{h^2 |u|_2^2}{18}.$$

On this basis, the two configurations can be compared. The dimension in case (A) is twice as great; in other words, it effectively works with squares of side $h/\sqrt{2}$ instead of h. With this change in (15), the constant becomes $\frac{1}{9}$, and *configuration (B) is better by a ratio of 12: 9*. The coefficient of x^2 in the error, with the number of free parameters equalized, will be smaller in case (B) by a factor of $\sqrt{\frac{3}{4}}$. By symmetry this is true also of the y^2 coefficient. For the twist term xy, the two configurations turn out to be equally effective, and there is nothing to choose.

These calculations are confirmed by numerical experiments, reported in the engineering literature, which favor the configuration (B). For elements of higher order a computer could determine the constants $K_{\alpha\beta}$ by solving a finite element problem, for which the true solution is $u = x^\beta$. It will at the same time be computing the leading terms in the truncation error for the finite difference scheme which arises on a regular mesh (see Sections 1.3 and 3.4).

Theorem 3.5 has also an important theoretical consequence.

COROLLARY

To achieve approximation of order h^{k-s} to the sth derivative, a trial space S^h on a regular mesh must be of degree at least $k - 1$. Therefore, the finite element method converges, in case of a differential equation of order $2m$, only if $k > m$; this is the constant strain condition, that all polynomials of degree m must be present in S^h.

This corollary is a converse, on a regular mesh, to Theorem 3.2. To prove it, suppose that S^h were only of degree $l - 1, l < k$. Let x^α be of degree l and not in S^h. Then by Theorem 3.4,

$$h^{s-l} \min |x^\alpha - v^h|_s \longrightarrow \text{constant} \neq 0.$$

Therefore, the order of approximation to x^α is only $l - s$, not $k - s$, and the corollary is proved. Clearly the theorem is much stronger than its corollary; the former applies to all u, while for the latter it is only necessary to consider approximation of the lowest-degree polynomial which is not in the trial space.

The conclusion is rather remarkable: everything depends on the presence of polynomials. It implies that *piecewise polynomials are the best trial functions, not only for their convenience but also for their approximation properties.*

From the beginning, the finite element method has worked with subspaces of the optimal kind.

The result of the corollary can be proved directly [S10], and because of its importance we shall do so in the simplest case. We show that for approximation of order h, the constant function 1 must lie in the trial space. Suppose, for example, that the roof function were replaced by the cosine (Fig. 3.5).

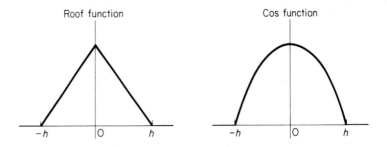

Fig. 3.5   A non-polynomial shape function.

The trial space contains all combinations $v^h = \sum q_j \operatorname{Cos}[\pi(x - jh)/2h]$, the notation Cos implying that the cosine is truncated outside its central arch $|\theta| \leq \pi/2$. It might appear that, in the limit as $h \to 0$, there would be little difference in comparison with piecewise linear approximation.

Suppose, however, that we attempt to approximate $u \equiv 1$ on a unit interval. In the linear case, this function lies in the space. In the new case it does not, and the error function E^h will be periodic of period h. We ignore boundary conditions, or rather assume them to be periodic, since in any case they have only a secondary effect on approximation in the interior of the interval. Thus the error is

$$\min_{S^h} \| 1 - v^h \|_0^2 = \int_0^1 (E^h(x))^2 \, dx.$$

This integral covers $1/h$ periods, and over each period its value is $K_0 h$, where K_0 is a constant *independent of h*. If h is changed, the error function (just as in Fig. 3.3) is simply scaled to its new period. Therefore, the approximation error is a constant K_0 (as predicted in Theorem 3.5) and *does not decrease with h*. The Cosine space has no approximation properties whatever.

Of course, this adverse conclusion does not apply to the ordinary cosines, which are among the most valuable trial functions in the Ritz method. In a sense they are of infinite accuracy, $k = \infty$, since they take advantage of each additional degree of smoothness of the solution u to be approximated. Since each cosine is nonzero over the whole interval, this case does not fall in the finite element framework and the restriction that the space must contain polynomials to be effective in approximation is no longer in force.

We conclude this section with some historical remarks on the case of a regular mesh, in other words, on approximation theory in the abstract finite element method. It is natural to approach this problem by use of the Fourier transform. The mean-square norms $\int \int | D^\alpha u |^2 \, dx$ can be converted by Parseval's formula to $\int \int | \xi^\alpha \hat{u} |^2 \, d\xi$, and the condition that φ generate polynomials produces zeros in its Fourier transform. The roof function, for example, generates all linear polynomials; its transform is $\hat{\varphi}(\xi) = (\sin \xi/2 / \xi/2)^2$, which has zeros of order $k = 2$ at all the points $\xi = \pm 2\pi, \pm 4\pi, \dots$. For the B-spline of arbitrary degree $k - 1$, the exponent in $\hat{\varphi}$ simply becomes k; it is the result of convolving the *box function* with itself k times. The connection between polynomials of degree $k - 1$ in x and zeros of order k at $\xi = 2n\pi$ was discovered by Schoenberg in his first paper on splines [S2] and rediscovered three times in the finite element context [G5, B2, F6].

In the papers [S5, S10] and in Aubin's book [4], the Fourier analysis of the abstract finite element method in n dimensions is pursued very seriously. The corollary stated earlier, that on a regular mesh S^h must be of degree $k - 1$ to achieve approximation of order h^k, was first proved by Fourier methods—together with the existence of the superfunction ψ referred to in Section 3.1. We confess to a profound regret that this Fourier analysis cannot be included in full detail, but it is only realistic for us to select those results which are unique to the case of a regular mesh and at the same time important to the general theory—the asymptotic theorem 3.5 and its corollary, the finite difference aspect of $KQ = F$ described in Section 3.4, and the discussion of condition number in Chapter 5.

Permit us to return to the main result of this section, that in the element e_i the difference between u and its interpolate satisfies

$$(16) \qquad\qquad | u - u_I |_{m, e_i} \leq C_m h_i^{k-m} | u |_{k, e_i}.$$

What happens if there is a singularity in the solution u, which prevents it from belonging to the space $\mathcal{K}^k$? With a regular mesh, the order of convergence will definitely be reduced; if u possesses only r derivatives in the mean-square sense, the error in energy will decrease like $h^{2(r-m)}$ instead of $h^{2(k-m)}$. And the pointwise error in the strains is visibly worse. The question is whether any improvement is possible by "grading" the mesh, in other words, by varying the mesh width h to produce a finer mesh near the singularity.

A useful rule of thumb, when it is inconvenient to introduce special singular trial functions, is suggested by formula (16): *The grading should be done so as to keep $h_i^{k-m} | u |_{k, e_i}$ roughly the same from one element to the next.* In one dimension, with an x^α singularity at the origin, this means that $h^{k-m+1/2} x^{\alpha-k}$ should be approximately constant; the larger x is, the larger $h = \Delta x$ is allowed to be. It appears that this rule has a remarkable consequence: *the same order of accuracy can be achieved for a singular as for a regular solution u,* *by properly grading the mesh.* In other words, suppose that for an irregular *n*-dimensional mesh with N elements we compute an average mesh size $\bar{h}$ from the formula $N\bar{h}^n = \text{vol}\,\Omega$. Then the correct grading can achieve $|u - u_I|_{m,\Omega} = O(\bar{h}^{k-m})$, even though u is singular and $|u|_{k,\Omega} = \infty$.

3.3. CURVED ELEMENTS AND ISOPARAMETRIC TRANSFORMATIONS

The basic idea is simple. Suppose it is intended to use a standard polynomial element, for example one of those defined on triangles or rectangles in Section 1.8. Suppose also that the regions into which Ω is subdivided are not of the proper shape; they may have one or more curved sides, or they may be nonrectangular quadrilaterals. By changing to a new ξ–η coordinate system, the elements can be given the correct shape. The element stiffness matrices are then evaluated by integrations in the new variables, over triangles or rectangles, and minimization leads to a finite element solution $u^h(\xi, \eta)$ which can be transformed back to x and y.†

There are several points to watch. First, since a typical two-dimensional element integral is transformed by

(17)
$$\iint_{e_i} p(x, y)(v_x)^2 \, dx \, dy$$
$$\longrightarrow \iint_{E_i} p(x(\xi, \eta), y(\xi, \eta))(v_\xi \xi_x + v_\eta \eta_x)^2 J(\xi, \eta) \, d\xi \, d\eta,$$

the coordinate change and its derivatives must be easily computable. Furthermore, the change must not distort the element excessively, or the Jacobian determinant $J = x_\xi y_\eta - x_\eta y_\xi$ may vanish within the region of integration; it is surprisingly easy for this to happen. Excessive distortion will also destroy the accuracy built into the polynomial element. Polynomials in the new variables do not correspond to polynomials in the old, and the requirement in order to preserve the approximation theory is that the coordinate changes should be uniformly smooth. Finally, in order that conforming elements in ξ–η shall be conforming in x–y, there is a *global* continuity condition on the coordinate change: If the energy involves mth derivatives, then the coordinate change must be of class $\mathbb{C}^{m-1}$ between elements. For the present we discuss only the case $m = 1$, arising from second-order differential equations, where the mapping must be continuous between elements: A point common to e_i and e_j must not split into two separate points when $e_i \rightarrow E_i$, $e_j \rightarrow E_j$.

How are these properties, especially the requirement of computability,

---

†The isoparametric technique is equally important in three dimensions. It is simpler to discuss examples in the plane, but there is no difference in the theory.

to be satisfied? *The isoparametric technique consists in choosing piecewise polynomials to define the coordinate transformations $x(\xi, \eta)$ and $y(\xi, \eta)$.* Strictly speaking, *isoparametric* means that the *same polynomial elements* are chosen for coordinate change as for the trial functions themselves; *subparametric* means that a subset of lower-degree polynomials is used. In either case we require continuity between elements and nonvanishing of the Jacobian.

The fundamental isoparametric example is the bilinear transformation from a square to a quadrilateral (Fig. 3.6). The coordinate change between

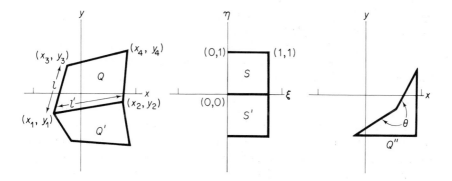

Fig. 3.6 Isoparametric mappings to quadrilaterals.

S and Q is given by

$$
(18) \quad
\begin{aligned}
x(\xi, \eta) &= x_1 + (x_2 - x_1)\xi + (x_3 - x_1)\eta + (x_4 - x_3 - x_2 + x_1)\xi\eta, \\
y(\xi, \eta) &= y_1 + (y_2 - y_1)\xi + (y_3 - y_1)\eta + (y_4 - y_3 - y_2 + y_1)\xi\eta.
\end{aligned}
$$

Each edge of the square S goes linearly into the corresponding edge of Q. For example, if $\eta = 0$ and ξ varies from 0 to 1, then the point (x, y) moves linearly from one corner (x_1, y_1) to the next corner (x_2, y_2). On this boundary, the location of the other vertices (x_3, y_3) and (x_4, y_4) has absolutely no effect. This guarantees that any conforming element in ξ and η—say a bilinear or bicubic Hermite element, for which the mapping is isoparametric or sub-parametric, respectively—will still be conforming in x and y.

We must check that the mapping (18) is invertible, in other words, that each point (x, y) in Q corresponds to one and only one pair (ξ, η) in S. Solving the equations (18) for ξ and η in terms of x and y will introduce complicated square roots and lead nowhere. Instead, since it is already verified that the boundaries of S and Q correspond, we have only to show that the Jacobian is nonvanishing inside S:

$$
J = x_\xi y_\eta - x_\eta y_\xi =
\begin{vmatrix}
x_2 - x_1 + A\eta & x_3 - x_1 + A\xi \\
y_2 - y_1 + B\eta & y_3 - y_1 + B\xi
\end{vmatrix},
$$

where $A = x_4 - x_3 - x_2 + x_1$ and $B = y_4 - y_3 - y_2 + y_1$. The Jacobian is actually *linear*, and not bilinear, since the coefficient of $\xi\eta$ in this 2×2 determinant is $AB - AB = 0$. Therefore, *if J has the same sign at all four corners of S, it cannot vanish inside*. At the typical corner $\xi = 0, \eta = 0$, the Jacobian is

$$J(0,0) = (x_2 - x_1)(y_3 - y_1) - (y_2 - y_1)(x_3 - x_1).$$

By the cross-product formula this equals $ll' \sin\theta$, where the lengths l and l' and the angle θ are shown in the figure. Therefore, J is positive at this corner provided the interior angle θ is less than π. The same is true at every other corner. Consequently J *is nonvanishing if and only if the quadrilateral Q is convex*: All its angles must be less than π. Otherwise, as for Q'' in the figure, J will change sign somewhere inside S. In this case the coordinate change is illegal.

Notice that even though polynomials in x and y do not generally transform into polynomials in ξ and η, *the linear polynomials 1, x, and y are special*. The coordinate transformation itself expresses x and y as bilinear functions of ξ and η, and of course constant functions remain constant. Since these three polynomials lie in the trial space, the convergence condition is guaranteed to hold—all solutions $u = \alpha + \beta x + \gamma y$ of constant strain are reproduced exactly in S^h. This is always the case for isoparametric mappings, and convergence is assured. The subparametric case is even better: If the trial space contains all biquadratics or bicubics in ξ and η, then it contains all quadratics or cubics in x and y, and S^h has degree 2 or 3, respectively. Therefore, assuming that the angles of Q are bounded away from 0 and π, approximation is possible to the full degree and the error in the strain is $O(h^{k-1})$, as it should be.

The most important example for triangles is one with a curved edge, such as might arise at the boundary of Γ (Fig. 3.7). The simplest curve possible is one of second degree, and a natural choice of element is the quadratic. On the ξ-η triangle, the trial function $v^h = a_1 + a_2\xi + a_3\eta + a_4\xi^2 + a_5\xi\eta +$

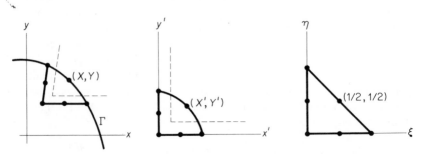

Fig. 3.7 Mappings of a curved triangle.

$a_6\eta^2$ is determined by its values at the six nodes of the figure, three at the vertices and three at the midpoints.

We imagine two mappings of the original curved triangle in the x–y plane. First a simple linear map normalizes the curved triangle, putting the two straight edges onto the coordinate axes in the x'–y' plane. The Jacobian is a constant, and this step is simply for convenience. The important step is to connect x'–y' to ξ–η, associating an arbitrarily specified boundary point (X', Y') with the midpoint $(\frac{1}{2}, \frac{1}{2})$. In practice, (X, Y) and therefore (X', Y') will often be chosen halfway along the curve, but this is not required. The mapping is bilinear:

(19)
$$x' = \xi + (4X' - 2)\xi\eta,$$
$$y' = \eta + (4Y' - 2)\xi\eta.$$

It is easy to verify that the straight edges are preserved, and continuity with adjacent elements is assured. The Jacobian is simply

$$J = \begin{vmatrix} 1 + (4X' - 2)\eta & (4X' - 2)\xi \\ (4Y' - 2)\eta & 1 + (4Y' - 2)\xi \end{vmatrix}$$
$$= 1 + (4X' - 2)\eta + (4Y' - 2)\xi.$$

Again the Jacobian is linear. It equals 1 at $\xi = 0$, $\eta = 0$, and will be nonzero in the triangle if and only if it is positive at the other two vertices. This condition, first shown to us by Mitchell, is simply

$$\text{At } (0, 1): \quad 4X' - 2 > -1, \quad \text{or} \quad X' > \tfrac{1}{4},$$
$$\text{At } (1, 0): \quad 4Y' - 2 > -1, \quad \text{or} \quad Y' > \tfrac{1}{4}.$$

Therefore, (X', Y') may lie anywhere in the quadrant formed by the dashed lines, and correspondingly (X, Y) should lie in the sector shown in the figure. Notice that even if the original triangle were straight, the point (X, Y) must lie in the middle half of its edge or shifting it to the midpoint would produce a vanishing Jacobian. (Of course in this case there would be no reason to shift it; on a straight triangle we could use quadratic elements in the x–y variables, even with arbitrarily placed midedge nodes. The mapping to ξ–η is really intended for the case when a curved edge needs to be straightened.)

In this example the curved side was a parabola. In the general isoparametric case, either with triangles or quadrilaterals, the mappings $x(\xi, \eta)$, $y(\xi, \eta)$ are given by the same type of polynomial elements as are used for displacements, and all edges may be polynomials of degree $k - 1$. The constraints will be the same as those for the element itself—when the unknowns include some derivatives at a node, this means that the corresponding derivatives of the boundary curves must be continuous there. The Lagrange case is therefore

much the simplest for isoparametric transformations, since the only unknowns are function values and the only constraint is the continuity between
elements, which is required in any case. In fact, it is especially simple if, as
in the serendipity rectangular element in Fig. 3.8, there are no internal nodes.

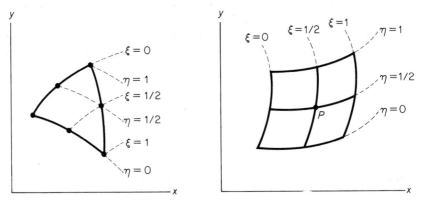

Fig. 3.8   Common isoparametric elements.

The mapping between the boundaries then determines the entire coordinate
change, which is otherwise *highly sensitive to the movement of internal nodes.*

We must emphasize that the whole isoparametric technique depends on
the use of *numerical integration* (in the ξ–η variables) to compute the entries
of K and F. From the change of variables in the element integral (17), it is
obvious even for an isotropic material ($p \equiv$ constant) that the mathematical
equivalent of variable material properties is introduced by the functions
ξ_x, η_x, and $J(\xi, \eta)$. In general the first two are rational functions and the last
is a polynomial, with smoothness depending on the distortion of the element.
We shall establish in Section 4.3 how the numerical integration error affects
the final result and what order of accuracy is required.

Here we deal with the question of approximation: How closely can isoparametric elements match the true solution $u(x, y)$? The answer must depend
on the size of the derivatives in the coordinate change F:

$$F(\xi, \eta) = (x(\xi, \eta), y(\xi, \eta)).$$

Let $\hat{u}(\xi, \eta)$ denote the solution $u(x, y)$ transformed to the new coordinates.
Then if S^h is of degree $k - 1$ in ξ and η, the interpolate $\hat{u}_I$ satisfies

$$(20) \qquad \sum_{|\alpha| \leq s} \int_{E_i} |D^\alpha \hat{u} - D^\alpha \hat{u}_I|^2 \, d\xi \, d\eta \leq C^2 h_i^{2(k-s)} \sum_{|\beta| \leq k} \int_{E_i} |D^\beta \hat{u}|^2 \, d\xi \, d\eta.$$

Changing to the x–y variables, each derivative of $\hat{u}$ is transformed into a sum of terms, for example,

$$\frac{\partial}{\partial \xi} \hat{u} \longrightarrow u_x x_\xi + u_y y_\xi,$$

$$\frac{\partial^2}{\partial \xi^2} \hat{u} \longrightarrow u_x x_{\xi\xi} + u_y y_{\xi\xi} + u_{xx} x_\xi^2 + u_{yy} y_\xi^2.$$

In general a derivative of order $|\beta| \leq k$ will be bounded by

$$| D^\beta \hat{u}(\xi, \eta)| \leq \| F \|_k \sum_{|\gamma|=1}^{k} | D^\gamma u(x, y)|,$$

where the constant $\| F \|_k$ is computed from powers of the derivatives x_ξ, y_ξ, x_η, . . . within the element up to order k. Ciarlet and Raviart [C7], whose argument we are following, write out this expression in detail. The reciprocal of the Jacobian J also enters the transformation of the integral:

$$\int_{E_i} | D^\beta \hat{u}|^2 \, d\xi \, d\eta \leq \frac{\| F \|_k^2}{\min\limits_{E_i} |J(\xi, \eta)|} \int_{e_i} \sum_1^k | D^\gamma u|^2 \, dx \, dy.$$

This holds for all $|\beta| \leq k$, and therefore

(21)
$$\| \hat{u} \|_{k, E_i}^2 \leq \frac{\| F \|_k^2}{\min |J|} \| u \|_{k, e_i}^2.$$

This yields an upper bound for the right side of (20).

For the left side we want a lower bound, and therefore the argument which led to (21) is reversed:

(22)
$$\| u - u_I \|_{s, e_i}^2 \leq \| F^{-1} \|_s^2 \max_{E_i} |J| \| \hat{u} - \hat{u}_I \|_{s, E_i}^2.$$

This time the factor $\| F^{-1} \|_s$ depends on powers of derivatives up to order s of the *inverse* mapping F^{-1}, taking x, y to ξ, η. If the Jacobian determinant J is nonvanishing, as we assume, then derivatives of F^{-1} can be bounded in terms of derivatives of F. For the first derivatives, this is expressed by the identity for the Jacobian matrices

$$\begin{pmatrix} \xi_x & \xi_y \\ \eta_x & \eta_y \end{pmatrix} = \begin{pmatrix} x_\xi & x_\eta \\ y_\xi & y_\eta \end{pmatrix}^{-1}.$$

For higher derivatives this identity is differentiated, and the essential question again becomes the boundedness of the derivatives of F.

Now we assemble the results. Substituting (21) and (22), the original inequality $\|\hat{u} - \hat{u}_I\|_s \leq Ch_i^{k-s}\|\hat{u}\|_k$ on the element E_i is converted to a similar inequality on e_i:

$$(23) \qquad\qquad \|u - u_I\|_{s,e_i} \leq C'h_i^{k-s}\|u\|_{k,e_i},$$

where the constant is now

$$C' = C\left(\frac{\max|J|}{\min|J|}\right)^{1/2}\|F\|_k\|F^{-1}\|_s.$$

This is the fundamental result. The order of approximation is the same with isoparametrics as with the standard polynomials, *provided C' remains bounded*. Note that h_i is still the diameter of the element in the ξ–η plane, even though the approximation inequality (23) is now in x–y. We may, however, suppose the diameters to be the same in the two coordinate systems. (A change of scale in ξ–η leaves the inequality unchanged, as it must; if h_i becomes αh_i, then the norms of F and F^{-1} are multiplied by α^{-k} and α^s, respectively.) This will be a convenient normalization, since with equal diameters the isoparametric approximation problem reduces exactly to the boundedness of F and its derivatives, and the nonvanishing of the Jacobian J, uniformly as $h \rightarrow 0$.

There is an important modification in the quadrilateral case. It is needed even for the map defined by

$$x(\xi, \eta) = \xi, \qquad y(\xi, \eta) = \eta - \frac{\xi\eta}{2h}.$$

This transformation takes the three points $(0, 0)$, $(h, 0)$ and $(0, h)$ into themselves and moves the fourth corner (h, h) of the square to a new point $(h, h/2)$. In other words, this is a completely typical map, transforming the square in the ξ–η plane into a quadrilateral of comparable size and shape. Nevertheless, the cross derivative $y_{\xi\eta}$, which enters the norm $\|F\|_k$ in the approximation inequality, is of order $1/h$. This would destroy the order of approximation. Ciarlet and Raviart were able, however, to make use of the presence of the twist term $\xi\eta$ in the trial space; bilinear interpolation reproduces it exactly, in addition to the linear terms $a_1 + a_2\xi + a_3\eta$ of the triangular case. Their result is that for quadrilaterals, the cross derivatives $x_{\xi\eta}$ and $y_{\xi\eta}$ do not appear in the factor $\|F\|_k$, and the expected order of approximation is possible. A similar conclusion holds for biquadratics and bicubics.

The consequences of the inequality (23) are summarized in the following

Theorem 3.6

Suppose the ξ–η trial space is of degree $k - 1$, and the element transformations F into x–y are uniform as $h \rightarrow 0$. Then if u belongs to $\mathcal{H}^k(\Omega)$, its interpo-

late u_I in S^h satisfies

$$\left(\int_\Omega |u - u_I|^2 \, dx \, dy\right)^{1/2} \leq C' h^k \|u\|_k,$$

$$\left(\int_\Omega |\text{grad}(u - u_I)|^2 \, dx \, dy\right)^{1/2} \leq C' h^{k-1} \|u\|_k.$$

The constant C' is given in (23). *The rate of convergence of the finite element solution u^h, for a second-order differential equation, is therefore the full $h^{2(k-1)}$ in strain energy.*

The inequalities in the theorem apply only to the function and its first derivatives, that is, to the cases $s = 0$ and $s = 1$ of (23). Approximation of order h^{k-s} holds also for higher derivatives within each element, but when the elements are joined the interpolate u_I is no more than continuous, and belongs only to $\mathfrak{IC}^1$ over the whole domain Ω.

The uniformity required by the theorem is a rather severe condition, and amounts essentially to this:

1. The Jacobians should be bounded away from zero, forcing all angles to be bounded away from 0 and π, and
2. The edges of the elements in the x–y plane should be polynomials with uniformly bounded derivatives, i.e., the curvatures and the higher derivatives along the edges must remain bounded. In particular, the edges should be displaced only by $0(h^2)$ from straight lines. Then a well-chosen map F will have $\|F\|_k \leq$ constant.

Computations by Fried [F14] have demonstrated the necessity of condition 2. He increased the curvature of the edges until their displacement from the original square was of the same order as the size of the square itself. On a unit square this still implies a bounded curvature, but scaled down to size h the curvature (being a second derivative) is of order $1/h$. The numerical results were correspondingly poor. Practical problems will fall somewhere between these overly distorted elements and the nearly straight ones required by the uniformity hypothesis in the theorem.

On the happy side, suppose that a curved element occurs at the boundary of Ω; its one curved side follows the true boundary Γ. (It will normally be a polynomial approximation to Γ, interpolating the true boundary at specified nodes.) Then *if Γ is smooth, the uniformity condition on this edge will automatically be satisfied.* The interpolating polynomial edge will vary only by $O(h^2)$ from a straight line, and the curvatures will be bounded in terms of the curvature of Γ. Therefore, *the isoparametric technique, for second-order equations, allows essential boundary conditions to be handled with no loss* *in accuracy or in simplicity over natural boundary conditions.* The same remarks apply to an internal boundary.

The improvement which this technique allows, in comparison with approximating the boundary by a polygon, is enormous. In his experiments with cubic triangular elements, for example, Zlamal [Z10] found a difference of an order of magnitude in displacements, and even more in strains. There can be no doubt of the technique's success in second-order problems.†

For fourth-order equations—plate and shell problems, for example—the position is much less happy. The theory insists that the coordinate change be $\mathbb{C}^1$: Its first derivatives should be continuous between elements, or the trial functions are nonconforming. (Convergence is still possible for nonconforming elements, as we prove in the next chapter. However, it appears that for plate elements even this hope is dashed, because they cannot pass the required *patch test*.) A $\mathbb{C}^1$ coordinate change is theoretically possible, but the extra continuity requirement is extremely restrictive (Fig. 3.9). Once

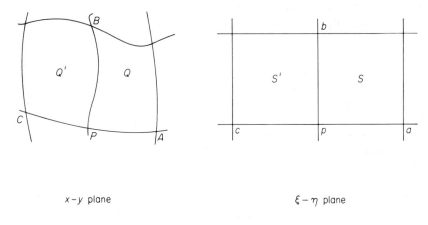

$x-y$ plane $\xi-\eta$ plane

Fig. 3.9   The constraints on a C^1 mapping.

two directions are known at a point—at P, the tangents to PA and PB are established when S is mapped to Q—all other directions are completely determined. [For any $\mathbb{C}^1$ function $f(x, y)$, all directional derivatives can be computed from the gradient, that is, from the derivatives f_x and f_y in two directions.] In the figure, this means that the tangent to PC is determined. In fact, the curve CPA must have a continuous tangent at P, since cpa is straight. It follows that *a general quadrilateral or triangular mesh, even one with straight edges, cannot be carried into a regular mesh by a $\mathbb{C}^1$ change of*

---

†Gordon and Hall recently proposed to map Ω all at once, rather than element by element, into a square. For this they use *blending functions*, a variant of the standard finite elements. If Ω is itself not too unlike a square (the mapping of a circular region would create artificial singularities at the corners) this should save time as compared to the isoparametric method on each element.

coordinates. The mapping in the figure could be carried out by bicubic Hermite elements, for example, allowing the curved edges to be cubics; but their tangents, and in the Hermite case also the cross-derivatives $x_{\xi\eta}$ and $y_{\xi\eta}$, must be continuous at the vertices. This is a severe, and almost an unacceptable, theoretical limitation on the isoparametric technique for fourth-order problems.

The message seems to be this, that it is a mistake to work with fourth-order equations rather than second-order systems. Analytically the elimination of unknowns may be of overwhelming importance, but not numerically. The construction of $\mathcal{C}^1$ polynomials on straight-sided elements is already difficult, and curved elements—especially when we come to consider numerical integration—are infinitely worse.

3.4. ERROR ESTIMATES

In this section we put the previous approximation theorems to use, in order to achieve the main goal of our whole analysis—to estimate the finite element error $u - u^h$. The function u is the solution to an n-dimensional elliptic boundary-value problem of order $2m$, and u^h is its Ritz approximation computed in a finite element space S^h. On a regular mesh, the finite element equations $KQ = F$ become a system of difference equations, and we shall find at the same time the order of accuracy of these difference equations.

The basic question is how well the subspaces S^h approximate the full admissible space $\mathcal{H}^m_E$. In the energy norm, nothing else matters: u^h is as close as possible to u, and the error in energy must be of the optimal order $h^{2(k-m)}$:

$$(24) \qquad a(u - u^h, u - u^h) \leq C^2 h^{2(k-m)} \|u\|_k^2.$$

This is the simplest and still the most fundamental error estimate. Since the expression $a(v, v)$ for strain energy is positive definite (in other words, since the problem is elliptic: $a(v, v) \geq \sigma \|v\|_m^2$), (24) is equivalent to

$$(24') \qquad \|u - u^h\|_m \leq c' h^{k-m} \|u\|_k.$$

In displacement, or more generally in the sth derivative, the first question is whether $u - u^h$ is again of the optimal order h^{k-s}. This would imply that everything is determined by the degree $k - 1$ of the finite elements, but it is not quite true. If we fix S^h and increase the order $2m$ of the problem, then eventually the trial functions will no longer be admissible and the Ritz method will collapse. Therefore, the order of accuracy must somehow depend on m as well as on k and s.

The correct order can be determined by an elegant variational argument due to Aubin and Nitsche. The argument has come to be known as the

Nitsche trick; it was extended by Schultz, and is now the standard approach to errors in displacement. In Section 1.6, we proved an h^2 error with linear elements; the general case is going to be more technical, but the results (25–26) are very simple. The upper limit $2(k - m)$ on the rate of convergence was found by the first author [S10].

THEOREM 3.7

Suppose that the finite element space S^h is of degree $k - 1$ and that the strain energy has smooth coefficients and satisfies the ellipticity condition $\sigma \|v\|_m^2 \leq a(v, v) \leq K \|v\|_m^2$. Then the finite element approximation u^h differs from the true solution u by

$$(25) \qquad \|u - u^h\|_s \leq C h^{k-s} \|u\|_k \qquad \text{if } s \geq 2m - k,$$

$$(26) \qquad \|u - u^h\|_s \leq C h^{2(k-m)} \|u\|_k \qquad \text{if } s \leq 2m - k.$$

These exponents are optimal, so that the order of accuracy never exceeds $2(k - m)$ in any norm; in almost all realistic cases the order is $k - s$.

Proof. Nitsche's trick is to introduce an *auxiliary problem* $Lw = g$, whose variational form (equation of virtual work) is

$$(27) \qquad\qquad a(w, v) = (g, v) \qquad \text{for all } v \text{ in } \mathcal{JC}_E^m.$$

From the theory of partial differential equations, there exists a unique solution w which is $2m$ derivatives smoother than the data $g \colon \|w\|_{2m-s} \leq c \|g\|_{-s}$.

We choose $v = u - u^h$ in (27):

$$(28) \qquad \begin{aligned} |(g, u - u^h)| &= |a(w, u - u^h)| = |a(w - v^h, u - u^h)| \\ &\leq K \|w - v^h\|_m \|u - u^h\|_m. \end{aligned}$$

This holds for any v^h in S^h, since $a(v^h, u - u^h) = 0$ by the fundamental Ritz theorem 1.1. Now suppose that v^h is the closest approximation to w in the $\mathcal{JC}^m$ norm. [Or, what is effectively the same, let v^h be the finite element solution to the auxiliary problem (27) and therefore the best approximation to w in the energy. Note that only approximation in energy is used in the proof, never a direct approximation in the $\mathcal{JC}^s$ norm.] According to the approximation theorem 3.3,

$$(29) \qquad \|w - v^h\|_m \leq \begin{cases} c h^{m-s} \|w\|_{2m-s} & \text{if } k \geq 2m - s, \\ c h^{k-m} \|w\|_k & \text{if } k \leq 2m - s. \end{cases}$$

In the first case k was reduced to $2m - s$ before the approximation theorem was applied; if the subspace is complete through degree $k - 1$, then it is certainly complete through any lower degree.

Substituting (24') and (29) into (28),

$$|(g, u - u^h)| \leq Kc \begin{Bmatrix} h^{m-s} \\ h^{k-m} \end{Bmatrix} \| w \|_{2m-s} \, c' h^{k-m} \| u \|_k$$

$$\leq C \begin{Bmatrix} h^{k-s} \\ h^{2(k-m)} \end{Bmatrix} \| g \|_{-s} \| u \|_k.$$

By duality in the definition (equation 1.58) of negative norms,

$$\| u - u^h \|_s = \max_g \frac{|(g, u - u^h)|}{\| g \|_{-s}} \leq C \begin{Bmatrix} h^{k-s} \\ h^{2(k-m)} \end{Bmatrix} \| u \|_k.$$

The proof is complete. It is most straightforward for the displacement, $s = 0$, because in this case the auxiliary data g is exactly $u - u^h$; this choice was made in Section 1.6.

A similar result holds with inhomogeneous essential boundary conditions [S6]. The error estimates have also been extended to problems of *forced vibration*, in which the basic ellipticity condition $a(v, v) \geq \sigma \| v \|_m^2$ is violated by the addition of a new zero-order term. Effectively, the original differential equation $Lu = f$ is altered to $Lu - \alpha u = f$: if α falls in between two eigenvalues of L, then $L - \alpha$ is no longer positive definite—but the equation still has a solution. Schultz has proved [S4] that the rate of convergence is unchanged. ∕

Remarks. The case $s = m$, corresponding to error in energy, always yields the correct power h^{k-m}. The error in higher derivatives, $s > m$, is not included in the theorem as it stands, since w lies outside $\mathcal{H}^m$ and (29) is nonsense. Nevertheless the exponent $k - s$ will be correct within each element, provided the subspaces S^h satisfy an *inverse hypothesis:* each differentiation of a trial function v^h increases its maximum by at most a factor c/h. This hypothesis is almost the same as the uniformity condition (2), except that there it was a case of c/h_i; therefore the inverse hypothesis is fulfilled if all elements are of comparable size. Otherwise a factor $(h_{max}/h_{min})^{s-m}$ appears in the error estimate for derivatives of order $s > m$.

The theorem and proof are valid without change in case $s < 0$. Surprisingly, *the rate of convergence in negative norms is not only of academic interest.* The reason is this: the error in the -1-norm is a bound for the average error over the domain:

$$\| u - u^h \|_{-1} = \max \frac{\left| \int_\Omega (u - u^h) v \, dx \right|}{\| v \|_1} \geq \frac{\left| \int (u - u^h) \, dx \right|}{(\text{vol } \Omega)^{1/2}},$$

if we choose $v \equiv 1$. Therefore, in the usual case $k > 2m$, Theorem 3.7 has the following consequence for $s = -1$: *The average error is much smaller than the typical displacement error at a point.* More precisely,

$$\int |u - u^h|^2\,dx \sim h^{2k} \qquad \text{but} \qquad \left| \int (u - u^h)\,dx \right|^2 \sim h^{2(k+1)}.$$

This must mean that *the error alternates rapidly in sign.* In fact, it does so within each element, and the practical problem is to discover even approximately the "special points" where these sign changes occur. Near such points the displacements u^h will be of exceptional accuracy.

(We had imagined that $-u'' = 1$, with linear elements, would provide the perfect example: u^h coincides with the interpolate u_I, so there is exceptional accuracy at the nodes.† Looking more closely, however, the average error is only of the same order h^2 as the error at a typical point. In fact, the linear interpolate never goes above the true (quadratic) solution, and the error is completely one-sided. The explanation is that the condition $k > 2m$ for the h^{k+1} phenomenon does not hold: $k = 2m = 2$. The nodes are exceptional, but it is because the trial functions solve the homogeneous differential equation $-u'' = 0$ [H6, T4]; this is not an example of rapid alternations in sign.)

The stresses are in a similar position. In second-order problems they are in error by h^{k-1} at typical points, and h^k on average; therefore, these errors also must alternate in sign, and there must exist exceptional *stress points.* Their presence was noticed in actual computations by Barlow, and it appears for quadratics on triangles that *the midpoints of the edges are exceptional.* The accuracy at these midpoints is better than at the vertices, where even after averaging the results from adjacent elements, the stress approximations are not good. Since the midpoints are also nodes for quadratic elements, the situation is extremely favorable; it is only spoiled by errors due to change in domain, which do not necessarily alternate in sign.

We believe that the stress points can be located in the following way. The leading term in the error is governed by the problem of approximating polynomials P_k of degree k, in energy, by the trial functions in S^h. On a regular

---

†The phenomenon of "super-convergence" at the nodes was recently made clearer by an elegant observation of Dupont and Douglas. Let $G_0(x)$ be the fundamental solution corresponding to the point x_0; G_0 is the response to a point load $f_0 = \delta(x - x_0)$. Then

$$|u(x_0) - u^h(x_0)| \le C \|u - u^h\|_m \|G_0 - v^h\|_m \quad \text{for any } v^h \text{ in } S^h.$$

Proof: $u(x_0) - u^h(x_0) = (u - u^h, f_0) = a(u - u^h, G_0) = a(u - u^h, G_0 - v^h)$. The most interesting case occurs if x_0 is a node, since the approximation of G_0 is likely to be especially good; in the one-dimensional case $-u'' = f$, G_0 is linear with a change in slope at x_0, and can be reproduced identically by v^h. This confirms the infinite accuracy in this special case. Normally, the term $\|G_0 - v^h\|$ will add some finite power of h to the h^{k-m} coming from $\|u - u^h\|_m$. The whole question of pointwise convergence, and of these increases in the power of h (super-convergence) at special points, is now being studied very intensively.

mesh this problem can be solved exactly. Then stress points are identified by the property that the true stresses (derivatives of P_k) coincide with their approximations (derivatives of a lower-degree polynomial.) With first-degree elements in one dimension, equality occurs at the midpoints of the intervals; that is where the slope of a quadratic matches the slope of its linear interpolate. (Equivalently, that is where the error function in Fig. 3.3 has a horizontal tangent.) Symmetry suggests that the midpoints are also exceptional for elements of higher degree. (The exceptional points for displacement lie elsewhere, and in the simplest case they were the zeros of the second Legendre polynomial. They seem more sensitive to the boundary conditions than the stress points.) In two dimensions, the results may depend on the choice of P_k, and exceptional points for one stress component need not be exceptional for the others. The midpoints of an edge seem likely to be exceptional for derivatives along the edge but not for stresses in the direction of the normal. This is still an area for research; eventually these special points will be more completely understood.

Three additional problems arise on a regular mesh:

1. To interpret $KQ = F$ as a system of finite difference equations, with corresponding local truncation errors.

2. To prove that the exponents in Theorem 3.7 are optimal.

3. To show that for smooth solutions, the same rates of convergence apply not only in the mean, but at every individual point.

We shall not attempt a technical discussion, particularly of problem 3. Roughly speaking, once the behavior of the truncation errors is established, the central question is the stability of the difference operator K. This property is difficult to establish in the maximum norm, but with the inverse hypothesis mentioned above it is almost certain to hold. We have verified it for constant-coefficient model problems, and the error at every point of the domain is then of the correct order. More precise results have been achieved by Nitsche, Bramble, and Ciarlet and Raviart, but the general problem is unsolved.

For problem 1 suppose first that there is only one node (and one associated unknown) for every mesh square; this will be the case for linear elements on right triangles, bilinear elements on squares, and splines. Then $KQ = F$ will look exactly like a conventional difference equation. This fact has led to innumerable discussions about the relation between finite elements and finite differences. It is clear that not all difference equations can be produced by an appropriate choice of element; the matrix K must be symmetric and positive definite, and even under these restrictions a corresponding element may not exist. On the other hand, a sufficiently tolerant reader may be willing to regard all finite element equations—even on an irregular mesh, with many nodal unknowns—as finite difference equations. We agree with this view.

The system $KQ = F$ is in general a novel kind of coupled difference equation, which in principle could have been devised without the variational principle as an intermediary. Historically, of course, that almost never happened; the finite element method leads systematically to a special class of equations—*the intersection between all possible difference equations and all possible Ritz–Galerkin equations*—which are astonishingly successful in computations.

If the original differential equation $Lu = f$ has constant coefficients the order of the local truncation error may be tested as follows: Let $f(x)$ be a pure exponential $e^{i\xi x}$, so that u can be found explicitly and substituted into the difference equation. The truncation error will have the form $e^{i\xi x}E(h, \xi)$. With $-u'' = f$, for example, the solution is $u = e^{i\xi x}/\xi^2$, and the truncation error of $h^{-2}(-u_{j+1} + 2u_j - u_{j-1}) = f_j$ is

$$\left(\frac{-e^{i\xi h} + 2 - e^{-i\xi h}}{h^2\xi^2} - 1\right)e^{i\xi x} = \left(\frac{2 - h^2\xi^2 - 2\cos h\xi}{h^2\xi^2}\right)e^{i\xi x} = Ee^{i\xi x}.$$

In general the coefficient E can be computed in terms of the Fourier transforms of the trial functions. (Here we are right at the heart of the abstract finite element method; this technique would be impossible on an irregular mesh.) Expanding E as a power series in h, the leading exponent is exactly the order of accuracy of the difference equation. We have computed this exponent and verified that it is the smaller of k and $2(k - m)$; the rate of convergence given in Theorem 3.7 is correct.

Suppose, finally, that there are M unknowns associated with each mesh square, so that the finite element equation $KQ = F$ becomes a coupled system of M difference equations. The unknowns may be the values of u^h at distinct nodes, or function values and derivatives at a multiple node. This introduced no complications when the error was estimated by a variational argument, in Theorem 3.7; that result depends only on the order of approximation achieved by S^h, and any additional facts about the subspace were irrelevant. The difference equation aspect, however, becomes much more subtle when $M > 1$.

Briefly, the problem is that *the truncation errors in the difference equations are not all of the expected order*. Therefore, it will not do simply to estimate these errors and then, applying stability to invert the matrix K, to convert them into estimates of the error $u - u^h$. The point is that u^h is given by a special combination of the trial functions, and if other combinations make little or no contribution to the problem of approximation, then also their contribution to u^h turns out to be small. We recall that in the abstract method, $\Phi_1, \ldots, \Phi_M$ generate approximation of order k if and only if it is possible to construct from them a single function ψ with the property (5) required in Theorem 3.2, that is, a function which *by itself* is adequate for approximation.

We may regard S^h as generated by this superfunction ψ and $M - 1$ more or less useless functions. Forming the corresponding combinations of the difference equations $KQ = F$, we can rewrite our finite element system as a set of difference equations having a very special form: one equation of the system is an accurate analogue of the original differential equation, and the other $M - 1$ (coming from the functions in S^h which are useless for approximation) are completely inconsistent. (This construction becomes very technical for large M, and Fourier analysis is the only possible tool.) The final conclusion is that the Ritz method attaches almost all weight to the one difference equation which corresponds to ψ, and that the order of accuracy of this equation agrees with the exponent $\min(k, 2(k - m))$ in Theorem 3.7. This is the order of accuracy (in displacement) of the finite element method.

4 VARIATIONAL CRIMES

4.1. VIOLATIONS OF THE RAYLEIGH–RITZ CODE

The one cardinal rule in the Ritz theory is that the trial functions should be admissible in the variational principle. In our notation, each v^h should belong to the space $\mathcal{H}^m_E$, and u^h should minimize $I(v^h)$. This rule is simple to state, but it is broken every day, and for good reason. In fact, there are three conditions to the rule, and all three present computational difficulties—perhaps not insuperable difficulties, but serious ones:

1. The trial functions should possess m derivatives in the mean-square sense, and therefore be of class $\mathcal{C}^{m-1}$ across element boundaries.
2. Essential boundary conditions should be respected.
3. The functional $I(v^h) = q^T K q - 2q^T F$ should be computed exactly.

Our goal is to analyze the consequences of violating these conditions.

The first is violated by nonconforming elements, and we show in the next section that in this case *convergence may or may not occur*. It is by no means automatic (or even probable) that the discrete problem is consistent with the continuous one. Instead, there is a *patch test* to be applied to the trial functions, which determines whether or not they consistently reproduce states of constant strain. If so, convergence takes place within each element.

Essential boundary conditions are fulfilled, at least along an approximate boundary, by the isoparametric or subparametric elements. However, there are many circumstances in which these elements are not available, either because they are too complicated for the program at hand or because the problem itself is too complicated—a full fourth-order system of shell equations, for example. In such cases the rule may be partly satisfied, in the following way: The essential boundary conditions may be imposed *at the boundary nodes*. Between the nodes, a polynomial trial function is incapable of following a general boundary and thus the essential conditions are only *interpolated*. We show in Section 4.4 that convergence still occurs.

Condition 3 is the most violated of all, since it is inconvenient to compute $I(v)$ exactly and very easy to compute it approximately. The approximations enter in two ways: The integrals over each element are computed by numerical integration, and the domain of integration Ω is itself altered to become a union of simple element shapes. In both cases $I(v)$ is replaced by a new functional $I_*(v)$. Therefore, the mathematical problem is to determine the dependence of the minimizing function u^h on the functional itself: If the convex paraboloid $I(v)$ is deformed, how far does the minimum move?

Each of these three possible violations requires a careful analysis, if we are to justify the finite element method as it is actually used. Before we start, it may be useful to call attention specifically to two ideas which appear over and over throughout this chapter; they are common to the analysis of all three problems. The first idea is to consider always the vanishing of the first variation—in physical terms, the equation of virtual work. Even when the rules are violated, the minimizing function u^h_* still satisfies

$$a_*(u^h_*, v^h) = (f, v^h)_* \qquad \text{for all } v^h \text{ in } S^h.$$

In the Ritz case this is compared with

$$a(u, v) = (f, v) \qquad \text{for all } v \text{ in } \mathcal{3C}^m_E.$$

If $S^h \subset \mathcal{3C}^m_E$, and if $I = I_*$ so that the asterisk can be removed, then one equation can be subtracted from the other; as in Theorem 1.1,

$$a(u^h - u, v^h) = 0 \qquad \text{for all } v^h \text{ in } S^h.$$

In our case this expression is *not* zero, and the whole problem is to show that the quantity which does appear is small.

The second idea applies specifically to finite elements; it consists in using the special properties of polynomials. We have noted already how the patch test applies to polynomial solutions. The situation is similar in numerical integration, where accuracy depends on the degree of the polynomials which are integrated exactly. And we note one property which will be useful in analyzing changes of domain: Polynomials cannot vary tremendously in the region between the given domain Ω and its approximation Ω^h.

We emphasize that the analysis does not depend on estimating the perturbation in the stiffness matrix, from K to $\tilde{K}$ or K^*. This change may be very large—the equations may seem completely altered—while still the piecewise polynomials of degree m are properly dealt with. In abstract terms, these polynomials become dense in the admissible space as $h \longrightarrow 0$, and their behavior is therefore decisive.

4.2. NONCONFORMING ELEMENTS AND THE PATCH TEST

A number of frequently used elements are nonconforming—their derivatives of order $m - 1$ have discontinuities at the element boundaries—and nevertheless they work quite well. Or rather, they sometimes work well and sometimes not. It is a risk that is taken most often in fourth-order problems, where the elements should lie in $\mathcal{C}^1$; matching the slopes between elements can be difficult for the normal displacement w of a plate under bending, and extremely difficult for shells. Therefore, the technique has been to compute the energies within each separate element in the usual way and then simply to add together the results. This has the effect of replacing the true functional $I(v)$ by a sum of element integrals

$$I_*(v) = \sum_e [a_e(v, v) - 2(f, v)_e] = a_*(v, v) - 2(f, v)_*.$$

The difference between I and I_* is that the singularities on the element boundaries are ignored in I_*, whereas $I(v) = \infty$ for nonconforming elements.

The Ritz approximation u_*^h is the (probably nonconforming) function in the trial space which minimizes $I_*(v^h)$. This property of u_*^h is expressed as usual by the vanishing of the first variation of I_*:

(1) $a_*(u_*^h, v^h) = (f, v^h)_*$ for all v^h in S^h.

This is the approximate equation of virtual work. It is identical to the usual equation, except that again the integrals are computed an element at a time and then summed, ignoring discontinuities between elements.

Our goal is to find conditions under which this finite element approximation u_*^h, in spite of its illegitimate construction, converges to u. This question was almost completely obscure until Irons [B9] had a simple but brilliant idea, now known as the *patch test*. Suppose that an arbitrary patch of elements is in a state of constant strain: $u(x, y) = P_m(x, y)$, a polynomial of degree m. Then since this polynomial is present in S^h—even on nonconforming elements the constant strain condition is imposed, that the degree $k - 1$ of the subspace must be at least m—a true Ritz solution u^h would coincide identically with P_m. (At the boundary of the patch, the conditions imposed are chosen to be consistent with constant strain; e.g. $u^h = P_m$ is imposed on the displacements at the patch boundary.) Then *the test is to see whether*, in spite of shifting from I to I_* by ignoring the inter-element boundaries, *the* *finite element solution u^h_* is still identical with P_m.* We may and shall assume that the problem has constant coefficients, since variations over an element have only an $O(h)$ effect on u^h_*.

There is a celebrated example in [B9] in which this test is passed for elements of shape (B), but is violated by about 1.5% for those of shape (A). (This is in the large-scale strain energy; the pointwise stress errors reached 25%.) The element in question is a cubic, with v, v_x, and v_y as parameters at the vertices. The tenth unknown is eliminated by a constraint on the coefficients; the authors disapproved of equating the coefficients of x^2y and xy^2, because this constraint can break down on a right triangle—all nine nodal parameters of $xy(1 - x - y)$ are zero on the standard triangle, and the constraint is satisfied, but the polynomial is not identically zero. Therefore the authors chose a condition which would be invariant under rotation, and never singular.

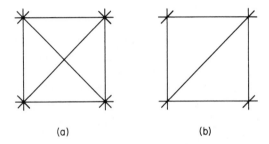

(a)   (b)

Fig. 4.1   Success and failure in the patch test.

With irregular triangles, the odds are strongly against success in the patch test. Nevertheless, an error of a few per cent may be acceptable, particularly when it is committed in the direction opposite to the true Ritz error. As explained in Section 1.10, the latter always comes from too stiff a structure, $a(u^h, u^h) \leq a(u, u)$; the nonconforming solution u^h_* is more relaxed, and the displacements as well as the strains are frequently overestimated.

One approach to the theory is to look directly at the discrete system $KQ = F$ as a finite difference equation, forgetting that it is improperly derived. If this equation is consistent and stable, then u^h_* must converge to u. For a regular mesh, this is exactly the content of the patch test: *It is a test for consistency.*† Normally the nonconforming elements will be short only one derivative, so that v^h lies in $\mathbb{C}^{m-2}$, and consistency is ensured for the lower-

---

†Remember that for difference equations, consistency is checked by looking at the first few terms of a Taylor series—in other words, by considering polynomials up to a certain degree. The patch test does exactly the same for finite elements.

order terms in the differential equation. Then the patch test in case (B) above confirmed consistency for the highest-order difference quotient, while in case (A) the equation $KQ = F$ proved to be the analogue of a *wrong differential equation*, with a new leading term. This link between consistency and the patch test can be established by Fourier methods; we omit the details, because it is effectively limited to a regular mesh and requires a good deal of preparation.

The second approach to the nonconforming theory is variational and therefore more general. We assert that the variational meaning of success in the patch test is this: *For each polynomial P_m and each (nonconforming) basis function φ_j,*

$$(2) \qquad\qquad a_*(P_m, \varphi_j) = a(P_m, \varphi_j).$$

This equality holds if the patch test is passed, and vice versa. Notice that the right side is well defined; the main terms are mth derivatives of P_m, which are constants, multiplied by mth derivatives of the nonconforming functions φ_j, which are δ-functions when the derivative is normal to the element boundaries. The boundary contributions are therefore finite; they are computable from the jumps in the normal derivative of order $m - 1$, as we shall show by example. These contributions *need not be zero*; the patch test may fail, and in this case convergence will not occur.

To prove the equivalence of (2) with the patch test, consider a patch outside of which φ_j vanishes identically. We assume that the true solution is $u = P_m$, and note that the corresponding load f (if any) will certainly have no δ-functions on the element boundaries:

$$(f, \varphi_j)_* = (f, \varphi_j).$$

Introducing the two equations of virtual work, this is the same as

$$a_*(u^h_*, \varphi_j) = a(P_m, \varphi_j).$$

If the patch test is passed, so that the approximate solution is exactly $u^h_* = P_m$, then obviously (2) holds. Conversely, suppose that (2) holds for every φ_j. Then $a_*(u^h_*, \varphi_j) = a_*(P_m, \varphi_j)$, and necessarily $u^h_* = P_m$; the patch test is passed.

We must mention one technical difficulty in this argument: Ordinarily the first variation might not vanish in the direction of φ_j, which lies outside $\mathcal{K}^m$. At the smooth solution $u = P_m$, however, we are able to show (based on an integration by parts [S7]) that it does vanish: $a(u, \varphi_j) = (f, \varphi_j)$, and the argument given above is valid. In short, (2) is what the patch test tests.

Perhaps the simplest instance in which the patch test can be verified occurs with Wilson's rectangular elements [W5]. Beginning with the standard bilinear functions—on the square $-1 \leq x, y \leq 1$ there are the four basis functions $(1 \pm x)(1 \pm y)/4$—he adds two new ones. Within the square these are $\varphi = 1 - x^2$ and $\psi = 1 - y^2$, and outside they are defined to be zero; therefore, *they are not continuous at the boundary.* Since they vanish outside the element, the solution of the final linear system admits static condensation, and the effect of φ and ψ should be to permit an improved representation within each element. Because of the discontinuity, however, the patch test is required.

Suppose for simplicity that the differential equation is $-\Delta u = f$. The energy inner product is $a(u, v) = \iint u_x v_x + u_y v_y$, and if $P = a + bx + cy$ is an arbitrary polynomial of degree $m = 1$, then

$$a_*(P, \varphi) = \int_{-1}^{1} \int_{-1}^{1} b(-2x)\, dx\, dy = 0.$$

On the other hand, Green's formula yields

$$a(P, \varphi) = \int \int (-\Delta P)\varphi - \int \varphi \frac{\partial P}{\partial n}\, ds.$$

Choosing a domain large enough to contain the given square in its interior, φ will vanish on the boundary and the line integral is zero. Since $-\Delta P = -P_{xx} - P_{yy} = 0$ for any linear polynomial P, we conclude that $a(P, \varphi) = 0 = a_*(P, \varphi)$ and the patch test is passed.

The patch test is evidently a very simple rule: The true value of $a(P_m, \varphi_j)$ is zero, as we have just seen, and therefore *the integral of every strain $D^m \varphi_j$—computed in the nonconforming way, by ignoring element boundaries—is also required to be zero.* This can be checked analytically; the patch test can be run without a computer. The size of these integrals, in case they don't vanish, decides the degree of inconsistency of the nonconforming equations.

There is another way to achieve the same result, using Green's theorem over a single square:

$$a(P, \varphi) - a_*(P, \varphi) = -\int \varphi \frac{\partial P}{\partial n}\, ds.$$

Since φ vanishes on the vertical sides $x = \pm 1$, there remains only an integral along the bottom and one back along the top. $\partial P/\partial n$ equals $-c$ in one case, and $+c$ in the other, so that

$$\int \varphi\, \partial P/\partial n\, ds = \int_{-1}^{1} (1 - x^2)(-c)\, dx + \int_{1}^{-1} (1 - x^2)c(-dx) = 0.$$

Thus the element again passes the test. It does *not* pass the test for a quadrilateral of arbitrary shape, where it is constructed isoparametrically: $\varphi = 1 - \xi^2$, $\psi = 1 - \eta^2$, and x, y are bilinear functions of ξ, η mapping the standard square into the given quadrilateral. In fact the patch test can be so strongly violated, even for quadrilaterals of reasonable shape, that the element is useless. To pass the test with isoparametrics, Taylor altered the non-conforming φ and ψ and then also modified the numerical quadrature; two wrongs *do* make a right, in California.

There is a second non-conforming element which is equally simple and useful. It is composed of piecewise linear functions on triangles. Rather than placing nodes at the vertices, however, which produces continuity across each interelement edge, the nodes are placed instead at the *midpoints of the edges*. Interelement continuity is therefore lost (except at these midpoints) and we have a larger trial space—roughly three times the dimension of the standard Courant space, because this is the ratio of edges to vertices. This larger space has made it possible to impose the side condition $div \, v^h = 0$ and still to retain enough degrees of freedom for approximation.

For this space to succeed, it must pass the patch test. We impose the test exactly as above, by computing $\int \varphi \, \partial P/\partial n \, ds$ along every edge. In fact, *it is the jump in φ which we integrate*, to compute the effect of going along an edge in one direction and then back along the other side. The jump in φ is a linear function, since φ is linear in each of the two triangles which meet at the edge. The jump is zero at the midpoint, where φ is continuous. Finally, $\partial P/\partial n$ is a constant because P is linear. Since the integral of a linear function is zero, given that the function vanishes at the midpoint of the range of integration, we conclude that each trial function passes the patch test.

Notice that the mesh was not even required to be regular! Témam has established a Poincaré inequality $\| v^h \|_0^2 \leq C a_*(v^h, v^h)$ for this element.

We intend to prove that the finite element solutions based on Wilson's nonconforming element converge to u. The rate of convergence will be only minimal—$O(h^2)$ in energy—although this may not give a fair description of its accuracy for large h. (An essential feature of finite elements is their success on a coarse mesh; even some elements which fail the patch test and are nonconvergent give very satisfactory results for realistic h.) If the element also passes the test for polynomials of higher degree P_n, this rate of convergence in energy would be increased to $h^{2(n-m+1)}$; but it doesn't.

Our plan is to begin with a general error estimate, applicable to any nonconforming element, and thereby to isolate the quantity which is decisive in determining the error. This is the quantity Δ, defined by

$$\Delta = \max_{v^h \text{ in } S^h} \frac{|a_*(u, v^h) - a(u, v^h)|}{|v^h|_*}, \qquad \text{where } |v^h|_* = [a_*(v^h, v^h)]^{1/2}.$$

Then, for the particular element studied above, we shall estimate Δ and deduce the rate of convergence.

The error bound on which everything is based is the following:

(3) $$|u - u_*^h|_* \leq \Delta + \min_{w^h \text{ in } S^h} |u - w^h|_*.$$

for the error in eigenvalues, which corresponds to the error in energy for static problems. This agrees with our prediction. Note that it is the nonconformity which is responsible; the element has $k = 4$, and approximation theory would have permitted $O(h^{2(k-m)}) = O(h^4)$.

It appears that element 3 passes the test even on an irregular mesh, which is rather remarkable. In fact, the vanishing of boundary integrals which the test requires would seem in general to be a rather lucky chance even on a regular mesh, and one must expect that not only nonconforming but *nonconvergent* elements will continue to achieve engineering accuracy in the future.

The list also contained a number of numerically integrated elements, but we prefer *not to regard numerical integration as producing nonconforming elements*. The effect of this integration is indeed to change the true functional $I(v)$ to a new one, but the difference between the two does not consist of boundary integrals. Therefore this effect, and the error it introduces into the finite element approximation u^h, will be analyzed separately.

4.3. NUMERICAL INTEGRATION

Numerical integration has become an increasingly important part of the finite element technique. In the early stages, one of the crucial advantages of finite elements appeared to lie in precisely the opposite direction, that the integration of polynomials over triangles and rectangles could be based on exact formulas. At the present stage, it might seem that the unique simplicity of polynomials is no longer essential, and that rational functions, or still more general shape functions, are equally convenient. In fact, nothing could be further from the truth: The key to the success of numerical integration in the finite element method is the presence of polynomials.

The fundamental question is this : What degree of accuracy in the integration formula is required for convergence? *It is not required that every polynomial which appears be integrated exactly.* The integrand in the energy $a(v, v)$ involves the *squares* of polynomials, and a formula which is exact to this degree may simply cost too much. It is very important to control properly the fraction of computer time which is spent on numerical integration.

Mathematically, we are again faced with a change in the functional $I(v)$; this is the effect of numerical quadrature. Suppose the true functional is

$$I(v) = a(v, v) - 2b(f, v) = \iint [p(x, y)(v_x^2 + v_y^2) - 2fv]\, dx\, dy.$$

[Notice the new notation $b(f, v)$ for the linear term.] Then typically

$$I^*(v) = a^*(v, v) - 2b^*(f, v) = \sum w_i[p(\xi_i)(v_x^2(\xi_i) + v_y^2(\xi_i)) - 2f(\xi_i)v(\xi_i)].$$

Each element region contributes a certain number of evaluation points $\xi_i = (x_i, y_i)$, with weights w_i which depend on the size and shape of the element and on the rule adopted for numerical integration. The rule is called *exact to degree q*, if the integral of every polynomial P_q is given correctly by $\sum w_i P_q(\xi_i)$.

Suppose that we minimize I^* over all trial functions v^h. Then the minimizing function $\tilde{u}^h = \sum \tilde{Q}_j \varphi_j$ is determined by an approximate finite element system $\tilde{K}\tilde{Q} = \tilde{F}$, in which the stiffness matrix and load vector are computed by numerical rather than exact integrations. *This is the system which* (apart from roundoff error) *the computer actually solves.* Our goal is to estimate the difference $u^h - \tilde{u}^h$, and we repeat the main point: It is not necessary that the energies I and I^* be close in order for $u^h - \tilde{u}^h$ to be small.

We shall summarize the main theorem before giving examples and proof. *The essential condition on the quadrature formula is identical with the patch test for nonconforming elements: $\tilde{u}^h$ converges to u^h in strain energy ($\| u^h - \tilde{u}^h \|_m \to 0$) if and only if for all polynomials of degree m and all trial functions,*

$$(6) \qquad\qquad a^*(P_m, v^h) = a(P_m, v^h) + O(h).$$

There is also a supplementary condition of positive definiteness, requiring that the approximate strain energy a^* should be elliptic on the subspaces S^h: $a^*(v^h, v^h) \geq \theta \| v^h \|_m^2$. The term $O(h)$ enters the condition only because, if the material properties vary within an element, this will not be treated exactly by the quadrature. This is a second-order effect.

The test (6) actually applies only to the leading terms in the strain energy, involving mth derivatives. Since such derivatives of P_m are constants, the inner product $a(P_m, v^h)$ involves only integrals of mth derivatives of v^h, and the convergence condition reduces to this: *the mth derivatives of every trial function should be integrated exactly.* If the trial functions are polynomials of degree $k - 1$, this means that the quadrature formula should be correct at least through degree $k - m - 1$. Rectangular elements are much more demanding; there are terms in the trial functions which add nothing to the degree of approximation, but whose derivatives must nevertheless be integrated correctly. With bilinear elements in second-order problems, for example, the twist term xy has linear derivatives; therefore the quadrature formula must be correct for these terms, and not only for the constant strains which come from the linear terms $a + bx + cy$.

In practice the test (6) will often hold for all polynomials P_n of some higher degree $n > m$. In this case the accuracy is more than just minimal: the error in the strains is of order h^{n-m+1}. Every additional degree of accuracy in the quadrature scheme contributes an additional power of h to the error estimate. In other words, if the trial functions are of degree $k - 1$, and the quadrature is exact to degree q, the error is of order $q - k + m + 2$. If there are any terms of degree higher than $k - 1$ in the trial functions, as there always are for rectangular elements, this will injure the order of the error. In all cases the correct test is expressed by $a^*(P_n, v^h) = a(P_n, v^h)$; it is complete polynomials of degree $n - m$, multiplied by trial strains $D^m v^h$, which have to be integrated correctly.†

No attempt will be made to devise new quadrature formulas, but it is astonishing that even on triangles and rectangles this classical problem has never been completely solved. Irons [15] has shown what can still be done in this direction, by achieving a given accuracy q with far fewer points than the standard product-Gauss rules. The Russian school of Sobolev, Liusternik, and others has made a profound study of "cubature formulas" over regular regions, and has discovered some remarkable formulas: the reader may be interested in a 14-point formula on a cube of edge $\sqrt{2}$, which is accurate to degree $q = 5$:

$$\frac{2\sqrt{5}}{361} \left(\frac{121}{8} \sum_1^8 f(\xi_i) + 40 \sum_1^6 f(\eta_i) \right),$$

where the ξ_i are the vertices of a similarly placed cube of edge $\sqrt{38/33}$ and the η_i are the vertices of a regular octahedron touching a circumscribed sphere of radius $\sqrt{19/60}$! We repeat, however, that numerical integration for finite elements must take account of the higher-degree terms which often occur in trial functions on rectangles, even though they may contribute nothing to approximation theory. On triangles the element is normally a complete polynomial of degree $k - 1$, or close to it, and then it is purely a question of correctly integrating complete polynomials of as high a degree as possible. We owe Table 4.1 on the following page to Cowper [C13], who has added several new quadrature rules over triangles to those reproduced in Zienkiewicz [22]. The formulas are symmetric in the area coordinates, so that if a sampling point $\xi_i = (\zeta_1, \zeta_2, \zeta_3)$ occurs, *so do all its permutations*. If the ζ_i are distinct, there are six sampling points; if two ζ_i coincide, there are three; the point $(\frac{1}{3}, \frac{1}{3}, \frac{1}{3})$ at the centroid, if used in the formula, is taken only once.

The isoparametric method could not exist without numerical integration, since the integrand is a rational function of the new coordinates ξ and η. At first it looks impossible that even numerical integration should succeed,

---

†It is interesting to compute the quadrature accuracy required in order that $u^h - \tilde{u}^h$ will be of the same order h^{k-m} in the strains as the basic approximation error $u - u^h$. If this exponent is to agree with $n - m + 1$, then $n = k - 1$; the mth derivatives of all polynomials of degree $k - 1$, multiplied by the mth derivatives of all trial polynomials v^h, must be integrated exactly. If the v^h themselves comprise all polynomials of degree $k - 1$, as is often true on triangles, and the material coefficients in the strain energy are constant, the leading terms in this energy—*the squares of mth derivatives*—must be computed exactly to maintain the full accuracy!

Table 4.1

w_i	ζ_1	ζ_2	ζ_3	Multiplicity
	3-point formula	degree of precision 2		
0.33333 33333 33333	0.66666 66666 66667	0.16666 66666 66667	0.16666 66666 66667	3
	3-point formula	degree of precision 2		
0.33333 33333 33333	0.50000 00000 00000	0.50000 00000 00000	0.00000 00000 00000	3
	4-point formula	degree of precision 3		
−0.56250 00000 00000	0.33333 33333 33333	0.33333 33333 33333	0.33333 33333 33333	1
0.52083 33333 33333	0.60000 00000 00000	0.20000 00000 00000	0.20000 00000 00000	3
	6-point formula	degree of precision 3		
0.16666 66666 66667	0.65902 76223 74092	0.23193 33685 53031	0.10903 90090 72877	6
	6-point formula	degree of precision 4		
0.10995 17436 55322	0.81684 75729 80459	0.09157 62135 09771	0.09157 62135 09771	3
0.22338 15896 78011	0.10810 30181 68070	0.44594 84909 15965	0.44594 84909 15965	3
	7-point formula	degree of precision 4		
0.37500 00000 00000	0.33333 33333 33333	0.33333 33333 33333	0.33333 33333 33333	1
0.10416 66666 66667	0.73671 24989 68435	0.23793 23664 72434	0.02535 51345 51932	6
	7-point formula	degree of precision 5		
0.22503 30000 03000 00	0.33333 33333 33333	0.33333 33333 33333	0.33333 33333 33333	1
0.12593 91805 44827	0.79742 69853 53087	0.10128 65073 23456	0.10128 65073 23456	3
0.13239 41527 88506	0.47014 20641 05115	0.47014 20641 05115	0.05971 58717 89770	3
	9-point formula	degree of precision 5		
0.20595 05047 60887	0.12494 95032 33232	0.43752 52483 83384	0.43752 52483 83384	3
0.06369 14142 86223	0.79711 26518 60071	0.16540 99273 89841	0.03747 74207 50088	6
	12-point formula	degree of precision 6		
0.05084 49063 70207	0.87382 19710 16996	0.06308 90144 91502	0.06308 90144 91502	3
0.11678 62757 26379	0.50142 65096 58179	0.24928 67451 70910	0.24928 67451 70911	3
0.08285 10756 18374	0.63650 24991 21399	0.31035 24510 33785	0.05314 50498 44816	6
	13-point formula	degree of precision 7		
−0.14957 00444 67670	0.33333 33333 33333	0.33333 33333 43333	0.33333 33333 33333	1
0.17561 52574 33204	0.47930 80678 41923	0.26034 59660 79038	0.26034 59660 79038	3
0.05334 72356 08839	0.86973 97941 95568	0.06513 01029 02216	0.06513 01029 02216	3
0.07711 37608 90257	0.63844 41885 69809	0.31286 54960 04875	0.4869 03154 253160	6

since it is never exact for rational functions. The entries $a^*(\varphi_j, \varphi_k)$ of $\tilde{K}$ will be completely different from the entries $K_{jk} = a(\varphi_j, \varphi_k)$, and no perturbation argument is possible. Nevertheless, we compute $a(P_m, v^h) - a^*(P_m, v^h)$ and apply the test; the crucial point is that the test involves only one trial function—not both φ_j and φ_k at the same time—and this will save us.

A typical transformation was given in Section 3.3, with $m = 1$; $\partial P_m/\partial x$ is a constant c, and

$$\int\!\!\int_{e_i} p(x, y)\frac{\partial P_m}{\partial x}\frac{\partial v^h}{\partial x}\, dx\, dy$$

$$\longrightarrow c\int\!\!\int_{E_i} p(x(\xi, \eta), y(\xi, \eta))\left(\frac{\partial v^h}{\partial \xi}\frac{\partial \xi}{\partial x} + \frac{\partial v^h}{\partial \eta}\frac{\partial \eta}{\partial x}\right)J\, d\xi\, d\eta.$$

It is crucial to observe the form of the transformation matrix:

$$\begin{pmatrix} \xi_x & \xi_y \\ \eta_x & \eta_y \end{pmatrix} = \begin{pmatrix} x_\xi & x_\eta \\ y_\xi & y_\eta \end{pmatrix}^{-1} = \frac{1}{J}\begin{pmatrix} y_\eta & -x_\eta \\ -y_\xi & x_\xi \end{pmatrix}.$$

Evidently $\xi_x = y_\eta/J$ and $\eta_x = -y_\xi/J$. With this substitution, the integral becomes

(7)
$$c\int\!\!\int_{E_i} p(\xi, \eta)(v_\xi^h y_\eta - v_\eta^h y_\xi)\, d\xi\, d\eta.$$

The rational functions have disappeared, and convergence will occur if this integral is computed correctly. (For fourth order equations the rational functions do *not* disappear, and we cannot justify numerical integration unless the coordinate changes satisfy the smoothness conditions $\|F\|_k \leq C$ described on p. 163. In that case, the elements are only slightly distorted, and J is apparently no worse than the variable coefficient p; the integration error *is of the same order as for constant coefficients without isoparametrics.*)

In the case of bilinear functions on quadrilaterals, all the derivatives $v_\xi^h, y_\eta, \ldots$ are linear, and it would appear that the quadrature must be correct for quadratic polynomials. It just happens, however, that the second-degree terms cancel in the particular combination $K = v_\xi^h y_\eta - v_\eta^h y_\xi$, so that exactness to first degree is actually sufficient for convergence. In practice the quadrature will be more accurate than that (it will probably have to be, just to achieve positive definiteness of a^*) and the benefits come in establishing more than a minimal rate of convergence. Suppose the conditions of Section 3.3 are satisfied—the Jacobian stays away from zero, and the coordinate transformations $x(\xi, \eta)$ and $y(\xi, \eta)$ have bounded coefficients. Then just as in the x–y plane, each additional degree of polynomials in ξ, η contributes one more order of approximation. Therefore we expect the integration error to improve at the same rate; since first-order quadrature is sufficient for convergence in the bilinear case, the error in strains should be $O(h^q)$ if the quadrature is exact to degree q.

One final isoparametric remark: since the coordinate transformations $x(\xi, \eta)$ and $y(\xi, \eta)$ have the same form as the shape function $v^h(\xi, \eta)$, it follows that the combination K has the same form as the Jacobian J. Therefore, our rule that K must be correctly integrated coincides with the rule [22], based on Irons' intuition, that the volume of each element (the integral of J) must be correctly computed by the quadrature. In three dimensions this requires a higher order of exactness: K and J will involve products of three rather than two derivatives. For subparametric elements the two rules diverge; the stiffness matrix depends on K, the mass matrix and load vector on J. To repeat, these rules apply to highly distorted isoparametrics; for small distortions J is smooth and can be discounted in the convergence test.

We begin now on the theory, which is entirely based on the following simple identity.

LEMMA 4.1

Suppose that u^h and $\tilde{u}^h$ minimize the functionals $I(v^h)$ and $I^*(v^h)$, respectively, so that the equations of virtual work (the Euler equations $\delta I = \delta I^* = 0$) become

$$a(u^h, v^h) = b(f, v^h) \quad \text{and} \quad a^*(\tilde{u}^h, v^h) = b^*(f, v^h) \qquad \text{for all } v^h \text{ in } S^h.$$

Then

$$(8) \qquad a^*(u^h - \tilde{u}^h, u^h - \tilde{u}^h) = (a^* - a)(u^h, u^h - \tilde{u}^h) - (b^* - b)(f, u^h - \tilde{u}^h).$$

Proof. The left side of the identity is

$$a^*(u^h, u^h - \tilde{u}^h) - a^*(\tilde{u}^h, u^h - \tilde{u}^h)$$
$$= (a^* - a)(u^h, u^h - \tilde{u}^h) + a(u^h, u^h - \tilde{u}^h) - a^*(\tilde{u}^h, u^h - \tilde{u}^h).$$

With $v^h = u^h - \tilde{u}^h$ in the equations of virtual work, the last two terms yield $(b - b^*)(f, u^h - \tilde{u}^h)$, and the proof is complete. There is a similar identity, with a, b, and u^h replaced by a^*, b^*, and $\tilde{u}^h$, but it is not so useful. Notice also that the terms in a and b actually cancel in (8), but it is important to keep them and to work with the differences $a^* - a$ and $b^* - b$.

Our principal theorem follows immediately from the identity.

THEOREM 4.1

Suppose that the approximate strain energy is positive definite, $a^*(v^h, v^h) \geq \theta \|v^h\|_m^2$, and that

$$(9) \qquad |(a^* - a)(u^h, v^h)| + |(b^* - b)(f, v^h)| \leq Ch^p \|v^h\|_m.$$

Then the error in strains due to approximate integration is of order

(10) $$\| u^h - \tilde{u}^h \|_m \leq \theta^{-1} C h^p.$$

Proof. The left side of the identity (8) is bounded below by $\theta \| u^h - \tilde{u}^h \|_m^2$, and the right side is bounded above by (9), with $v^h = u^h - \tilde{u}^h$. Cancelling the common factor, the result is (10).

The lemma and theorem are not limited only to numerical quadrature. They apply also to *a change in the coefficients of the original differential equation*; in other words, they describe the manner in which the solution, to the continuous as well as the discrete problem, depends on the coefficients and on the inhomogeneous term. Consider a one-dimensional problem $-(pu')' + qu = f$, and suppose that p, q, and f are changed to $\tilde{p}, \tilde{q}$, and $\tilde{f}$. Then the identity, applied on the whole space $\mathcal{H}_E^1$ instead of the subspace S^h, becomes

(11)
$$\int \tilde{p}(u' - \tilde{u}')^2 + \tilde{q}(u - \tilde{u})^2$$
$$= \int (\tilde{p} - p)u'(u' - \tilde{u}') + (\tilde{q} - q)u(u - \tilde{u}) - (\tilde{f} - f)(u - \tilde{u}).$$

The left side, which is a^*, will be positive definite if $\tilde{p} > 0$ and $\tilde{q} \geq 0$. On the right side, each term is less than $\| u - \tilde{u} \|_1$ times a constant multiple of the perturbation. This yields a simple bound, not the most precise one possible, on the resulting perturbation in the solution.

COROLLARY

Suppose that the coefficients and inhomogeneous term are perturbed by less than ϵ:

$$\max_x (|p - \tilde{p}|, |q - \tilde{q}|, |f - \tilde{f}|) < \epsilon.$$

Then the solution is also perturbed by $O(\epsilon)$:

(12) $$\| u - \tilde{u} \|_1 < C \frac{\epsilon}{\theta}.$$

In terms of finite elements this has the following interpretation. Suppose that p, q, and f are replaced by their interpolates in the finite element space. This is a perturbation of order h^k. If the resulting finite element problem is solved exactly (there will be products of *three* polynomials in the element integrals), then by the corollary $\| u^h - \tilde{u}^h \|_1 = O(h^k)$. Thus interpolation is a possible alternative to numerical integration, and it has had the lion's share of attention in the numerical analysis literature; Douglas and Dupont

[D8] have successfully explored even nonlinear parabolic problems. In engineering calculations, however, direct numerical integration has consistently been preferred; with isoparametric or shell elements, there is effectively no choice. A rough operation count suggests that also in other problems direct quadrature is the more efficient, and we therefore intend to concentrate on this technique.

The simplest example is, as usual, the most illuminating. Therefore, we begin with the equation $-u'' = f$, and apply numerical integration to $I(v) = \int (v')^2 - 2fv$. Consider first the requirement of positive definiteness:

$$(13) \qquad a^*(v^h, v^h) = \sum w_i(v_x^h(\xi_i))^2 \geq \theta \int (v_x^h)^2 \, dx.$$

The interval is divided into elements, in other words into subintervals, and a standard quadrature rule is applied over each element. If v^h is a polynomial of degree $k - 1$, and the quadrature weights w_i are positive, then the definiteness requirement amounts to this: *There must be at least $k - 1$ integration points ξ_i in the subintervals.* Otherwise, there will exist in each element a nonzero polynomial of degree $k - 2$ which vanishes at every ξ_i. Joining these polynomials and integrating once, we have constructed a trial function v^h which contradicts (13):

$$\int (v_x^h)^2 \, dx > 0 \quad \text{but} \quad v_x^h(\xi_i) = 0.$$

If there are negative weights, then even more integration points will be needed for definiteness. We do not expect such formulas to be popular; they are also vulnerable to roundoff errors.

In two dimensions, with bilinear trial functions on rectangles, the definiteness condition would fail if we were to choose the *midpoint rule* for quadrature:

$$\int_{-h/2}^{h/2} \int_{-h/2}^{h/2} g(x, y) \, dx \, dy \sim h^2 g(0, 0).$$

This is the one-point Gauss rule on a square, and it is exact for the polynomials $g = 1, x, y, xy$. However, *it is indefinite;* for the trial function $v^h = xy$ numerical integration of $(v_x^h)^2 + (v_y^h)^2$ gives zero. If we set $v^h = +1$ or -1 in a checkerboard pattern over the whole set of nodes in Ω, the result is a high frequency oscillation (pure twist, with the smallest wavelength $2h$ which the mesh can accept) whose numerical energy is exactly zero. This is reflected in the discrete approximation to the Laplacian which would arise from this midpoint rule; a typical row of the stiffness matrix $\tilde{K}$ gives

$$4\tilde{u}_{j,k} - \tilde{u}_{j+1,\,k+1} - \tilde{u}_{j-1,\,k+1} - \tilde{u}_{j+1,\,k-1} - \tilde{u}_{j-1,\,k-1}.$$

This is the 5-point scheme rotated through 45°, and seems dangerously unstable; in fact, our oscillating twist gives a solution to the homogeneous equation $\tilde{K}\tilde{Q} = 0$.†

A similar one-point rule at the centroids of triangles, with linear trial functions, will *not* be indefinite. The strains are constant, and cannot vanish at the centroid without being identically zero. In fact, the Laplacian gives rise to the normal 5-point scheme.

The basic test for definiteness is to determine whether or not there are trial functions which, under numerical quadrature, give up all their strain energy. In practice this is decided from the rank of the element stiffness matrices; if the only zero eigenvalues come from rigid body motions, the quadrature is all right. If there are additional zero eigenvalues, the quadrature might still be acceptable: it has to be checked whether or not the offending polynomials in separate elements can be fitted together—as in the case of the twist described above—into a trial function v^h which has too little energy over the whole domain. For example, four-point Gauss integration (2×2) does not satisfy our stability condition for the nine-parameter biquadratic. With Gauss points ($\pm\xi, \pm\xi$) on the square centered at the origin, $(x^2 - \xi^2)(y^2 - \xi^2)$ has zero strain energy, and this pattern can be translated to give trouble over the whole domain. (The matrix $\tilde{K}$ may not be actually singular, if this pattern does not fit the boundary conditions (say $v^h = 0$) of the problem. In this case one could live dangerously and try such a four-point integration, even though $\tilde{K}$ is much nearer to singularity than the theory demands.)

The question of definiteness becomes rather delicate for the important eight-parameter element, obtained from the biquadratic by eliminating the x^2y^2 term and the node at the center of each mesh square. Since the trial functions no longer include $(x^2 - \xi^2)(y^2 - \xi^2)$, the four Gauss points are apparently enough for a stable finite element approximation to Laplace's equation. Taylor has pointed out, however, that for plane elasticity with *two* dependent variables, the situation is different: the combination $u = x(y^2 - \xi^2)$, $v = -y(x^2 - \xi^2)$ lies in the trial space, and its strain energy vanishes for the 2×2 rule. But Taylor has also demonstrated that *this pattern cannot be continued into a neighboring element;* the rank of a single element matrix is too low, but after assembly the global stiffness matrix is perfectly nonsingular and the numerical integration is all right.

So much for definiteness, which is a question of having enough integration points. We turn now to accuracy, which is decided by the polynomials for which the quadrature is exact. There are two ways to develop the theory. One is to compute directly the exponent p which appears in equation (9) of the

†A recent paper by V. Girault analyzes the discrete system which results from this quadrature rule. It turns out to be surprisingly useful in some contexts, even though it violates our hypothesis of positive definiteness.

theorem; we take this approach in the next two paragraphs, writing down explicit bounds (14–15) for the errors in numerical integration. Then in the final paragraphs of this section we describe a simpler and neater argument, which leads immediately to the connection between the exactness of the quadrature rule and the order of the resulting error $u^h - \tilde{u}^h$.

First approach. For a quadrature exact to degree q, the error in computing $\int g(x)\,dx$ numerically will be bounded by $C h^{q+1} \int |g^{(q+1)}(x)|\,dx$. This is the exact counterpart of the Approximation Theorem 3.3, and is proved in the same way. Applied to the expressions (9) which appear in Theorem 4.1, this becomes

$$(14) \qquad |(a^* - a)(u^h, u^h - \tilde{u}^h)| \le C h^{q+1} \sum_i \int_{e_i} \left| \left(\frac{d}{dx}\right)^{q+1} [p(x)u_x^h(u_x^h - \tilde{u}_x^h) \right.$$
$$\left. + q(x)u^h(u^h - \tilde{u}^h)] \right| dx.$$

$$(15) \qquad |(b^* - b)(f, u^h - \tilde{u}^h)| \le C h^{q+1} \sum_i \int_{e_i} \left| \left(\frac{d}{dx}\right)^{q+1} [f(u^h - \tilde{u}^h)] \right| dx.$$

Suppose the data f is smooth, as well as the variable coefficient $p(x)$ we have now allowed. Then u is also smooth, and so is its finite element approximation u^h. Therefore the only uncontrolled terms on the right sides are $u^h - \tilde{u}^h$ and its derivative. Every differentiation of these trial functions can introduce a factor h^{-1}.† It would appear that $q + 1$ differentiations could completely cancel the factor h^{q+1}, and destroy the proof of convergence. *Here is the point at which it is essential for the trial functions to be polynomials.* Since $u_x^h - \tilde{u}_x^h$ is of degree $k - 2$, this is the maximum number of factors h^{-1} which can appear; further differentiations would annihilate the polynomial. Therefore the expression (14) is of order $h^{(q+1)-(k-2)} \|u^h - \tilde{u}^h\|_1$. The same is true of the other expression (15); the first differentiation gives $u_x^h - \tilde{u}_x^h$, and we are back to the same argument. (We shall return to this point, and show in more detail why $b^* - b$ is of the same order as $a^* - a$ if the same quadrature rule is applied.) We conclude that $p = q - k + 3$ in Theorem 4.1, and therefore that the effect of numerical quadrature is bounded by

$$\|u^h - \tilde{u}^h\|_1 \le C h^{q-k+3}.$$

This coincides with the exponent $q - k + m + 2$ given earlier. For N-point Gauss quadrature, the degree of accuracy is $q = 2N - 1$, and the resulting error is $O(h^{2N-k+2})$. Therefore the use of $k - 1$ Gauss points should be completely successful in one dimension: it is enough to satisfy the requirement of definiteness, and it leads to an error of order h^k. This is even of lower order than the approximation error in the strains.

---

†More precisely, $|v^h|_{s+1} \le C h^{-1} |v^h|_s$ for all s.

In two or three dimensions the principle is the same. For certain poly-nonials of degree $q + 1$, say $x^\alpha y^\beta$, the quadrature will be inexact. This gives rise to corresponding terms in the error $a^* - a$, of the form

$$Ch^{q+1} \sum_i \int \int \left| \left(\frac{\partial}{\partial x} \right)^\alpha \left(\frac{\partial}{\partial y} \right)^\beta [D^m u^h \cdot D^m v^h] \right|, \qquad v^h = u^h - \tilde{u}^h.$$

The strain $D^m u^h$ is smooth, but every differentiation of $D^m v^h$ can introduce a factor h^{-1}. The convergence condition is that at most q of these factors ap-pear. Therefore the full $q + 1$ differentiations $(\partial/\partial x)^\alpha (\partial/\partial y)^\beta$ must anni-hilate $D^m v^h$ for any trial function v^h. In other words, $D^m v^h$ must not include any of the terms $x^\alpha y^\beta$ for which the quadrature is inexact. This is precisely the convergence criterion (6) given earlier, that the mth derivatives of every trial function should be integrated exactly.

Second approach. Suppose that the integral of any polynomial of degree $n - m$, multiplied by any mth derivative $D^m v^h$ of any trial function, is com-puted exactly by the quadrature. We want to show, following [S7], that the exponent p in Theorem 4.1 is $p = n - m + 1$; then the error in the strains due to numerical integration will be of this order h^p. To do so we consider the two quantities $a - a^*$ and $b - b^*$ which appear in the inequality (9).

A typical term in $(a - a^*)(u^h, v^h)$, with material properties expressed by a coefficient $c(x, y)$, looks like

(16a) $\displaystyle \int \int c(x, y) D^m u^h D^m v^h \, dx \, dy - \sum w_i c(\xi_i) D^m u^h(\xi_i) D^m v^h(\xi_i).$

The essential point, coming from our condition $a(P_n, v^h) = a^*(P_n, v^h)$ on the exactness of the quadrature, is that this term can be rewritten as

(16b) $\displaystyle \int \int (c D^m u^h - P_{n-m}) D^m v^h \, dx \, dy - \sum w_i (c D^m u^h - P_{n-m})(\xi_i) D^m v^h(\xi_i).$

Since numerical integration is carried out an element at a time, we can con-centrate on a specific element and choose P_{n-m} close to $c D^m u^h$. The difference between the two, according to our mean-square approximation theory, will be of order h^{n-m+1}—provided $c D^m u^h$ is sufficiently smooth, which will be the case if the original problem is smooth. Summing the results over the separate elements, the error (16a) is of the right order h^{n-m+1}.

A variation of this argument yields the same bound for any derivatives below order m in the energy, and also for the term $b - b^*$. We shall describe the steps involved in estimating this error in the latter term, assuming for simplicity that the mth derivatives of the trial functions v^h consist of all polynomials of some degree t. Then our exactness condition on the quadrature amounts to correct integration of all polynomials of degree $n - m + t$.

Therefore

$$(b - b^*)(f, v^h) = \int\int f v^h \, dx \, dy - \sum w_i(f v^h)(\xi_i)$$

$$= \int\int (f v^h - P_{n-m+t}) \, dx \, dy - \sum w_i(f v^h - P_{n-m-t})(\xi_i).$$

With the right choice of P in each element, these quantities are of order $h^{n-m+t+1}$, multiplied by an integral of the absolute value of derivatives up to this order of $f v^h$. But v^h can be differentiated only $t + m$ times, after which (being a polynomial) it disappears. Therefore, assuming f is smooth enough to allow $n - m + t + 1$ differentiations, the error $b - b^*$ is of order

$$h^{n-m+t+1} \|v^h\|_{t+m} \leq h^{n-m+1} \|v^h\|_m.$$

At the last step, the removal of t differentiations from v^h was paid for, as in the footnote on p. 190, by a factor h^{-t}.

Thus both $a - a^*$ and $b - b^*$ are of the correct order h^{n-m+1}, and by Theorem 4.1 the strains are in error to this order. This is the main result of the section: *If $a(P_n, v^h) = a^*(P_n, v^h)$, then $\|u^h - \tilde{u}^h\|_m = O(h^{n-m+1})$.* Ciarlet and Raviart have been able to show, even allowing for the use of iso-parametrics [6], that the error in displacement shows the usual improvement over the error in strains. (Their proof is a subtle variation of Nitsche's trick.) Therefore the displacement error is of order h^{n+1}, and the theory of numerical quadrature is in a satisfactory state: $n = m$ is necessary for convergence, and $n = k - 1$ is sufficient to reduce the errors due to quadrature to the same level as the errors due to approximation by polynomial trial functions.

4.4. APPROXIMATION OF DOMAIN AND BOUNDARY CONDITIONS

At the same time that the admissible functions in $\mathcal{H}_E^m$ are being approximated by piecewise polynomials, some other and quite different approximations are being made in the finite element method. In the first place, the domain itself may be changed: Ω is replaced by a nearby polygon Ω^h, or in the isoparametric method by a domain whose boundary is a piecewise polynomial. It would be extremely difficult to handle an arbitrary domain in any other way. Second, the boundary conditions themselves are subject to approximation; if the problem specifies that $u = g(x, y)$ on Γ, or that $u_n + \alpha u = b(x, y)$, then these functions g and b will almost inevitably be interpolated at the nodes on the boundary Γ (or on its approximation). We want to estimate the errors involved.

This section discusses four problems in detail:

1. Change of domain with a homogeneous Dirichlet condition, $u = 0$ on Γ.

2. Change of domain with a homogeneous Neumann condition, $u_n = 0$ on Γ.

3. Approximation of an inhomogeneous Dirichlet condition, $u = g(x, y)$ on Γ.

4. Approximation of an arbitrary inhomogeneous Neumann condition, $u_n + \alpha(x, y)u = b(x, y)$ on Γ.

In each case we work with a second-order equation in two dimensions, say Poisson's equation $-\Delta u = f$. In large part this is for convenience and simplicity of exposition: very little is altered if there are several unknowns and three space dimensions. For a pure Dirichlet or a pure Neumann problem of higher order, say a plate with its edges either clamped or free, the order of the error in strain energy is the same as the one demonstrated below.

A new and quite different situation arises when an essential and a natural boundary condition are combined. This occurs in a *simply supported plate*: the biharmonic equation $\Delta^2 u = f$ accounts for the loading in the interior, and Poisson's ratio v enters the natural boundary condition:

$$(17) \qquad u = 0 \quad \text{and} \quad v\Delta u + (1 - v)u_{nn} = 0 \qquad \text{on } \Gamma.$$

It is no longer possible to guarantee, in this physically important and mathematically well-posed problem, *that if the polygon Ω^h is close to Ω, then the solution on the polygon is close to the solution u on Ω.* This applies both to the exact solution U^h on Ω^h and to the finite element approximation u^h. More is required than simple convergence of the boundaries.

The difficulty is easy to see when Ω^h is a polygon. On each edge, the condition $U^h = 0$ immediately forces the tangential derivative U^h_{tt} to vanish. Therefore, the second boundary condition in (17) is equivalent on a straight edge to $\Delta U^h = 0$, and the dependence on v has disappeared. Babuska [B1] proposes introducing $V^h = \Delta U^h$ as a new unknown, yielding two second-order equations:

$$\Delta U^h = V^h \quad \text{and} \quad \Delta V^h = f \quad \text{in } \Omega^h, \qquad U^h = V^h = 0 \quad \text{on } \Gamma^h.$$

For such a second-order system convergence *is* guaranteed, and U^h, V^h approach the solutions of

$$\Delta U = V \quad \text{and} \quad \Delta V = f \quad \text{in } \Omega, \qquad U = V = 0 \quad \text{on } \Gamma.$$

This is precisely the problem of the clamped plate with $v = 1$, and therefore *the limiting function is independent of the Poisson's ratio actually specified in the boundary conditions.* Convergence occurs, but nearly always to the wrong answer. Corresponding difficulties with finite element calculations are reported in [R1] and discussed in [B11]. We would anticipate success in the isoparametric method, on the other hand, if the approximation of Γ is at least piecewise quadratic; the curvature of the boundary converges in this case. Alternatively, suppose the essential condition $u = 0$ were replaced at boundary nodes by $u^h = \partial u^h/\partial t = 0$, using the cubic trial space Z_3 (see Section 1.9) and taking the tangent to the true boundary Γ. Then convergence is expected even on a polygon. At this writing, however, the required theory does not exist.

We turn to the four problems listed above. In each case the analysis is rather technical, but the conclusions are straightforward.

1. Suppose that Ω is replaced by an inscribed polygon Ω^h, and the trial functions are made to vanish on the straight edges of Γ^h. Then imagining that they are defined to vanish everywhere outside Γ^h, they are admissible in the variational problem; they vanish on the true boundary Γ, and the trial space S^h is a bona fide subspace of $\mathcal{H}_0^1(\Omega)$. Therefore, the fundamental Ritz theorem 1.1 assures that u^h minimizes the error in strain energy:

$$(18) \qquad \iint_\Omega (u - u^h)_x^2 + (u - u^h)_y^2 = \min_{S^h} \iint_\Omega (u - v^h)_x^2 + (u - v^h)_y^2.$$

Since every v^h vanishes outside Ω^h, the integral over the skin $\Omega - \Omega^h$ is fixed; it is just the integral of $u_x^2 + u_y^2$. Therefore, u^h is minimizing over Ω^h as well as Ω:

$$(19) \qquad \iint_{\Omega^h} (u - u^h)_x^2 + (u - u^h)_y^2 = \min_{S^h} \iint_{\Omega^h} (u - v^h)_x^2 + (u - v^h)_y^2.$$

The question is simply to estimate (18) and (19). Up to now we have chosen v^h to be the interpolate u_I. If S^h is Courant's piecewise linear space, with nodes at the vertices of the triangles, this is still a good choice. Since it interpolates $u = 0$ at boundary nodes, it vanishes along the boundary of the polygon and lies in S^h. The standard approximation theorem 3.3 yields an h^2 error in energy. No doubt the approximation is very poor near the boundary.

Suppose a more refined element is used, for example a piecewise quadratic (Fig. 4.2). As usual, the nodes are placed at the vertices and midpoints of the edges. Each trial function will vanish along Γ^h provided it is zero at all boundary nodes, but *this excludes u_I from the trial space S^h*. The true solution u vanishes at the vertices P and Q but not at the midpoint M, so the interpolate is not zero along the boundary. Since M is a distance $O(h^2)$ from the true

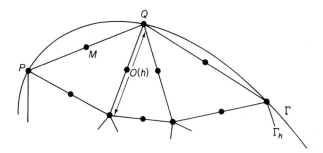

Fig. 4.2   Polygonal approximation of the boundary.

boundary Γ, the mean-value theorem yields

$$(20) \qquad\qquad |u(M)| \leq Ch^2 \max_{\Omega} |\operatorname{grad} u|.$$

It is no longer permissible to choose $v^h = u_I$ in (19). Instead, a convenient v^h will be the piecewise quadratic u_I^* which interpolates u at *interior nodes only* and vanishes on Γ^h. The error in strain energy over Ω^h for this choice, using the triangle inequality, is

$$\iint_{\Omega^h} (u - u_I^*)_x^2 + (u - u_I^*)_y^2 = |u - u_I + u_I - u_I^*|_1^2$$

$$\leq (|u - u_I|_1 + |u_I - u_I^*|_1)^2$$

$$\leq 2|u - u_I|_1^2 + 2|u_I - u_I^*|_1^2.$$

The term $|u - u_I|_1^2$ is completely familiar; it is $O(h^4)$ by Theorem 3.3. The new term $u_I - u_I^*$ is a piecewise quadratic which vanishes over all interior triangles: the only nodes at which $u_I \neq u_I^*$ are the boundary midpoints M_i; at these points $u_I^* = 0$ and $u_I = u = O(h^2)$. The number of such boundary triangles is $O(1/h)$, and

$$|u_I - u_I^*|_1^2 = \sum \iint_{e_i} |u(M_i)|^2 (\varphi_x^2 + \varphi_y^2)\, dx\, dy$$

$$\leq O\!\left(\frac{1}{h}\right) c^2 h^4 \max |\operatorname{grad} u|^2 \iint (\varphi_x^2 + \varphi_y^2)\, dx\, dy.$$

Here φ is the quadratic with value 1 at the midpoint M_i and zero at the other five nodes of the triangle. Its first derivatives are of order $1/h$, and the area of the triangle is of order h^2, so that finally

$$|u_I - u_I^*|_1^2 = O(h^3).$$

THEOREM 4.2

The error in energy over Ω^h, *produced by polygonal approximation of the domain, satisfies*

(21)
$$|u - u^h|_1^2 \leq |u - u_I^*|_1^2 = O(h^3).$$

The error in function values is $u - u^h = O(h^2)$.

The latter estimate (which we shall not prove) is suggested by the continuous problem on the approximate domain: $-\Delta U^h = f$ on Ω^h, $U^h = 0$ on Γ^h. Since $\Delta(u - U^h) = -f + f = 0$, the maximum principle applies, and $u - U^h$ attains its maximum on Γ^h. But $U^h = 0$ and $u = O(h^2)$ on this boundary, so that $u - U^h$ is of order h^2 everywhere. The theorem asserts the same result for $u - u^h$ [B20].

It follows that if we look only at the exponent of h, and approximate Ω by a polygon, then there is no point in going beyond quadratic shape functions in computing stresses. Furthermore, even linear polynomials will be correct to the best possible order h^2 in the displacements. Probably this is an instance in which the exponents of h are not sufficient to indicate the true accuracy; the actual computed error with linear elements may be excessive.

A similar estimate applies to three-dimensional and to higher-order Dirichlet problems. It is most unusual that the exponent should be odd; the average error in the derivatives of $u - u^h$ is of the *fractional* order $h^{3/2}$. The estimate itself is nevertheless correct, and in fact the exact solution U^h to the equation on the polygon [which is closer to u than u^h is, because it is minimizing over the space $\mathcal{K}_0^1(\Omega^h)$, which contains S^h] is also in error to order $h^{3/2}$ [B10]. For a rectangular approximation of the boundary, which is much more crude, the order becomes $h^{1/2}$; the computational results would be completely unsatisfactory. That is the reason for triangular elements.

The explanation for the fractional exponent $\frac{3}{2}$ appears to be this: There is a *boundary layer*, a couple of elements thick, within which the derivatives are in error by $O(h)$. It is easy to verify that the angle θ in the figure is of this order, so that the true derivative of u along the chord is $O(h)$ rather than zero. This layer alone accounts for the $O(h^3)$ error in energy. Beyond the boundary layer, the error is quite different: it is of the optimal order h^2 in the displacement, that is, $u - u^h = O(h^2)$, and it is so smooth that the first derivatives are also of order h^2. The error is much better behaved in the interior than in the boundary, illustrating the smoothing property common to all elliptic problems. In the finite element approximation, $u - u^h$ oscillates from zero to $O(h^2)$ and back along each chord PMQ; this singular behavior is damped out rapidly in the direction normal to the boundary. The damping is very visible in computations and has been explicitly verified. It represents a kind of

St. Venant principle, but one which applies to the geometry rather than, as is more customary, to the boundary data. The rule is the same: *Away from the boundary, it is the average which matters and not the local short-wavelength oscillations.*

The situation is not unlike that of mapping the unit circle onto a circle with a wavy boundary. Suppose the latter boundary is $R(\theta) = 1 + h^2 \cos \theta/h$. Then in the conformal map between these "circles," the point (r, θ) goes into $(r + h^2 r^{1/h} \cos \theta/h, \theta)$. All the activity lies inside a boundary layer $r \geq 1 - Ch$; for smaller values of r, the amplitude of the "wavy" perturbation term is exponentially small,

$$h^2 r^{1/h} \leq h^2 (1 - Ch)^{1/h} \sim h^2 e^{-C}.$$

The derivatives of the mapping are $O(h)$ near the boundary. In the interior they are virtually zero, because of $r^{1/h}$. Note that the two circles have the same area; if one were inscribed in the other, there would be an additional term rh^2, whose derivatives do not decay but are of lower order h^2 throughout the domain. In St. Venant terms, the average over the local oscillations is nonzero and is propagated. Furthermore, if the circle were scalloped instead of wavy, with $|\cos \theta/h|$ instead of $\cos \theta/h$, then the conformal map would possess weak singularities at the cusps. In the finite element approximation, however, these are smeared over and the mean error in derivatives is of order h at the boundary and h^2 inside.

There are a number of ways to recover the optimal convergence rate for quadratics. Coordinate change by isoparametric elements, and Mitchell's x–y element which has piecewise hyperbolas (!) as boundary, have already been mentioned. Another possibility is to compute a correction term to the polygonal approximation u^h [B26], or to modify the original functional $I(v)$ [N3]. In all these techniques it is not essential that Ω^h lie inside Ω; in fact, the error due to change of domain is partly self-cancelling, if Γ^h goes systematically in and out of Γ. This is St. Venant again, that the leading error term depends on the area of $\Omega - \Omega^h$ (with algebraic sign), and suggests a further possibility: to average the approximate solutions for inscribed and circumscribed polygons. All these proposals, except for the isoparametric one, are largely untested in practical problems.

2. Consider next the equation $-\Delta u + qu = f$, with natural boundary condition $u_n = 0$. In this case there is no question of imposing conditions on the wrong boundary; the trial functions are unrestricted at the boundary. However, a change of domain does enter, if it is inconvenient to carry out integrations over the true domain Ω. If the potential energy is computed instead over an approximate domain Ω^h, the effect is to introduce a new defi-

nition of the potential energy:

$$I^h(v) = a^h(v, v) - 2(f, v)^h$$

$$= \iint_{\Omega^h} (v_x^2 + v_y^2 + qv^2 - 2fv)\, dx\, dy.$$

Assuming that u^h minimizes the true functional I over S^h and that $\tilde{u}^h$ minimizes I^h, the problem is to estimate $e^h = u^h - \tilde{u}^h$.

Mathematically, this is essentially the same question which arises for Gaussian quadrature: the integrals are computed inexactly. Therefore, we apply a variant of the identity of Lemma 4.1; it follows directly from the vanishing of the first variations at u, u^h, and $\tilde{u}^h$ that

$$E^2 = a^h(e^h, e^h) = a^h(u^h - u, e^h) + (a^h - a)(u, e^h) + (f, e^h) - (f, e^h)^h.$$

The first term on the right, by the Schwarz inequality, is not larger than

$$[a^h(u^h - u, u^h - u)a^h(e^h, e^h)]^{1/2} \le C_1 h^{k-1} E.$$

The other terms on the right amount to an integral over the skin $\Omega - \Omega^h$:

$$B = \iint_{\Omega - \Omega^h} (-u_x e_x^h - u_y e_y^h - q u e^h + f e^h)\, dx\, dy.$$

Assuming that u_x, u_y, q, and f are bounded, the Schwarz inequality this time involves the area A of the skin:

$$|B| \le C_2 A^{1/2} \left[\iint_{\Omega - \Omega^h} (e_x^h)^2 + (e_y^h)^2 + (e^h)^2 \right]^{1/2}.$$

The remaining problem is to estimate this last integral in terms of E, in other words, to relate the size of e^h in the skin to its size in the interior. For arbitrary functions this would be impossible. The function $u^h - \tilde{u}^h$, however, is by no means arbitrary: in each triangle it agrees with some polynomial, and polynomials cannot suddenly explode—a bound in Ω^h implies a bound in Ω.

We denote by T a typical curved triangle at the boundary, and by T^h the inscribed straight triangle; their difference $T - T^h$ is one of the pieces that go to make up $\Omega - \Omega^h$. The trial functions are polynomials on each triangle T, and we are dealing with the error $e^h = u^h - \tilde{u}^h$ which is caused by integrating only over T^h. This error is itself a polynomial $P(x, y)$ on each T, and obeys the following lemma given by Berger.

LEMMA 2.2

Suppose that $\rho = \text{area}(T - T^h)/\text{area}(T^h)$. *Then there is a constant* c, *depending only on the degree of the polynomial* $P(x, y)$, *such that*

$$\iint_{T-T^h} P_x^2 + P_y^2 + P^2 \leq c\rho \iint_{T^h} P_x^2 + P_y^2 + P^2.$$

Summing over all boundary triangles and increasing the right side to include also the interior triangles, the lemma gives

$$\iint_{\Omega-\Omega^h} (e_x^h)^2 + (e_y^h)^2 + (e^h)^2 \leq c\rho E^2.$$

Substituting back into the estimate for $|B|$, the identity now reads

$$E^2 \leq [C_1 h^{k-1} + C_2(Ac\rho)^{1/2}]E.$$

In other words, the error in strain energy due to integration over Ω^h instead of Ω satisfies

$$E^2 \leq C_3(h^{2(k-1)} + A\rho).$$

The first term reflects the strain energy in $u - u^h$ and contributes nothing new. The second term is the interesting one, and it is purely geometrical. Suppose, for example, that Ω^h is a polygon. Then the area A is $O(h^2)$, and the ratio ρ to the area of the neighboring triangles is $O(h)$. Thus *the strain energy error in changing to a polygon is of the same order* h^3 *for natural as for essential boundary conditions.* We conjecture that there is again a boundary layer. For isoparametrics instead of subparametrics, A is $O(h^k)$ and ρ is $O(h^{k-1})$, and their product h^{2k-1} is dominated by the ordinary approximation error in strain energy of order h^{2k-2}. Therefore isoparametrics should be successful with curved regions, whether the boundary conditions are essential or natural.

3. The next problem is that of an inhomogeneous essential boundary condition, $u = g(x, y)$ on Γ. This condition is satisfied by every member v of the true admissible space $\mathcal{JC}_E^1$. Any two admissible functions therefore differ by a function v_0 in $\mathcal{JC}_0^1$; the boundary values of v_0 are zero.

Suppose that the situation is the same for the trial functions v^h in S^h: All trial functions assume the same values on the boundary Γ (not necessarily $v^h = g$, that is too much to expect from polynomials, but say $v^h = g^h$ instead). Then any two trial functions differ by a function v_0^h, which vanishes on Γ; these functions v_0^h form a space S_0^h, which is a subspace of the true homogeneous space $\mathcal{JC}_0^1$.

THEOREM 4.3

Suppose that u minimizes $I(v)$ over $\mathfrak{IC}_E^m$ and u^h minimizes over S^h. Then the vanishing of the first variation is expressed by

(22) $$a(u, v_0) = (f, v_0) \quad \text{for all } v_0 \text{ in } \mathfrak{IC}_0^1,$$

(23) $$a(u^h, v_0^h) = (f, v_0^h) \quad \text{for all } v_0^h \text{ in } S_0^h.$$

As in Theorem 1.1, u^h has the additional minimizing property

(24) $$a(u - u^h, u - u^h) = \min_{v^h \text{ in } S^h} a(u - v^h, u - v^h).$$

The proof is copied from Theorem 1.1. Since u and u^h are minimizing, any perturbations ϵv_0 and ϵv_0^h must increase the functional I:

$$I(u) \le I(u + \epsilon v_0) \quad \text{and} \quad I(u^h) \le I(u^h + \epsilon v_0^h).$$

Expanding, the coefficients of ϵ must vanish, and these are the virtual work equations (22) and (23).

To prove that u^h minimizes the error in strain energy, write

$$\begin{aligned} a(u - v^h, u - v^h) &= a(u - u^h + u^h - v^h, u - u^h + u^h - v^h) \\ &= a(u - u^h, u - u^h) + 2a(u - u^h, u^h - v^h) \\ &\quad + a(u^h - v^h, u^h - v^h). \end{aligned}$$

The middle term must vanish automatically, subtracting (23) from (22) and choosing $v_0^h = u^h - v^h$. The last term will be positive unless v^h happens to equal u^h; therefore, the minimum occurs at this point, and the theorem is proved.

COROLLARY

Suppose that Ω is a polygon, and the essential condition $u = g(x, y)$ on Γ is interpolated in the finite element method: at all boundary nodes, the trial functions satisfy $v^h(z_j) = g(z_j)$ [or more generally $D_j v^h(z_j) = D_j g(z_j)$]. Then

(25) $$a(u - u^h, u - u^h) \le a(u - u_I, u - u_I).$$

Therefore, the error estimate in the finite element method is reduced to the standard approximation estimate of $u - u_I$.

Proof. The conditions that must be checked are

a. u_I must lie in S^h, so that the choice $v^h = u_I$ is possible in (24). Since u_I satisfies the boundary conditions imposed in the corollary, that is, $u_I(z_j) = u(z_j) = g(z_j)$ at boundary nodes, this requirement is met.

b. All trial functions v^h must assume the same boundary values on Γ, so that their differences belong to $\mathcal{3C}_0^1$. In other words, the trial functions v^h should be determined on Γ by their values at the boundary nodes. Since Γ is composed of flat edges, and we assume that a conforming element is used, this will be the case.

We hesitated to give the usual estimate $h^{2(k-1)}$ for (25), since such an estimate requires that the solution be smooth (u in $\mathcal{3C}^k$). At the corners of polygonal domains, a singularity in the derivatives of u is almost automatic. For second-order equations, u normally just fails to lie in $\mathcal{3C}^{1+\pi/\alpha}$, where α is the largest interior angle. The first derivatives will actually be approximated to order $h^{\pi/\alpha}$; the strain energy error will be $h^{2\pi/\alpha}$. In Chapter 8 we consider how the usual order $h^{2(k-1)}$ can be preserved, by refining the mesh at the corners or (better) by introducing special trial functions with the right singularities.

Suppose that a nonpolygonal Ω is replaced by a polygonal Ω^h—either directly in the x–y plane, or else in the ξ–η plane after an isoparametric transformation. If the computations are done on Ω^h, then again the boundary nodes completely determine all the boundary values, and the corollary can be applied: The final error is dominated by the error in $u - u_I$ over the approximate domain, *plus the error studied earlier due to change in domain.*

There remains a further possibility: to work on a curved domain Ω in the given x–y coordinates and to interpolate $v^h = g$ at boundary nodes. Presumably the integrations would be done numerically, especially over the curved elements at the boundary, although the experiments in [C14] were carried out by exact algebraic operations. In this method the boundary values vary from one trial function to another, except of course at the nodes themselves; the differences between the trial functions are small but not zero on Γ. The Ritz rules are therefore violated, and the theoretical questions which arise are:

a. How large can these differences on the boundary become?

b. Is the error $u - u^h$ affected by this worst possible behavior on Γ, or does u^h still yield an optimal approximation to u from S^h?

These questions are relevant even with homogeneous boundary conditions, $u = g = 0$ on Γ. In this case the trial functions are zero at the boundary nodes; question a asks how large they can be on the rest of Γ. Surprisingly, the answer is independent of their degree. The maximum size of v^h on Γ is the same whether v^h is linear, vanishing only at the vertices of the boundary triangles, or quadratic or cubic, vanishing at one or two additional nodes on each piece of Γ. It is even the same for the cubic space Z_3, with values and true tangential derivatives set to zero at boundary vertices. The answer is: *There exists a particular trial function V^h of unit energy whose mean value on*

Γ *is of order* $h^{3/2}$:

$$\iint_\Omega (V_x^h)^2 + (V_y^h)^2 = 1, \qquad ch^3 \le \int_\Gamma |V^h|^2 \, ds \le Ch^3.$$

To answer question b, that is, to find a bound on $u - u^h$, it will be necessary to extend the classical Ritz theory. It is true that the first variations still vanish at the minimizing functions u and u^h:

$$a(u, v) = (f, v) \quad \text{for } v \text{ in } \mathcal{K}_0^1, \qquad a(u^h, v^h) = (f, v^h) \quad \text{for } v^h \text{ in } S^h.$$

However, even with $g = 0$, S^h is not a subspace of $\mathcal{K}_0^1$; in the inhomogeneous case $g \ne 0$, S_0^h is not contained in $\mathcal{K}_0^1$. Nevertheless, we may still appeal to Green's formula: Since $-\Delta u = f$,

$$a(u, v^h) = \iint_\Omega u_x v_x^h + u_y v_y^h = \iint_\Omega f v^h + \int_\Gamma u_n v^h \, ds.$$

Subtracting $a(u^h, v^h) = (f, v^h)$, the boundary term remains:

$$(26) \qquad a(u - u^h, v^h) = \int_\Gamma u_n v^h \, ds \qquad \text{for all } v^h \text{ in } S^h.$$

It is this term which controls the error due to violating the essential boundary conditions.

We propose to show that question *b* usually has the worst possible answer: The order of error will be decided by the trial function V^h which is largest on Γ. In fact, with $v^h = V^h$ in (26), that is easy to see. Since V^h has unit energy, the left side is bounded by the square root of $a(u - u^h, u - u^h)$; this is the Schwarz inequality. On the right side, V^h is of average value $h^{3/2}$, and unless there is some special cancellation (see below!) we must expect the integral of $u_n V^h$ to be of the same order. Therefore,

$$a(u - u^h, u - u^h)^{1/2} \sim h^{3/2}.$$

This would mean that interpolating the boundary conditions and integrating over Ω is no better, even for an element of high degree, than replacing Ω by a polygon Ω^h. (At least, the exponent of h is no better: we have no estimate of the constants involved.) One explanation is this: Each polynomial which interpolates $u = 0$ at boundary nodes will vanish along a curve which runs close to the true boundary, but the two curves may differ by as much as $O(h^2)$—the same distance as for a polygon Ω^h.

It is shown in [B10] that $h^{3/2}$ is also an *upper* bound for the error due to interpolating the boundary conditions. From the manipulations in that paper, a second rather surprising fact emerges: Even though the accuracy is limited by the existence of an undesirable V^h of order $h^{3/2}$ on Γ, nevertheless *the true displacement error $u - u^h$ is actually of order h^3 on Γ.* In other words, the boundary values of the Ritz solution will look deceptively good. It must be the normal derivatives at the boundary which are most in error.

We consider now the possibility of cancellation in the integral $\int u_n v^h \, ds$ on the right side of (26). If the boundary values v^h happen to oscillate around zero, then even though their average modulus $|v^h|$ may be as large as $h^{3/2}$, the integral itself will tend to be an order of magnitude smaller. *This oscillation does occur for some elements, but not for all.* It occurs for quadratics, for example, when the condition $u = 0$ (or $u = g$) is interpolated at boundary vertices, and at the points on Γ which lie halfway between. The leading term in v^h is a cubic $s(s - h/2)(s - h)$ in the arc length variables on Γ, vanishing at boundary nodes, and this cubic has average value zero. In other words, v^h oscillates around zero. Correspondingly, Scott has shown that the error in energy due to interpolating $u = g$ is improved to $O(h^{5/2})$. (This does not occur for cubics [B10], unless the nodal points are specially placed on Γ to give v^h an average value of zero.) Since the ordinary approximation error for quadratics is $O(h^2)$ in energy, the boundary error now appears perfectly acceptable. We don't know whether this alternative to isoparametrics for boundary triangles, namely integration over curved elements directly in the $x-y$ plane, has a useful future.

4. The final boundary problem on our list arises from an inhomogeneous natural condition

$$u_n + \alpha(x, y)u = b(x, y) \quad \text{on } \Gamma, \qquad \alpha \geq 0.$$

For Poisson's equation, this leads to the appearance of boundary integrals in the potential energy:

$$I(v) = \iint_\Omega v_x^2 + v_y^2 + \int_\Gamma \alpha v^2 - 2 \iint_\Omega fv - 2 \int_\Gamma bv.$$

Strictly speaking, there are three questions to consider in relation to these boundary integrals: polynomial approximation on Γ, numerical integration, and change of domain. The last two arise because, when Γ is curved, it will be nearly impossible to compute the boundary integrals exactly. Some numerical procedure will be adopted, and in case the trial functions are defined on Ω^h rather than Ω, the integrals will be shifted to Γ^h. We propose to omit a detailed analysis of these errors on Γ^h, because we are confident that they are no larger than those already studied on the interior Ω^h. A proof would be very desirable.

The novel question is that of polynomial approximation on Γ. It arises in the usual way: The method minimizes the error in energy,

$$a(u - u^h, u - u^h) = \min \iint_\Omega (u - v^h)_x^2 + (u - v^h)_y^2 + \int_\Gamma \alpha(u - v^h)^2 \, ds.$$

For a finite element space of degree $k - 1$, and the choice $v^h = u_I$, the integral over Ω is of order $h^{2(k-1)}$. Fortunately, the integral over Γ is even of higher order; *the rate of convergence is not reduced by the presence of boundary integrals.* This is obvious if Γ is composed of straight lines; the restriction of the trial functions to Γ yields a complete polynomial of degree $k - 1$ in the boundary variable s, and the integral over Γ is of order h^{2k}. Nitsche [6] has obtained the same result for a curved boundary.

With isoparametric elements the boundary in the ξ–η plane is straight, and all the boundary integrals can be computed directly by numerical integration. In fact, this seems to be the main conclusion of the boundary theory for second-order problems: *The isoparametric technique establishes a local change of coordinates into normal and tangential directions, which is more accurate and convenient than was ever achieved with finite differences.*

5 STABILITY

5.1. INDEPENDENCE OF THE BASIS

In one sense, there should be no problem of stability. The solution of an elliptic variational problem depends continuously on the data; if the load f and all prescribed boundary displacements and forces are small, then the strain energy in u is small. In other words, the problem is well posed. Furthermore, *regardless of the choice of the subspace S^h*, the strain energy in the Ritz approximation u^h is automatically bounded by that in u; the Ritz method projects u onto S^h, and this can only reduce the energy (corollary to Theorem 1.1). Therefore, the approximate problems are *uniformly well posed*, and it should always be possible to construct a "stable" procedure for the numerical calculation of u^h.

The difficulty is that to achieve total numerical stability—conceding nothing as the mesh size is reduced to zero—the required algorithm may simply be too fancy. For the standard five-point difference scheme, one could systematically use all the constraints on the coefficient matrix which come from consistency with the Laplace operator: The sums along each row of the matrix, as well as the first moments, vanish at all stages of the Gauss elimination process. For finite elements of irregular shape, however, the corresponding constraints will be extremely difficult to use. Therefore, we shall analyze the standard elimination algorithm, accepting an increase in roundoff error as $h \longrightarrow 0$ but intending that the numerical stability should be as foolproof as possible; unnecessary numerical instabilities will not be accepted.

It is known that the key to stability lies in the *uniform linear independence of the basis elements φ_j*. Even though u^h is completely independent of the choice of basis, the roundoff which enters its computed value $\bar{u}^h$ does depend on this choice. (Mikhlin refers to the "strong minimality" of the basis, and Soviet authors are generally careful to consider numerical methods from this point of view, as well as for stability with respect to coefficient changes in the differential equation.) To quantify the linear independence of the basis, the standard procedure is to consider the *Gram matrix* (in physical terms, the *mass matrix*) whose entries are the inner products of the basis elements:

$$M_{jk} = (\varphi_j, \varphi_k) = \int_\Omega \varphi_j(x)\varphi_k(x)\, dx_1 \ldots dx_n.$$

In many cases we shall prefer, because the Ritz method operates always with the energy $a(v, v)$ which is intrinsic to the problem, to work with the stiffness matrix: $K_{jk} = a(\varphi_j, \varphi_k)$. Both matrices are Hermitian and positive definite.

As a first measure of the independence of the basis, we propose the *condition number*

$$\kappa(M) = \frac{\lambda_{\max}(M)}{\lambda_{\min}(M)}.$$

If the basis were orthonormal, M would be the identity matrix and $\kappa = 1$. This is not the case for finite elements, but the important fact on a regular mesh is that *the finite element basis functions are uniformly linearly independent*: $\kappa(M) \leq$ constant. In other words, the eigenvalues of M are all of the same order. As Schultz observed, this makes piecewise polynomials infinitely more stable for least-squares approximation (which is just the Ritz method applied to the differential equation $u = f$ of order zero) than the sequence $1, x, y,$ $x^2, \ldots$ of ordinary polynomials. The condition number for this sequence, whose mass matrix is the Hilbert matrix (1.5), would increase exponentially.

There are applications in which a more realistic measure of independence is provided by the *optimal condition number*

$$c(M) = \min_D \kappa(DMD).$$

Here D may be any positive diagonal matrix, corresponding to a *rescaling* of the basis elements; $c = 1$ if the original basis is only orthogonal rather than orthonormal. With irregular elements, some trial functions may be much smaller than others, and this rescaling could make a significant difference to the condition number. We view rescaling in the following way: *When the condition number of M or of K is improved by rescaling, that suggests that the numerical difficulties which are thereby cured were never likely to propagate throughout the whole problem.* Rescaling cures *local* difficulties; if a particular φ_j is badly out of scale, say the diagonal entry K_{jj} is too small, then roundoff will destroy confidence in the computed value of the weight Q_j. The effect which is measured depends on the choice of norm; such a roundoff error would not mean a large mistake in the energy, because φ_j is small, but at the particular node z_j the approximation may be comparatively poor.

The standard rule in scaling a sparse positive definite matrix is *to keep all the diagonal entries equal*: in the finite element case, which is primarily governed by the stiffness matrix, that means that the strain energies K_{jj} in the basis elements are equal. This rule yields a diagonal scaling matrix D which is nearly optimal [V3]. The question of scaling arises even on a regular mesh for finite elements of Hermite type, in which both function values and derivatives appear among the unknowns Q_j. Perhaps a natural procedure is to keep the unknowns dimensionally correct, by using the rotation $\theta = hv'_j$ rather than v'_j itself.

Fried [F17] has observed that some problems can be successfully rescaled, while others cannot. Take, for example, a two-point boundary-value problem, say $-u'' = f$ with piecewise linear elements. If the subintervals are all of length h, except that the first is of length h/c, then the stiffness matrices with a natural condition $u'(0) = 0$ or an essential condition $u(0) = 0$ are proportional respectively to

$$K_{\text{nat}} = \begin{pmatrix} c & -c & & \\ -c & 1+c & -1 & \\ & -1 & 2 & \\ & & & \ddots \\ & & & & \ddots \end{pmatrix} \quad \text{or} \quad K_{\text{ess}} = \begin{pmatrix} 1+c & -1 & \\ -1 & 2 & \cdot \\ & \cdot & \cdot \end{pmatrix}$$

The largest eigenvalue grows with c, whereas the smallest is of the usual order N^{-2}, $N =$ order of the matrix. Therefore, the condition deteriorates as $c \to \infty$.

Scaling produces diagonal entries all equal to 2, and leaves the tail of the matrices unchanged:

$$DK_{\text{nat}}D = \begin{pmatrix} 2 & -2\sqrt{c/1+c} & & \\ -2\sqrt{c/1+c} & 2 & -\sqrt{2/1+c} & \\ & -\sqrt{2/1+c} & 2 & -1 \\ & & -1 & \cdot \end{pmatrix},$$

$$DK_{\text{ess}}D = \begin{pmatrix} 2 & -\sqrt{2/1+c} & \\ -\sqrt{2/1+c} & 2 & -1 \\ & -1 & \cdot \end{pmatrix}.$$

The largest eigenvalues are now bounded; the difficult question is always the size of $\lambda_{\min}$. In this case there is no problem with the second matrix; c has very little effect on $\lambda_{\min}$, which is again of order N^{-2}. The first matrix, however, begins with a block of order 2 which is nearly singular for large c, and the smallest eigenvalue for the whole matrix is necessarily below the eigenvalues for this block. Thus $\lambda_{min} \to 0$ as $c \to \infty$. We conclude that rescaling tends to be successful with an essential condition but not with a natural condition.

This is the numerical analogue of a physical situation which is well understood: A stiff system connected to earth by a soft spring is extremely unstable, whereas a firm connection (essential condition) is stable. It should be emphasized that while ill-conditioning from numerical sources is to be watched and degenerate elements are to be avoided, there will arise cases of *physical ill-conditioning* which cannot be altered—except perhaps by a change from stiffness method to force method, with stresses as unknowns. This situation occurs when there is a sharp change in the stiffness of the medium, or when Poisson's ratio approaches the limit $\nu = \frac{1}{2}$ of incompressibility [F18]. For shells there will be difficulties with large stiffness in the thickness direction, or with extremely thin shells. Roughly speaking, extensional modes can involve the ratio $(r/th)^2$ [20], whereas bending modes bring out the fourth-order (plate-like) aspect of the problem and the roundoff is proportional to h^{-4}.

To conclude this introduction, we must clarify the connection between the condition number of a matrix and its sensitivity to perturbations. The extreme case occurs when the given linear system $KQ = F$ has a load vector F which coincides with a unit eigenvector of K, in particular with the eigenvector V_{max} corresponding to λ_{max}. The solution is then $Q = V_{max}/\lambda_{max}$. Suppose this load vector is slightly perturbed by the eigenvector at the other extreme, so that $\tilde{F} = V_{max} + \epsilon V_{min}$. Then the solution becomes $\tilde{Q} = V_{max}/\lambda_{max} + \epsilon V_{min}/\lambda_{min}$. The relative change in the solution is therefore

$$\frac{|Q - \tilde{Q}|}{|Q|} \sim \epsilon \frac{\lambda_{max}}{\lambda_{min}} = \kappa\epsilon.$$

Thus a perturbation in F of order ϵ is amplified to a perturbation in Q of order $\kappa\epsilon$.

It is easy to see that this case is extreme, and that always

$$\frac{|\delta Q|}{|Q|} \leq \kappa \frac{|\delta F|}{|F|}.$$

Proof. $|F| = |KQ| \leq \lambda_{max}|Q|$, and $|\delta F| = |K \delta Q| \geq \lambda_{min}|\delta Q|$.

The consequences of this simple inequality, and a parallel one with $|\delta K|/|K|$ on the right side [20], are very far-reaching. If the condition number is $\kappa \sim 10^{-s}$, *as many as s significant digits may be lost during the solution of* $KQ = F$. If this is close to the number of digits carried by the computer, it may require double precision to protect the accuracy of the computed result.

It has been objected that the condition number might have nothing to do with the load vector in a particular problem. The constant κ may be, in fact it must be, at least partly pessimistic. Irons [12] has proposed several alternative numbers, which the computer can form as elimination proceeds, and which automatically take into account the scaling and the data f of the *given* problem. We accept that for on-the-spot decisions—terminating a specific calculation or changing to double precision—computable quantities of this kind are the best. However, our goal here is to find some a priori measure of sensitivity, and for this purpose the condition number is very satisfactory. The rule to which it leads, that the roundoff error will increase proportionately with h^{-2m}, is definitely obeyed in normal computations. We emphasize that *the dependence is not on the total number of elements in the domain; it is on the number of elements per side.* In other words, there is no significant dependence on the number of spatial dimensions.

5.2.   THE CONDITION NUMBER

Our goal is to estimate the ratio $\kappa = \lambda_N(K)/\lambda_1(K)$ between the maximum and minimum eigenvalues of the stiffness matrix.

THEOREM 5.1.

For each variational problem and each choice of finite element there is a constant c such that

$$(1) \qquad\qquad \kappa \leq c h_{\min}^{-2m}.$$

The constant depends inversely on the smallest eigenvalue λ_1 of the given continuous problem, and it increases if the geometry of the elements becomes degenerate.

We give two proofs, both correct but rather informal. The first applies only to the special case of a regular mesh, and illustrates how Fourier (or Toeplitz) analysis permits a quite precise computation of the eigenvalues and condition number. The second applies to finite elements of arbitrary shape.

To begin, suppose that the mesh is regular and that the coefficients in the given problem $Lu = f$ are constants. The element stiffness matrices k_i, and also the mass matrices m_i, will all be identical. This means that after assembly, K and M are essentially *Toeplitz matrices.* The entries of a Toeplitz matrix are constant along each diagonal: K_{ij} depends only on the difference $j - i$ between the column and row indices. (This corresponds in the continuous case to an integral operator $\int K(s - t) f(t)\, dt$, in other words to a *convolution*, in which the kernel depends only on $s - t$.) Although this property may be lost at boundaries, particularly with natural boundary conditions, we want to show how calculations with Toeplitz matrices are still useful and easy.

Suppose we compare two stiffness matrices, both arising from $-u'' = f$ with piecewise linear elements. The first is of order N, and it is constrained by $u(0) = u'(\pi) = 0$; it has been the basic example throughout the book. The second is formed with no boundaries whatsoever, extending the interval $[0, \pi]$ all the way to $(-\infty, \infty)$ by adding on more and more element matrices:

$$K = \frac{1}{h} \begin{pmatrix} 2 & -1 & & & \\ -1 & \cdot & \cdot & & \\ & \cdot & \cdot & \cdot & \\ & & \cdot & 2 & -1 \\ & & & -1 & 1 \end{pmatrix}, \qquad K_\infty = \frac{1}{h} \begin{pmatrix} \cdot & \cdot & \cdot & & \\ -1 & 2 & -1 & & \\ & -1 & 2 & -1 & \\ & & \cdot & \cdot & \cdot \end{pmatrix}.$$

Since K is formed by constraining K_∞—eliminating all elements outside $[0, \pi]$ and imposing the essential condition $Q_0 = 0$—the extreme eigenvalues of K are enclosed by those of K_∞:

$$\lambda_{\min}(K_\infty) \le \lambda_{\min}(K) \le \lambda_{\max}(K) \le \lambda_{\max}(K_\infty).$$

The same will be true for the mass matrices, and therefore linear independence of the basis can be tested first on an infinite interval.

It is simple to work with the Toeplitz matrix K_∞, which can also be described as a discrete convolution matrix. Its eigenvectors are pure exponentials, just like the eigenfunctions of any constant-coefficient differential equation. We try the vector whose components are $v_j = e^{ij\theta}$, and apply the matrix to it:

$$K_\infty v_j = \frac{1}{h}(-e^{-i\theta} + 2 - e^{i\theta})v_j.$$

Therefore, the eigenvalue is

$$\lambda(\theta) = \frac{1}{h}(-e^{-i\theta} + 2 - e^{i\theta}) = \frac{2(1 - \cos\theta)}{h}.$$

Since this number ranges between 0 and $4/h$, we conclude that

$$0 \le \lambda_{\min}(K) \le \lambda_{\max}(K) \le \frac{4}{h}.$$

In this case the result coincides with the conclusion of Gerschgorin's theorem (1.4): Every eigenvalue lies in the circle with center given by the main diagonal $K_{ii} = 2/h$ and radius given by the absolute row sums $\sum_{j \ne i}|K_{ij}| = 2/h$. In general, Gerschgorin will not be nearly so good: Even for the bilinear elements below, with off-diagonal entries all of the same sign, it is not precise.

The same technique applies to the mass matrices, which have $h/6$, $4h/6$, $h/6$ along each row. The eigenvalues in the infinite case are

$$\mu(\theta) = \frac{h}{6}e^{-i\theta} + \frac{4h}{6} + \frac{h}{6}e^{i\theta} = \frac{2h}{3} + \frac{h}{3}\cos\theta.$$

Since $\cos\theta$ ranges from -1 to 1, the eigenvalues of M_∞ lie between $h/3$ and h. Notice the way in which the mass matrices are better: We obtain not only the correct upper bound on the largest eigenvalue, but also a good *lower* bound on $\lambda_{\min}(M)$. The condition number of M is given very accurately by

$$\kappa(M) \le \kappa(M_\infty) = \frac{h}{h/3} = 3.$$

Since this is independent of h, the piecewise linear roof functions are linearly independent uniformly as $h \to 0$. The zero lower bound for $\lambda_{\min}(K)$ arose from the rigid-body motions still allowed in K_∞—a constant function satisfies $-u'' = 0$ on the whole line, corresponding to a discrete eigenvector of $(\dots 111 \dots)$ which is eliminated from the finite matrix K only by the essential boundary condition. Therefore, we shall eventually have to argue by way of the mass matrices to achieve a rigorous estimate for $\kappa(K)$.

First, to give a better understanding of K_∞, we study two more examples:

1. Bilinear elements on a square mesh, for $-\Delta u = f$. There is one unknown at every node, and a typical row of K (multiplied by $3h$) shows an 8 on the main diagonal and eight -1's on the adjacent diagonals. The equation $KQ = F$ is

$$8Q_{jk} - \sum Q_{j'k'} = 3hF_{jk},$$

where (j', k') represent the eight nearest neighbors of the point (j, k) on a square grid. Again the eigenvectors v of K_∞ are pure exponentials, but now there are two frequencies: the components are $v_{jk} = e^{i(j\theta + k\varphi)}$. Calculating the eigenvalues,

$$3hK_\infty v_{jk} = (8 - e^{i\theta} - e^{-i\theta} - e^{i\varphi} - e^{-i\varphi} - e^{i(\theta+\varphi)} - e^{-i(\theta+\varphi)}$$
$$- e^{i(\theta-\varphi)} - e^{-i(\theta-\varphi)})v_{jk}.$$

The extreme eigenvalues come from maximizing and minimizing the expression in parentheses. Since it is linear in $\cos\theta$ and $\cos\varphi$, the extrema must occur where these equal ± 1, and we find

$$\lambda_{\min}(K_\infty) = 0, \quad \lambda_{\max}(K_\infty) = \frac{12}{3h} \quad \text{(at } \theta = \pi, \; \varphi = 0).$$

Gerschgorin's argument would give $\lambda_{\max} \le 16/3h$.

2. Cubic elements in one dimension for $u^{(iv)} = f$. There are now two unknowns at every node, and $KQ = F$ is a coupled system of two difference equations. Correspondingly, K and K_∞ are *block Toeplitz matrices*: there is a 2×2 block appearing over and over on the main diagonal, flanked on one side by another such block and on the other side by its transpose. From Section 1.7,

$$K_\infty = \begin{pmatrix} \ddots & \cdot & \cdot & & \\ & B^T & A & B & \\ & & B^T & A & B \\ & & & \cdot & \cdot & \cdot \end{pmatrix},$$

$$A = \frac{1}{h^3}\begin{pmatrix} 24 & 0 \\ 0 & 8h^2 \end{pmatrix}, \qquad B = \frac{1}{h^3}\begin{pmatrix} -12 & 6h \\ -6h & 2h^2 \end{pmatrix}.$$

In analogy with the previous examples, the eigenvalues now come from the block $\lambda(\theta) = B^T e^{-i\theta} + A + B e^{i\theta}$. This is itself a 2×2 matrix; *its* eigenvalues, maximized over the whole range $-\pi \leq \theta \leq \pi$, yield $\lambda_{\max}(K_\infty)$. Again the extreme case must occur at $\theta = 0$ or $\theta = \pi$, since the Rayleigh quotient $x^T \lambda(\theta) x / x^T x$ is linear in $\cos \theta$. We compute

$$\lambda(0) = \frac{1}{h^3}\begin{pmatrix} 0 & 0 \\ 0 & 12h^2 \end{pmatrix}, \qquad \lambda(\pi) = \frac{1}{h^3}\begin{pmatrix} 48 & 0 \\ 0 & 4h^2 \end{pmatrix}.$$

Therefore, $\lambda_{\max}(K_\infty)$ is the larger of $12/h$ and $48/h^3$. The result can be influenced by altering the relative scaling of the two types of basis functions, corresponding to displacement and slope.

The extreme eigenvalues of any block Toeplitz matrix K_∞ can be found in the same way. If the blocks are of order M (the number of unknowns per mesh square), then $\lambda(\theta)$ will be a matrix of this order; θ represents $\theta_1, \ldots, \theta_n$ in an n-dimensional problem. Since each square is connected only to its nearest neighbors in the nodal finite element method, the matrix λ will be linear in $\cos \theta_1, \ldots, \cos \theta_n$. Therefore, $\lambda_{\max}(K_\infty)$ can be computed by trying all possible ± 1 combinations for these cosines and evaluating the largest eigenvalues of the 2^n resulting matrices λ. The condition number for a pure Toeplitz matrix can thus be computed exactly by working only with matrices of order M (which is less than the order of the element matrices and is given for common elements in the tables of Section 1.8). For the mass matrices (also denoted by M!) we obtain $\kappa(M) \leq \kappa(M_\infty)$, and this is a constant depending only on the element. *Finite elements are uniformly independent.*

For the stiffness matrix there is a good upper bound on $\lambda_{\max}$, but the lower bound is useless. At this point we have to abandon exact computations and return to inequalities. According to Rayleigh's principle, the smallest eigenvalue of K is

$$(2) \qquad \lambda_1(K) = \min \frac{x^T K x}{x^T x} \geq \min \frac{x^T K x}{x^T M x} \min \frac{x^T M x}{x^T x} = \lambda_1(M^{-1}K)\lambda_1(M).$$

With the appearance of $M^{-1}K$, it becomes easy to use variational arguments. The eigenproblem $KQ = \lambda MQ$ is exactly the finite element analogue of the continuous problem $Lu = \lambda u$, and we show in Chapter 6 that *the fundamental eigenvalue $\lambda_1(M^{-1}K)$ in the discrete problem always equals or exceeds the fundamental eigenvalue $\lambda_1(L)$ in the continuous problem.* This is a lower bound which is independent of h.

In the example $Lu = -u''$, with boundary conditions $u(0) = u'(\pi) = 0$, the lowest eigenvalue is $\lambda_1(L) = \frac{1}{4}$. Since we have already proved that $\lambda_1(M)$ $\geq \lambda_1(M_\infty) \geq h/3$, it follows that $\lambda_1(K) \geq h/12$. The exact value for the smallest eigenvalue of K is $\sin^2(h/4)/(h/4)$, and therefore about $h/4$. The estimate (2) misses by a factor of 3, essentially the condition number of M, because the true minimum of $x^T K x/x^T x$ occurs at a fundamental mode x whose components are all of one sign. This corresponds more closely to $\lambda_{\max}(M)$ than to the $\lambda_{\min}(M)$ which appears in (2).

The essential point, however, is that the exponent of h in the condition number is correctly determined:

$$\kappa(K) = \frac{\lambda_{\max}(K)}{\lambda_{\min}(K)} \leq \frac{4/h}{h/12} = \frac{48}{h^2}.$$

For an equation of order $2m$, in a regular domain, the same Toeplitz matrix approach [combined with (2)] yields a condition number of order h^{-2m}. The constant involves $\lambda_1(L)$, as it should; if the physical problem is poorly conditioned, that must be reflected in its finite element analogue.

Second proof of Theorem 5.1.

The remaining problem is to extend this h^{-2m} estimate of the condition number to the case of irregular elements. The Toeplitz structure falls apart, and the whole argument must be based on the individual element matrices. For this second proof we follow Fried [F14]. To find an upper bound for $\lambda_{\max}(K)$, we recall that the global stiffness matrix is assembled from the element matrices k_i by

$$(3) \qquad\qquad x^T K x = \sum_i x_i^T k_i x_i.$$

The vector x and matrix K are of order N; the vector x_i and matrix k_i are of order d_i, the number of degrees of freedom within the ith element. In fact, x_i is formed from x just by striking out all except the appropriate d_i components; we may think of multiplying x by an *incidence matrix*, made up of 0's and 1's, to obtain x_i.

If Λ is the maximum eigenvalue of all the element stiffness matrices, so that $x_i^T k_i x_i \leq \Lambda\, x_i^T x_i$ by Rayleigh's principle, we conclude from (3) that

$$x^T K x \leq \Lambda \sum_i x_i^T x_i.$$

Now suppose that q is the maximum number of elements which meet at any node. Then no component of x will appear in more than q of the abbreviated vectors x_i, and $\sum x_i^T x_i \leq q x^T x$. Therefore,

$$x^T K x \leq \Lambda q x^T x \qquad \text{and} \qquad \lambda_{\max}(K) \leq \Lambda q.$$

To find a lower bound on $\lambda_{\min}(M)$, we proceed in exactly the same way:

$$x^T M x = \sum x_i^T m_i x_i \geq \theta \sum x_i^T x_i \geq \theta r x^T x,$$

where θ is the smallest eigenvalue of any of the element mass matrices and r is the minimum number of elements which meet at a node. (Often $r = 1$ or $r = 2$ because of elements at the boundary; certainly $r = 1$ if there are nodes internal to the elements.) Rayleigh's principle again gives

$$\lambda_1(M) = \min \frac{x^T M x}{x^T x} \geq \theta r.$$

Therefore, according to (2), the condition number of the stiffness matrix is less than

(4) $$\kappa(K) \leq \frac{\lambda_{\max}(K)}{\lambda_1(L)\lambda_1(M)} \leq \frac{\Lambda q}{\lambda_1(L)\theta r}.$$

If the geometry of the elements does not degenerate, so that the basis is uniform in the sense of Chapter 3, a direct calculation yields the expected $\kappa(K) = O(h^{-2m})$. If there is a degeneracy, so that triangles become very thin or rectangles approach triangles, it is reflected in the parameters Λ and θ. Since these are eigenvalues of small matrices, the effects of this degeneracy can be rigorously estimated. Fried [F14] has computed this dependence on the geometry for several examples—the estimate in Theorem 5.1 is sometimes pessimistic—and gives also a lower bound on the condition number.

He has also given a nice estimate for irregular meshes in one dimension, using the fact that $a(u^h, u^h) \leq a(u, u)$; the discrete structure is always stiffer than the continuous one. Suppose we apply a point load at the jth node, $f = \delta(x - z_j)$. Then it was observed in Section 1.10 that the load vector F has only one nonzero component and that

$$a(u^h, u^h) = Q^T K Q = F^T K^{-1} F = (K^{-1})_{jj}.$$

Therefore, $(K^{-1})_{jj}$ is less than the true energy $a(u, u)$, and is bounded *independently of the distribution of nodes.* [We assume function values to be the unknowns, and a second-order problem $-(pu')' + qu = f$.] Since K^{-1} is positive definite and thus $(K^{-1})_{ij}^2 \leq (K^{-1})_{ii}(K^{-1})_{jj}$, the largest entry must occur on the main diagonal. Therefore, along each row of K^{-1}, the sum of the absolute values is bounded by cN; N is the order of the matrix. We may take N^{-1} to be the average mesh width $\bar{h}$. Along each row of K, the corresponding sum is bounded by $c/h_{\min}$, since h_i appears in the denominator of the ith element matrix. Therefore, the condition number, even in the stronger sense of absolute row sums—corresponding to pointwise rather than mean-square bounds—is less than $C/\bar{h}\, h_{\min}$. With a regular mesh we recover C/h^2.

The important conclusion is this: *The roundoff error does not depend strongly on the degree of the polynomial element.* It depends principally on h and on the order and fundamental eigenvalue of the continuous problem. Therefore, the way to achieve numerical accuracy in the face of roundoff is to increase the degree of the trial functions. The condition number for cubics is only slightly worse than for linear elements, so that the roundoff errors for a given h are comparable. The discretization error, however, is an order of magnitude smaller for cubics. Therefore, at the crossover point where roundoff prohibits any further improvement coming from a decrease in h, the cubic element is much more accurate. This applies especially to the computation of stresses, where differentiation (or differencing) of displacements introduces an extra factor h^{-1} into the numerical error. Even in second-order problems the roundoff becomes significant, and an increase in the degree of the trial functions is the best way out.

6 EIGENVALUE PROBLEMS

6.1. VARIATIONAL FORMULATION AND THE MINMAX PRINCIPLE

Eigenvalue problems—which we shall write as $Lu = \lambda u$, or more generally as $Lu = \lambda Bu$—arise in a tremendous variety of applications. We mention the buckling of columns and shells, the vibration of elastic bodies, and multigroup diffusion in nuclear reactors. Fortunately the Rayleigh–Ritz idea is as useful for these problems as for steady-state equations $Lu = f$. In fact, the idea had its beginning in Rayleigh's description of the fundamental frequency as the minimum value of the *Rayleigh quotient*. Therefore, the step which has been taken in the last 15 years was completely natural and inevitable: to apply the new finite element ideas to this long-established variational form of the eigenvalue problem.

From a practical point of view, this means that piecewise polynomial functions can be substituted directly as trial functions into the Rayleigh quotient. The evaluation of this quotient becomes exactly the problem which has already been discussed and which large-scale computer systems have been developed to carry out: the assembly of the stiffness and mass matrices K and M. The next step, however, leads to a different and more difficult computational problem in linear algebra: instead of a linear system $KQ = F$, there arises a discrete eigenvalue problem, $KQ = \lambda MQ$. Fortunately, it is now known how the properties of the two matrices—symmetry, sparseness, positive definiteness of M—can be used to speed up the numerical algorithm. We shall discuss several effective numerical processes in Section 6.4.

From a theoretical point of view, the basic steps in establishing error bounds depend once again on the approximation properties of the finite element subspace S^h. Therefore, the approximation theorems of Chapter 3 can be put directly to use. Mathematically, the new step which is required is to deduce, starting from these approximation theorems, satisfactory bounds for the error in the eigenvalues and the eigenfunctions. This step is therefore our main goal.

We shall precede the general theory by a study of some specific examples, in order to illustrate the behavior of finite element approximations. Then we link the approximation theory to the eigenvalue problem, aiming to find bounds on $\lambda_l - \lambda_l^h$ and $u_l - u_l^h$, the errors in the lth eigenvalue and eigenfunction, which are sharp enough to give the right dependence on l as well as on h.

We begin with the eigenvalue problem

$$(1) \qquad Lu = -\frac{d}{dx}\left(p(x)\frac{du}{dx}\right) + q(x)u = \lambda u, \qquad 0 < x < \pi.$$

This is the same operator L which was studied in the first chapter; its simplicity lies in the fact that it is only one-dimensional. We shall again illustrate the distinction between natural and essential boundary conditions by introducing one of each:

$$(2) \qquad u(0) = 0, \qquad u'(\pi) = 0.$$

It is known that such a Sturm–Liouville problem has an infinite sequence of real eigenvalues

$$\lambda_1 \leq \lambda_2 \leq \cdots \leq \lambda_j \leq \cdots \longrightarrow \infty,$$

and an associated complete set of orthonormal eigenfunctions:

$$(3) \qquad (u_j, u_k) = \int_0^\pi u_j(x)u_k(x)\,dx = \delta_{jk}.$$

It follows immediately that these eigenfunctions are orthogonal also in the *energy inner product*

$$(4) \qquad (Lu_j, u_k) = (\lambda_j u_j, u_k) = \lambda_j \delta_{jk},$$

or, with the usual integration by parts,

$$(5) \qquad a(u_j, u_k) = \int_0^\pi (pu_j'u_k' + qu_ju_k)\,dx = \begin{cases} \lambda_j & j = k, \\ 0 & j \neq k. \end{cases}$$

In the case of constant p and q, which will serve as a model problem, the eigenfunctions are sinusoidal:

$$u_j(x) = \sqrt{\frac{\pi}{2}} \sin(j - \tfrac{1}{2})x, \qquad \lambda_j = p(j - \tfrac{1}{2})^2 + q.$$

The first step in the Rayleigh–Ritz method is to rewrite $Lu = \lambda u$ as a variational problem. There are two possibilities, corresponding to the Ritz minimization and the Galerkin weak form of steady-state equations $Lu = f$. Both lead to the same result. The first is to introduce the *Rayleigh quotient*, which is defined by

$$(6) \qquad\qquad R(v) = \frac{a(v, v)}{(v, v)} = \frac{\int p(v')^2 + qv^2}{\int v^2}.$$

We claim that the stationary (or critical) points of this functional $R(v)$, which are the points where the gradient of R vanishes, are exactly the eigen-functions of the problem. To understand this, let the trial function v be replaced by its eigenfunction expansion $\sum \alpha_j u_j$, $\alpha_j = (v, u_j)$:

$$(7) \qquad\qquad R(v) = \frac{a(\sum \alpha_j u_j, \sum \alpha_k u_k)}{(\sum \alpha_j u_j, \sum \alpha_k u_k)} = \frac{\sum \lambda_j \alpha_j^2}{\sum \alpha_j^2},$$

using the orthogonality conditions (3) and (5). At a stationary point of $R(v)$, the derivative with respect to each α_k must be zero:

$$\frac{\partial R}{\partial \alpha_k} = \frac{2\alpha_k \sum \alpha_j^2 (\lambda_j - \lambda_k)}{(\sum \alpha_j^2)^2} = 0.$$

If v is one of the eigenfunctions u_j, this is certainly satisfied: Every $\alpha_k = 0$ with the exception of the term with $k = j$, and this remaining survivor is cancelled by the vanishing of $\lambda_j - \lambda_k$. There are no stationary points other than the eigenfunctions. (For a repeated eigenvalue $\lambda_j = \lambda_{j+1}$, all combinations $v = \alpha_j u_j + \alpha_{j+1} u_{j+1}$ are eigenfunctions, and there is a whole plane of stationary points.) The value of $R(v)$ at a stationary point $v = u_j$ is easy to determine: it is exactly the eigenvalue λ_j.

$$(8) \qquad\qquad R(u_j) = \frac{a(u_j, u_j)}{(u_j, u_j)} = \lambda_j.$$

It is useful to try to visualize the graph of $R(v)$. The numerator $a(v, v)$ corresponds to a convex surface, an "egg cup," exactly as for $I(v)$ in the first chapter; the only difference is that the linear term $-2(f, v)$ is now absent, so that the bottom of the egg cup lies at the origin. There are two ways to consider the effect of the denominator (v, v). Since it makes the quotient homogeneous, $R(\alpha v) = R(v)$, one possibility is to consider only unit vectors v. In other words, we can fix the denominator to equal 1, and look at the cross section of the egg cup which is cut out by the right circular cylinder $(v, v) = 1$. The lowest point on this cross section will correspond to the fundamental frequency λ_1.

Alternatively, we can fix the numerator at the value 1. This means that the egg cup is cut by a horizontal plane, lying one unit above the base. The cross section is the ellipse $a(v, v) = 1$, or more precisely the ellipsoid $\sum \lambda_j \alpha_j^2 = 1$ in infinitely many dimensions. Its major axis is in the direction of the first eigenfunction u_1, since it is in this direction that the ellipsoid is farthest from the axis. In other words, with the numerator of $R(v)$ fixed at 1, the Rayleigh quotient is minimized when the denominator is largest. If we then consider the cross section of the ellipse normal to this major axis, that is, we fix $\alpha_1 = 0$ and look in a space of one lower dimension, then the major axis of *this* ellipse has length $\sqrt{1/\lambda_2}$. [And any other cross section of the original ellipsoid will have a major axis whose length falls in between, corresponding to the minmax principle described in (13) below: The smallest eigenvalue λ_1' under any one given constraint satisfies $\lambda_1 \leq \lambda_1' \leq \lambda_2$.]

It appears that in four dimensions, the cross section of an egg cup is the shell of a whole egg.

We have not yet specified which functions v are to be admitted in the variational, or Rayleigh quotient, characterization of eigenvalues and eigenfunctions. This is exactly the same question which arose for the steady-state problem in Section 1.3. In that case $I(v)$ was finally minimized over all functions in $\mathcal{3C}^1$ satisfying the essential boundary conditions, that is, over $\mathcal{3C}_E^1$. This space arose naturally through a completion process, starting with smooth functions that satisfied *all* boundary conditions and then including any v which was a limit of such v_N in the sense that $a(v - v_N, v - v_N) \rightarrow 0$. Here we shall do exactly the same thing, since it still leaves the surface $R(v)$ unchanged; all stationary points remain so in the completion process. [And if $p(x)$ were discontinuous, this process of filling in the holes on the surface of $R(v)$ actually fills in the eigenfunctions; because they are not smooth, they were initially excluded.]

The resulting admissible space is then $\mathcal{3C}_E^1$, precisely as before. This means that all the same finite element subspaces may be used in the discrete approximation, the difference being that we now look for stationary points of $R(v)$ rather than the minimum of $I(v)$.

It was mentioned earlier that there is a second way to reformulate the eigenvalue problem. This is to put the equation $Lu = \lambda u$ into its *weak form*, or *Galerkin form*, which is produced in the following way: We multiply $Lu = \lambda u$ by a function v and integrate by parts to obtain

$$(9) \qquad \int_0^\pi (pu'v' + quv)\, dx = \lambda \int_0^\pi uv\, dx.$$

The eigenvalue problem is then to find a scalar λ and a function u in $\mathcal{3C}_E^1$ such that (9) holds for all v in $\mathcal{3C}_E^1$. The boundary conditions on u come out correctly, since equation (9) is actually equivalent to

$$\int Lu(x)v(x)\, dx - pu'v \Big|_0^\pi = \lambda \int uv\, dx.$$

At $x = 0$, the integrated term $pu'v$ automatically vanishes. Equality of the remaining terms, for all v in $\mathfrak{IC}_E^1$, forces both the natural condition $u' = 0$ to hold at $x = \pi$ and the differential equation $Lu = \lambda u$ to hold over the interval.

The equation (9) is a specific case of the standard weak form for the eigenvalue problem: *Find a scalar λ, and a function u in the admissible space V, such that*

$$(10) \qquad\qquad a(u, v) = \lambda(u, v) \qquad \text{for all } v \text{ in } V.$$

This resembles the condition $a(u, v) = (f, v)$ for the vanishing of the first variation in the steady-state problem of minimizing $I(v)$. Indeed, the equation represents exactly the *vanishing of the first variation of R at the stationary point u*:

$$R(u + \epsilon v) = \frac{a(u + \epsilon v, u + \epsilon v)}{(u + \epsilon v, u + \epsilon v)} = \frac{a(u, u) + 2\epsilon a(u, v) + \cdots}{(u, u) + 2\epsilon(u, v) + \cdots}$$

$$= R(u) + 2\epsilon \frac{a(u, v)(u, u) - a(u, u)(u, v)}{(u, u)^2} + \cdots$$

$$= R(u) + 2\epsilon \frac{a(u, v) - \lambda(u, v)}{(u, u)} + O(\epsilon^2).$$

Thus the weak formulation and the stationary point formulation are equivalent.

In the case of λ_1 and u_1—that is, for the fundamental frequency and its associated normal mode—the stationary point is actually a minimum:

$$(11) \qquad\qquad \lambda_1 = \min_{v \text{ in } \mathfrak{IC}_E^1} R(v).$$

This is obvious if v is expanded as $\sum \alpha_j u_j$, since then $R(v) = \sum \lambda_j \alpha_j^2 / \sum \alpha_j^2 \geq \lambda_1$.

It is valuable to describe also the higher eigenfunctions in terms of a minimization, since convergence to a minimum is so much simpler to analyze than convergence to a stationary point. One possibility in this direction to constrain v to be orthogonal to the first $l - 1$ eigenfunctions: $\alpha_1 = (v, u_1) = 0, \ldots, \alpha_{l-1} = (v, u_{l-1}) = 0$. Under these constraints the minimum of the Rayleigh quotient becomes λ_l:

$$(12) \qquad\qquad \lambda_l = \min_{v \perp E_{l-1}} R(v).$$

Here E_{l-1} is the space spanned by the eigenfunctions $u_1, \ldots, u_{l-1}$.

There is an alternative formula for λ_l which does not depend on knowing $u_1, \ldots, u_{l-1}$. It will be of fundamental importance in what follows, and was discovered by Poincaré, Courant, and Fischer; we refer to it as the

Minmax Principle: *If $R(v)$ is maximized over an l-dimensional subspace S_l, then the minimum possible value for this maximum is λ_l:*

$$(13) \qquad\qquad \lambda_l = \min_{S_l} \max_{v \text{ in } S_l} R(v).$$

The minimum is taken over all l-dimensional subspaces of $\mathcal{JC}_E^1$. For the special subspace $S_l = E_l$, the maximum of $R(v)$ is exactly λ_l.

To prove the minmax formula (13), we have to show that for any other choice of S_l,

$$(14) \qquad\qquad \max_{v \text{ in } S_l} R(v) \geq \lambda_l.$$

The argument depends on selecting v^* in S_l so as to be orthogonal to E_{l-1}, that is, so as to satisfy the $l - 1$ equations $(v, u_i) = 0$, $1 \leq i < l$. There exists such a v^*, since we are imposing only $l - 1$ homogeneous constraints on an l-parameter space. Now it follows from (12), since $v^* \perp E_{l-1}$, that $\lambda_l \leq R(v^*)$. In other words, (14) holds, and the minmax formula (13) is proved.

One useful consequence of this formula is a rough estimate for the eigenvalue λ_l, based on a comparison with the constant-coefficient case. It is clear that for any v,

$$\int p_{\max}(v')^2 + q_{\max}v^2 \geq \int p(x)(v')^2 + q(x)v^2 \geq \int p_{\min}(v')^2 + q_{\min}v^2.$$

Dividing by $\int v^2$, this becomes a comparison of Rayleigh quotients, with the variable-coefficient case sandwiched between two problems whose eigenvalues are known explicitly. By the minmax principle, each eigenvalue λ_l of the middle problem must lie between the two known eigenvalues,

$$(15) \qquad p_{\max}(l - \tfrac{1}{2})^2 + q_{\max} \geq \lambda_l \geq p_{\min}(l - \tfrac{1}{2})^2 + q_{\min}.$$

In particular, λ_l is of order l^2 as $l \to \infty$.

Finally, we come to the main object of this section, which is to establish the *Rayleigh–Ritz principle for approximating the eigenvalues*. It begins either with the weak form $a(u, v) = \lambda(u, v)$ or with the description of the eigenvalues as critical (stationary) points of $R(v) = a(v, v)/(v, v)$. In either case, the idea is to work only within a finite-dimensional subspace S^h of the full admissible space $\mathcal{JC}_E^1$. In this subspace, we look for a pair λ^h and u^h such that

$$(16) \qquad\qquad a(u^h, v^h) = \lambda^h(u^h, v^h) \qquad \text{for all } v^h \text{ in } S^h.$$

In other words,

$$\int p(x)(u^h)'(v^h)' + q(x)u^hv^h = \lambda^h \int u^hv^h.$$

Alternatively, *the approximate eigenvectors are the critical points of $R(v^h)$ over the space S^h.*

* To see that these methods lead to the same approximations, choose a basis $\varphi_1, \ldots, \varphi_N$ for S^h. Then any v^h in S^h can be expanded as

$$(17) \qquad\qquad v^h = \sum q_j\varphi_j,$$

where q_j are the generalized coordinates (the nodal parameters of v_h, if S^h is a finite element space). Substituting into the Rayleigh quotient,

$$(18) \qquad R(v^h) = \frac{a(v^h, v^h)}{(v^h, v^h)} = \frac{\sum\sum q_jq_k \int (p(x)\varphi_j'\varphi_k' + q(x)\varphi_j\varphi_k)\, dx}{\sum\sum q_jq_k \int \varphi_j\varphi_k\, dx}.$$

The integrals in the numerator and denominator are by now familiar: *They are the entries in the stiffness matrix K^h and the mass matrix M^h.* Thus the Rayleigh quotient, in terms of the vector $q = (q_1, \ldots, q_N)$, is exactly

$$(19) \qquad\qquad R(v^h) = \frac{q^T K^h q}{q^T M^h q}.$$

The critical points of this discrete quotient are the solutions to the matrix eigenvalue problem

$$(20) \qquad\qquad K^hQ^h = \lambda^h M^h Q^h.$$

Therefore, this is the eigenproblem which has to be assembled and solved. The eigenvalues λ_l^h are expected to approximate the continuous eigenvalues λ_l, at least for small values of l, and the eigenvectors Q_l^h lead to a corresponding approximate eigenfunction

$$(21) \qquad\qquad u_l^h = \sum_1^N (Q_l^h)_j\varphi_j.$$

Thus the components of the discrete eigenvectors, in the matrix problem $KQ = \lambda MQ$, yield the nodal values of the finite element eigenfunctions.

The weak form of the eigenproblem leads directly to the same result. Suppose that in equation (16) we take $v^h = \varphi_k$:

$$a(\sum Q_j^h\varphi_j, \varphi_k) = \lambda^h(\sum Q_j^h\varphi_j, \varphi_k).$$

This is simply the kth row of the matrix equation $K^hQ^h = \lambda^h M^h Q^h$.

If the basis functions φ_j are orthonormal, then the mass matrix M^h is the identity and the discrete problem is to find the eigenvalues of K^h. However, this orthogonality condition on the φ_j is incompatible with a more important property of finite elements, that φ_j should vanish over all elements not containing the corresponding node z_j. Therefore, we must either accept $M^h \neq I$ or do violence to Rayleigh's idea by "lumping" the masses. We prefer the former, since numerical algorithms for the general problem $KQ = \lambda MQ$ are now appearing which are comparable in efficiency to those for $KQ = \lambda Q$. We emphasize that in all cases the mass matrix is symmetric and positive definite; it is the Gram matrix for the linearly independent vectors $\varphi_1, \ldots, \varphi_N$.

The fundamental frequency λ_1 is often the quantity of greatest significance, and we especially hope that λ_1^h provides a good approximation. Notice that λ_1^h *always lies above* λ_1, $\lambda_1^h \geq \lambda_1$, since λ_1^h is the minimum value of $R(v)$ over the subspace S^h and λ_1 is the minimum over the whole admissible space $\mathcal{K}_E^1$. It is natural to expect that if the true eigenfunction u_1 can be well approximated by the subspace S^h, then λ_1^h is automatically close to λ_1; this will be a key result of the theory.

The minmax principle applies equally well to the discrete problem, with the same proof, so that the approximate eigenvalues can be characterized by

(22)
$$\lambda_l^h = \min_{S_l} \max_{v^h \text{ in } S_l} R(v^h),$$

Here S_l *ranges over all l-dimensional subspace of the subspace* S^h. Of course, the definition makes sense only for $l \leq N$, since if the dimension of S^h is N, there exist only N approximate eigenvalues. Comparing the minmax principles (22) and (13), it follows immediately that *every eigenvalue is approximated from above:*

(23)
$$\lambda_l^h \geq \lambda_l \qquad \text{for all } l.$$

Every space S_l which is allowed in the minimization (22) is also allowed in (13), and therefore the minimum λ_l in (13) is at least as small as λ_l^h.

The minmax principle extends verbatim to the case $Lu = \lambda Bu$ as long as B is positive definite; the Rayleigh quotient is then $R(v) = (Lv, v)/(Bv, v)$. The mass matrix in the discrete problem becomes $M_{jk} = (B\varphi_j, \varphi_k)$.

6.2. SOME ELEMENTARY EXAMPLES

In this section we shall treat some specific examples in order to isolate the patterns which occur in the general theory of eigenvalue approximation.

We shall concentrate on the constant-coefficient problem

$$-p\frac{d^2u}{dx^2} + qu = \lambda u, \qquad 0 < x < \pi,$$

with the boundary conditions $u(0) = 0$, $u'(\pi) = 0$.

The first trial space to consider is made up of piecewise linear functions, with equally spaced nodes $x_j = jh$. Using the standard roof functions φ_j as a basis, the key matrices are

$$M^h = \frac{h}{6}\begin{bmatrix} 4 & 1 & & & \\ 1 & 4 & 1 & & \\ & & & & \\ & & 1 & 4 & 1 \\ & & & 1 & 2 \end{bmatrix}, \qquad K_1^h = \frac{1}{h}\begin{bmatrix} 2 & -1 & & & \\ -1 & 2 & -1 & & \\ & & & & \\ & & -1 & 2 & -1 \\ & & & -1 & 1 \end{bmatrix}$$

The stiffness matrix can be written as $K^h = pK_1^h + qM^h$. The optimal weights for the Rayleigh–Ritz eigenfunction

$$u^h(x) = \sum_{j=1}^{N} Q_j \varphi_j(x)$$

are the values of u^h at the nodes, and, in fact, the system $K^h Q^h = \lambda^h M^h Q^h$ is precisely the difference equation

(24)
$$\frac{p}{h}(2Q_j^h - Q_{j+1}^h - Q_{j-1}^h) + \frac{qh}{6}(4Q_j^h + Q_{j+1}^h + Q_{j-1}^h)$$
$$= \frac{\lambda^h h}{6}(4Q_j^h + Q_{j+1}^h + Q_{j-1}^h), \qquad 1 \le j \le N.$$

For this to be valid at the boundary point $j = N$ we put $Q_{N+1}^h = Q_{N-1}^h$. Recall that the Dirichlet condition $u(0) = 0$ gives $Q_0 = 0$.

The eigensystem (24) has a Toeplitz structure away from the boundaries, in the sense that the (i, j) entries of the stiffness and mass matrices depend only on the difference $i - j$. It is therefore reasonable to expect trigonometric eigenvectors, and indeed the components of the lth eigenvector are

$$(Q_l^h)_j = \sqrt{\frac{\pi}{2}} \sin((l - \tfrac{1}{2})jh).$$

To write out an expression for the associated eigenvalue λ_l^h, put

$$k_h(l) = 2h^{-2}(1 - \cos(l - \tfrac{1}{2})h), \qquad m_h(l) = \frac{2 + \cos(l - \tfrac{1}{2})h}{3}.$$

Then

(25)
$$\lambda_l^h = \frac{pk_h(l) + qm_h(l)}{m_h(l)}.$$

In this special problem the eigenfunctions are infinitely accurate at the nodes,

(26)
$$u_l^h(jh) = u_l(jh) = \sqrt{\frac{\pi}{2}} \sin((l - \tfrac{1}{2})jh).$$

It follows that u_l^h is equal to the interpolate of u_l, and the approximation theorems give

(27)
$$\|u_l - u_l^h\|_s \leq Ch^{2-s} \|u_l\|_2 = O(l^2 h^{2-s}).$$

Although the exactness of the Ritz eigenfunction u_l^h at the nodes is special, the estimate (27) is typical of the general case. *The errors in the Ritz eigenfunctions $u_l - u_l^h$ are of the same order as the errors in the approximate solution to steady-state problems $Lu = f$.*

Turning now to the approximate eigenvalue λ_l^h given by (25), we expand

$$k_h(l) = (l - \tfrac{1}{2})^2 - \frac{h^2}{12}(l - \tfrac{1}{2})^4 + O(l^6 h^4),$$

$$m_h(l) = 1 - \frac{h^2}{6}(l - \tfrac{1}{2})^2 + O(l^4 h^4).$$

Recalling that $\lambda_l = p(l - \tfrac{1}{2})^2 + q$, the eigenvalue error is

(28)
$$\lambda_l^h = \lambda_l + h^2 \frac{p}{12}(l - \tfrac{1}{2})^4 + O(l^6 h^4) = \lambda_l + O(h^2 l^4).$$

Observe that the eigenvalues are as accurate as the *energies* in the eigenfunctions. This is true in general:

$$\lambda_l^h - \lambda_l \leq C \|u_l^h - u_l\|_m^2$$

for $2m$th-order problems. The reason is that the Rayleigh quotient is flat near a critical point. Moderately accurate trial functions yield very accurate eigenvalues.

The eigenvalues are actually independent of the constant q, since the addition of the term qu to the operator simply *translates the spectrum* by a constant amount: $\lambda_l = p(l - \tfrac{1}{2})^2$ when $q = 0$ and $\lambda_l = p(l - \tfrac{1}{2})^2 + q$ in general. The crucial point is that it has the same effect on the Ritz–Galerkin approximation λ_l^h. It will be convenient to exploit this invariance of the error $\lambda_l^h - \lambda_l$, by adding a sufficiently large constant term to ensure throughout this chapter that $\lambda_1 > 0$.

In a *lumping process* the mass matrix M^h is replaced by a diagonal matrix, in this simple case by the identity I. In our example it happens that the eigenfunctions remain unchanged. The eigenvalues, on the other hand, are altered to

$$(29) \qquad \tilde{\lambda}_l^h = pk_h(l) + q = \lambda_l - \frac{ph^2}{12}(l - \tfrac{1}{2})^4 + O(l^6 h^4).$$

Thus $\tilde{\lambda}_l^h$ is a *lower bound* to λ_l and has the same $O(h^2)$ accuracy as λ_l^h.

Lumping has a very seductive physical interpretation in terms of the stiffness of the system $K^h Q^h = \lambda^h M^h Q^h$. From this point of view, the replacement of the mass matrix M^h with the identity matrix I has the effect of making the structure "softer" and hence reducing the magnitude of the approximate eigenvalues. Since the Rayleigh–Ritz approximations are necessarily upper bounds, $\lambda_l^h \geq \lambda_l$, the hope is that this reduction in magnitude will increase the accuracy of the approximation. Our concern, on the other hand, is that this violation of the Ritz rules may make the structure too soft and thereby adversely affect the accuracy. In the present case the damage has not been great, but a more typical example at the end of this section shows that a far more serious loss of accuracy is possible.

We turn to an example of a higher-degree piecewise polynomial approximation, namely the cubic Hermite space S^h on a uniform mesh. There are basis functions ψ_j corresponding to function values, and ω_j to slopes, so that†

$$v^h(x) = \sum_{j=1}^{N} u_j \psi_j(x) + \sum_{j=0}^{N-1} u_j' \omega_j(x).$$

The matrix eigenvalue problem for the optimal weights, with $p = 1$, $q = 0$, becomes

$$(30) \qquad \begin{aligned} &\frac{6}{5h}(2u_j - u_{j+1} - u_{j-1}) + \frac{1}{10}(u_{j+1}' - u_{j-1}') \\ &= \lambda^h\left[\frac{h}{70}(52u_j + 9u_{j+1} + 9u_{j-1}) - \frac{13h^2}{420}(u_{j+1}' - u_{j-1}')\right] \end{aligned}$$

$$(31) \qquad \begin{aligned} &\frac{h}{30}(8u_j' - u_{j+1}' - u_{j-1}') - \frac{1}{10}(u_{j+1} - u_{j-1}) \\ &= \lambda^h\left[\frac{h^3}{420}(8u_j' - 3u_{j+1}' - 3u_{j-1}') + \frac{13h^2}{420}(u_{j+1} - u_{j-1})\right]. \end{aligned}$$

Strictly speaking (30) and (31) hold only for $1 \leq j \leq N - 1$, but they remain

---

†For simplicity we shall require that all functions v^h in S^h satisfy the natural boundary condition at π as well as the essential boundary condition at 0.

valid at the boundaries if we put

$$u_0 = u_{-1} + u_1 = u_1' - u_{-1}' = u_N' = u_{N+1}' + u_{N-1}' = u_{N+1} - u_{N-1} = 0.$$

[Observe that (30) reduces to $0 = 0$ at $j = 0$, and (31) reduces to $0 = 0$ at $j = N$.]

As in the piecewise linear case, the approximate eigenfunctions for this simple problem happen to agree at the nodes with trigonometric polynomials:

$$(32) \qquad u_j = \sqrt{\frac{\pi}{2}} \sin((l - \tfrac{1}{2})jh), \qquad u_j' = \sqrt{\frac{\pi}{2}} (l - \tfrac{1}{2})\alpha_l \cos((l - \tfrac{1}{2})jh).$$

Indeed, with $v_l = l - \tfrac{1}{2}$, the substitution of (32) into (30)–(31) produces a 2×2 eigenvalue problem

$$(33) \quad
\begin{bmatrix}
\dfrac{12}{5h}(1 - \cos v_l h) & -\dfrac{\sin v_l h}{5} \\[2ex]
-\dfrac{\sin v_l h}{5} & \dfrac{h}{15}(4 - \cos v_l h)
\end{bmatrix}
\begin{bmatrix} 1 \\[1ex] v_l \alpha_l \end{bmatrix}
$$
$$
= \lambda_l^h
\begin{bmatrix}
\dfrac{h}{70}(52 + 18 \cos v_l h) & \dfrac{13}{210} h^2 \sin v_l h \\[2ex]
\dfrac{13}{210} h^2 \sin v_l h & h^3 \left(\dfrac{2}{105} - \dfrac{\cos v_l h}{70} \right)
\end{bmatrix}
\begin{bmatrix} 1 \\[1ex] v_l \alpha_l \end{bmatrix}.
$$

Observe that (33) gives two eigenvalues, $\lambda_{l,0}^h$ and $\lambda_{l,1}^h$, for each integer l, $l < N$. Since the system (30)–(31) is of order $2N - 2$, these are necessarily all the finite element eigenvalues. By a direct calculation,

$$(34) \qquad\qquad \lambda_{l,0}^h = \lambda_l + O(l^8 h^6).$$

The corresponding eigenfunctions are in error by $l^4 h^{4-s}$ in the sth derivative.

The second eigenvalues $\lambda_{l,1}^h$ are of order $O(h^{-2})$, and *they are not close to any eigenvalue of the differential equation*. At first notice this seems to be a serious disadvantage of using cubic Hermite elements in eigenvalue calculations, that *at least half of the eigenvalues are spurious and completely useless as approximations*. However, this phenomenon is quite typical of finite element approximations. A closer look at even the piecewise linear approximations shows that for each h there is an integer l_h such that only the first l_h eigenvalues are reasonable; moreover, $hl_h \to 0$ as $h \to 0$. This has important implications for the choice of a method to use in computing the eigenvalues of $KQ = \lambda MQ$; *only the leading eigenvalues are wanted*.

We close this section by noting that "lumping" can lead to a serious loss of accuracy [T8]. For example, a typical approach would be to replace

$$\frac{1}{70}(52u_j + 9u_{j+1} + 9u_{j-1}), \qquad \frac{1}{10}(u'_{j+1} - u'_{j-1})$$

in (30) with u_j and 0, respectively, and to replace

$$\frac{1}{420}(8u'_j - 3u'_{j+1} - 3u'_{j-1}), \qquad \frac{13}{420}(u_{j+1} - u_{j-1})$$

in (31) with $u'_j/210$ and 0. This leads to a problem $\tilde{K}\tilde{Q} = \tilde{\lambda}\tilde{Q}$, which appears simpler because M is replaced by the identity. However, a direct calculation shows that the $O(h^6)$ eigenvalue error is increased to $O(h^2)$.

6.3. EIGENVALUE AND EIGENFUNCTION ERRORS

In this section we give a general theory of Rayleigh–Ritz approximations to elliptic eigenvalue problems $Lu = \lambda u$. The setting is by now familiar: Integration by parts converts (Lv, v) into a symmetric form $a(v, v)$, which is defined for all v in the admissible space $\mathfrak{H}^m_E$. The eigenfunctions are the points u_l at which the Rayleigh quotient $R(v) = a(v, v)/(v, v)$ is stationary, and the corresponding eigenvalues are $\lambda_l = R(u_l)$. These eigenfunctions are orthogonal and, since L is symmetric, the eigenvalues are real.

On the subspace S^h, the Rayleigh quotient becomes

$$R(v^h) = \frac{a(v^h, v^h)}{(v^h, v^h)} = \frac{q^T K q}{q^T M q},$$

and the stationary points Q yield the approximations u^h_l and λ^h_l. These points are determined by the matrix eigenvalue problem $KQ = \lambda^h MQ$, and any two eigenvectors Q_i and Q_l satisfy the standard orthogonality relations

$$(35) \qquad\qquad Q_i^T M Q_l = \delta_{il}, \qquad Q_i^T K Q_l = \lambda^h_l \delta_{il}.$$

Translated in terms of the eigenfunctions $u^h_i = \sum(Q_i)_j \varphi_j$ and $u^h_l = \sum(Q_l)_j \varphi_j$, this means that

$$(36) \qquad\qquad (u^h_i, u^h_l) = \delta_{il}, \qquad a(u^h_i, u^h_l) = \lambda^h_l \delta_{il}.$$

Thus the approximations mirror the basic properties of the true solutions. They share also the minmax principle:

$$(37) \qquad\qquad \lambda^h_l = \min_{S_l \subset S^h} \max_{v^h \text{ in } S_l} R(v^h),$$

where S_l denotes any subspace of dimension l.

Now we begin on the eigenvalue estimates. Let P be the Rayleigh–Ritz projection, defined as follows: *If u is in $\mathcal{K}_E^m$, then Pu is its component in the subspace S^h* (with respect to the energy inner product):

$$(38) \qquad a(u - Pu, v^h) = 0 \qquad \text{for all } v^h \text{ in } S^h.$$

This means that in the energy norm $a(v, v)$, Pu is the closest function in S^h to the given u. In other words, if u were the solution to a steady-state problem $Lu = f$, Pu would be exactly its Ritz approximation u^h. This link to our previous work on approximation (Theorem 3.7) guarantees that

$$(39) \qquad \|u - Pu\|_s \leq C[h^{k-s} + h^{2(k-m)}] \|u\|_k.$$

Our strategy for estimating $\lambda_l^h - \lambda_l$ is this: We let E_l denote the subspace spanned by the true eigenfunctions $u_1, \ldots, u_l$, and choose $S_l = PE_l$ as the subspace of S^h in which to use the minmax principle. Thus S_l is spanned by the trial functions $Pu_1, \ldots, Pu_l$. These are not identical with the approximate eigenfunctions $u_1^h, \ldots, u_l^h$, but the essential point of the proof is that they are close.

LEMMA 6.1

Let e_l be the set of unit vectors in E_l, and let

$$(40) \qquad \sigma_l^h = \max_{u \text{ in } e_l} |2(u, u - Pu) - (u - Pu, u - Pu)|.$$

Then provided that $\sigma_l^h < 1$, the approximate eigenvalues are bounded above by

$$(41) \qquad \lambda_l^h \leq \frac{\lambda_l}{1 - \sigma_l^h}.$$

Proof. To apply the minmax principle we must make sure that $S_l = PE_l$ is l-dimensional. Certainly E_l itself is l-dimensional, and therefore the question is whether possibly $Pu^* = 0$ for some nonzero u^* in E_l. Normalizing u^* to be a unit vector, and therefore in e_l, $Pu^* = 0$ would imply that

$$\sigma_l^h \geq |2(u^*, u^* - Pu^*) - (u^* - Pu^*, u - Pu^*)| = |(u^*, u^*)| = 1.$$

Since this contradicts $\sigma_l^h < 1$, S_l must be l-dimensional.

Now from the minmax principle (37),

$$\lambda_l^h \leq \max_{v^h \text{ in } S_l} R(v^h) = \max_{u \text{ in } e_l} \frac{a(Pu, Pu)}{(Pu, Pu)}.$$

The numerator is bounded above by $a(Pu, Pu) \leq a(u, u)$, since P is a projection in the energy norm.† The denominator is bounded below by

$$(Pu, Pu) = (u, u) - 2(u, u - Pu) + (u - Pu, u - Pu) \geq 1 - \sigma_l^h.$$

Therefore, the lemma is proved:

$$\lambda_l^h \leq \max_{u \text{ in } e_l} \frac{a(u, u)}{1 - \sigma_l^h} = \frac{\lambda_l}{1 - \sigma_l^h}.$$

The problem is now to estimate σ_l^h, and for this we need the following identity.

LEMMA 6.2

If $u = \sum_1^l c_i u_i$ is in e_l, then

(42) $(u, u - Pu) = \sum c_i \lambda_i^{-1} a(u_i - Pu_i, u - Pu).$

Proof. Since u_i is a true eigenfunction,

$$(u_i, u - Pu) = \lambda_i^{-1} a(u_i, u - Pu),$$

by (10). Furthermore, $a(Pu_i, u - Pu) = 0$, if we take $v^h = Pu_i$ in the definition (38) of the projection P. Subtracting,

$$(u_i, u - Pu) = \lambda_i^{-1} a(u_i - Pu_i, u - Pu).$$

Multiplying by c_i and summing on i, the result is (42).

Now we have the basic error estimate for eigenvalues.

THEOREM 6.1

If S^h is a finite element space of degree $k - 1$, then there is a constant δ such that the approximate eigenvalues are bounded for small h by

(43) $\lambda_l \leq \lambda_l^h \leq \lambda_l + 2\delta h^{2(k-m)} \lambda_l^{k/m}.$

This is in agreement with the explicit bounds (28) and (34) for linear and cubic elements in one dimension.

---

†This is in the corollary to Theorem 1.1 and is easy to see directly:
$$a(v, v) = a(Pv, Pv) - 2a(v - Pv, Pv) + a(v - Pv, v - Pv).$$
The last term is nonnegative, and the next-to-last term is zero by the definition (38) of the projection P.

Proof. We want to estimate σ_l^h. Its first term is

$$2|(u, u - Pu)| = 2\left|\sum_1^l c_i \lambda_i^{-1} a(u_i - Pu_i, u - Pu)\right|$$

$$\leq 2K \|(I - P)\sum c_i \lambda_i^{-1} u_i\|_m \|(I - P)u\|_m$$

since $|a(v, w)| \leq K\|v\|_m\|w\|_m$. Applying the approximation theorem,

$$2|(u, u - Pu)| \leq 2KC^2 h^{2(k-m)}\left\|\sum_1^l c_i \lambda_i^{-1} u_i\right\|_k \|u\|_k$$

(44)

$$\leq C'h^{2(k-m)}\left\|\sum_1^l c_i \lambda_i^{(k/2m)-1} u_i\right\|_0 \left\|\sum_1^l c_i \lambda_i^{k/2m} u_i\right\|_0$$

$$\leq C'h^{2(k-m)}\lambda_l^{(k/m)-1}.$$

This holds for all functions $u = \sum c_i u_i$ in e_l, since $\sum c_i^2 = 1$; the extreme case occurs at $c_l = 1$, because it is multiplied by a power of the largest eigenvalue λ_l. The next-to-last step was more delicate and required the inequality $\|v\|_k \leq c\|L^{k/2m}v\|_0$. For $k = m$ this is exactly the ellipticity condition $\|v\|_m^2 \leq c^2 a(v, v)$, and the inequality holds for all k if the coefficients of the differential operator L are smooth.

The other term in σ_l^h is of higher order in h, since the approximation theorem with $s = 0$ yields

(45) $(u - Pu, u - Pu) \leq C^2[h^k + h^{2(k-m)}]^2 \|u\|_k^2.$

Therefore, if we increase the constant C' in (44) to a larger constant δ, that expression will exceed σ_l^h for all small h.

The theorem is now proved; if h is small enough to ensure that $\sigma_l^h \leq \frac{1}{2}$, then

$$\lambda_l^h \leq \lambda_l(1 - \sigma_l^h)^{-1} \leq \lambda_l(1 + 2\sigma_l^h) \leq \lambda_l + 2\delta h^{2(k-m)}\lambda_l^{k/m}.$$

Such error bounds for the Rayleigh–Ritz method have a long history, particularly in the Russian literature. Vainikko [7] has given a very complete theory, including eigenvalue and eigenfunction estimates even in the non-self-adjoint case. Combined with the approximation theorems for finite elements, his analysis establishes the same error bounds which we have proved for self-adjoint problems by means of the minmax principle. This principle must be replaced by a contour integration in the complex plane, followed by an application of the Galerkin estimates for static problems given in Section 2.3. The integral of $(L - zI)^{-1}$ around a true eigenvalue λ_l produces exactly $2\pi i u_l^T u_l$; Bramble and Osborn [B29], and Babuska and Fix, have computed the error in using finite elements instead.

The minmax argument, which we have preferred for self-adjoint problems because it is more elementary, has been applied to finite elements by a number of authors; the basic reference is [B16]. We have made the argument more precise, in order to determine not only the power $h^{2(k-m)}$ but also the correct dependence on l:

$$\lambda_l^h - \lambda_l \sim ch^{2(k-m)}\lambda_l^{k/m}.$$

The last factor means that *higher eigenvalues are progressively more difficult to compute*. This prediction has been tested experimentally, and the appearance of $\lambda_l^{k/m}$ is confirmed by the calculations described in [C14] with a quintic element, those in Section 8.4 for the eigenvalues of an L-shaped membrane, and the following numerical results which are reproduced from [L3]. The bicubic Hermite element was applied to a square plate which is simply supported either on all sides (SSSS) or on only two sides, with the other two left free (SFSF); the number of elements is fixed at 10 per side (Fig. 6.1). Since $k = 4$ and $m = 2$, the prediction is a relative error of

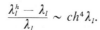

$$\frac{\lambda_l^h - \lambda_l}{\lambda_l} \sim ch^4\lambda_l.$$

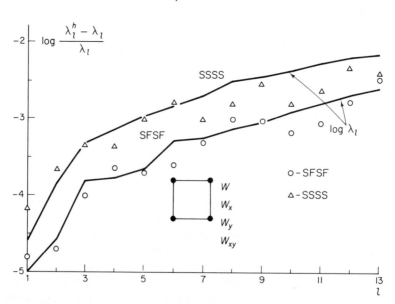

Fig. 6.1 Errors in the higher eigenvalues of a square plate.

The generalized eigenvalue problem $Lu = \lambda Bu$ yields to exactly the same analysis. Suppose that B is a symmetric operator of order $2m' < 2m$, corresponding to the quadratic form $(Bv, v) = b(v, v)$. Then the true Rayleigh quotient is $R(v) = a(v, v)/b(v, v)$, and the discrete one becomes $q^T K q/q^T M_b q$, where the mass matrix M_b is formed with respect to the new inner product: $(M_b)_{jk} = b(\varphi_j, \varphi_k)$. One change occurs in the last term of σ_l^h, where the bound (45) in the 0-norm is replaced by

$$b(u - Pu, u - Pu) \leq C^2 [h^{k-m'} + h^{2(k-m)}]^2 \|u\|_k^2.$$

However, the first term in σ_l^h remains larger, $2b(u, u - Pu) \sim h^{2(k-m)}$, and the final estimate in the theorem is unchanged.

The effect on the computed eigenvalues of other approximations in the finite element technique—change of domain or coefficients, numerical quadrature, nonconformity of elements—is exactly comparable to the effect on energy in static problems. We caution only that, for a change from Ω to a polygonal domain Ω^h, the error in energy must be measured over Ω. Therefore, this error is not the $O(h^3)$ established in Section 4.4 over the polygon; if all trial functions vanish over $\Omega - \Omega^h$, then the energy in this region is completely lost, with an effect which is proportional to its area, of order h^2.

We turn now to the eigenfunctions. We should like to prove that their errors are of the same magnitude as those in steady-state problems $Lu = f$. It must be expected, of course, that there will be some dependence on l; the accuracy will deteriorate for higher eigenfunctions. Again the arguments involve two simple identities, stated in the following lemmas, and a rather technical computation.

LEMMA 6.3

With the normalizations $(u_l, u_l) = (u_l^h, u_l^h) = 1$, we have

$$(46) \qquad a(u_l - u_l^h, u_l - u_l^h) = \lambda_l \|u_l - u_l^h\|_0^2 + \lambda_l^h - \lambda_l.$$

Proof. Observe that

$$a(u_l - u_l^h, u_l - u_l^h) = a(u_l, u_l) - 2a(u_l, u_l^h) + a(u_l^h, u_l^h)$$
$$= \lambda_l - 2\lambda_l(u_l, u_l^h) + \lambda_l^h$$
$$= \lambda_l[2 - 2(u_l, u_l^h)] + \lambda_l^h - \lambda_l.$$

But the quantity in brackets is exactly

$$2 - 2(u_l, u_l^h) = (u_l, u_l) - 2(u_l, u_l^h) + (u_l^h, u_l^h) = \|u_l - u_l^h\|_0^2.$$

With this identity, and the estimate already established for $\lambda_l^h - \lambda_l$, it is necessary only to bound the error in the norm $\|u_l - u_l^h\|_0$; then the error in energy will be an immediate consequence.

LEMMA 6.4

For all j and l,

$$(47) \qquad (\lambda_j^h - \lambda_l)(Pu_l, u_j^h) = \lambda_l(u_l - Pu_l, u_j^h).$$

Proof. Since the term $-\lambda_l(Pu_l, u_j^h)$ appears on both sides, we have only to show that

$$\lambda_j^h(Pu_l, u_j^h) = \lambda_l(u_l, u_j^h).$$

Because u_j^h and u_l are eigenfunctions, these two sides can be rewritten as $a(Pu_l, u_j^h)$ and $a(u_l, u_j^h)$, respectively. Now equality comes from the definition (38) of the projection P.

The expression (47) resembles a truncation error in the eigenvalue equation; it coincides with $a(Pu_l, u_j^h) - \lambda_l(Pu_l, u_j^h)$.

The set $u_1^h, \ldots, u_N^h$ forms an orthogonal basis for S^h, and in particular

$$(48) \qquad Pu_l = \sum_{j=1}^{N} (Pu_l, u_j^h)u_j^h.$$

The identities (47) and (48) can be interpreted in the following way. First, from (47) it is seen that the coefficient (Pu_l, u_j^h) is small if λ_j^h is not close to λ_l. Then from (48) it follows that Pu_l is close to u_l^h. This will be our strategy for estimating $u_l^h - Pu_l$ (and hence $u_l^h - u_l$), but to make the process rigorous it is convenient to consider separately the cases of distinct and repeated eigenvalues.

If λ_l is distinct from the other eigenvalues, then according to our eigenvalue bounds (43) there is a separation constant ρ such that for small h,

$$(49) \qquad \frac{\lambda_l}{|\lambda_j^h - \lambda_l|} \le \rho \qquad \text{for all } j.$$

Now come the computations. Writing β for the key coefficient (Pu_l, u_l^h) in (48), the size of the other terms is given by

$$
\begin{aligned}
(50) \qquad \|Pu_l - \beta u_l^h\|_0^2 &= \sum_{j \ne l} (Pu_l, u_j^h)^2 \\
&= \sum_{j \ne l} \left(\frac{\lambda_l}{\lambda_j^h - \lambda_l}\right)^2 (u_l - Pu_l, u_j^h)^2 \\
&\le \rho^2 \sum_{j \ne l} (u_l - Pu_l, u_j^h)^2 \\
&\le \rho^2 \|u_l - Pu_l\|_0^2,
\end{aligned}
$$

since the square of the norm is the sum of the squares of the components. Thus we have the crucial estimate

$$(51) \quad \|u_l - \beta u_l^h\|_0 \le \|u_l - Pu_l\|_0 + \|Pu_l - \beta u_l^h\|_0 \le (1 + \rho)\|u_l - Pu_l\|_0.$$

The eigenfunction error is now bounded in terms of the Ritz approximation error $u_l - Pu_l$; it follows from (39) that

(52)
$$\| u_l - \beta u_l^h \|_0 \leq C'(1 + p)[h^k + h^{2(k-m)}] \| u_l \|_k$$
$$\leq C''[h^k + h^{2(k-m)}] \lambda_l^{k/2m}.$$

This is the essential part of our theorem.

THEOREM 6.2

If S^h is a finite element space of degree $k - 1$ and λ_l is a distinct eigenvalue, then for small h,

(53)
$$\| u_l - u_l^h \|_0 \leq c[h^k + h^{2(k-m)}] \lambda_l^{k/2m},$$

(54)
$$a(u_l - u_l^h, u_l - u_l^h) \leq c' h^{2(k-m)} \lambda_l^{k/m}.$$

If λ_l is a repeated eigenvalue, then the orthonormal eigenfunctions u_j can be chosen so that these estimates still hold. The estimates are the best possible and are consistent with the special case (27) of linear elements.

Proof. The bound (53) is virtually the same as (52), which is already proved. It remains only to show that the factor β is close to 1, and this comes from fiddling about with the triangle inequality:

$$\| u_l \|_0 - \| u_l - \beta u_l^h \|_0 \leq \| \beta u_l^h \|_0 \leq \| u_l \|_0 + \| u_l - \beta u_l^h \|_0.$$

Recalling that u_l and u_l^h are unit vectors, and choosing their sign so that $\beta \geq 0$, this is the same as $| \beta - 1 | \leq \| u_l - \beta u_l^h \|_0$. Therefore,

$$\| u_l - u_l^h \|_0 \leq \| u_l - \beta u_l^h \|_0 + \| (\beta - 1) u_l^h \|_0 \leq 2 \| u_l - \beta u_l^h \|_0.$$

Now the right side is given by (52), and the first statement of the theorem is proved: $c = 2C''$. The error in energy (54) follows immediately from Lemma 6.3. This is the simplest argument we have seen for eigenfunctions.

The case of a repeated eigenvalue $\lambda_l = \lambda_{l+1} = \cdots = \lambda_{l+R}$ is more awkward but not essentially different. There is still a separation constant p between these eigenvalues and the approximations λ_j^h to the others, as in (49). The constant β becomes a matrix of order $R + 1$,

$$\beta_{ri} = (Pu_{l+r}, u_{l+i}^h), \qquad 0 \leq i, r \leq R.$$

The long computation (50) is now carried out with all of the terms $j = l, \ldots, l + R$ on the left side rather than the right side and leads to

$$\left\| \sum_0^R \beta_{ri} u_{l+i}^h - u_{l+r} \right\|_0 \leq C[h^k + h^{2(k-m)}] \lambda_l^{k/2m}.$$

Inverting β, this means that new eigenfunctions U_{l+r}—linear combinations of the old ones—can be chosen so that

$$\| u^h_{l+r} - U_{l+r} \|_0 \leq C[h^k + h^{2(k-m)}]\lambda_l^{k/2m}.$$

Since the u^h_{l+r} are known to be orthonormal, it is even possible to keep the U_{l+r} also orthonormal, without any harm to the estimate.

The same theorem and proof—with the inner product (u, v) replaced throughout by $b(u, v)$—apply to the generalized eigenvalue problem $Lu = \lambda Bu$.

6.4. COMPUTATIONAL TECHNIQUES

The Rayleigh–Ritz principle has led to the matrix eigenvalue problem $KQ = \lambda MQ$, which now remains to be solved. This is not a trivial problem, and it is scarcely discussed in a typical text on linear algebra. An efficient algorithm should take advantage of the fact that K and M are symmetric positive definite and also of the fact that they are sparse. The last property would be lost, for example, if we were to factor M into LL^T by a Cholesky elimination, and compute the eigenvalues of $L^{-1}K(L^{-1})^T$ by a standard algorithm (we would choose QR, or a Givens method which begins by reduction to a triangular matrix, rather than the older Jacobi method). This loss of sparseness will not be so serious for a small problem, which can be handled within the core of the computer; but for a large system it is inefficient.

We propose to find the eigenvalues—or more precisely the first few eigenvalues, since it would be useless to compute higher eigenvalues which have no physical significance—directly from the equation $KQ = \lambda MQ$. We shall reject lumping, since a diagonal M is not enormously better than a banded M.

There is also a technique of *economization*, which reduces the order of the system by working only with a small number of *master variables*. The dependence of the other *slave variables* is assumed a priori, and these degrees of freedom are thereby eliminated [22]. Fried has described this idea in the following terms: Placing a point load at a node z_j corresponding to a master variable, let Φ_j be the finite element solution in the static problem. Then these functions Φ_j are a basis (not local, as the original φ_j were) for the trial space in the economized problem. These functions can be expected to represent low-frequency modes fairly well, and those are the ones which are calculated. Nevertheless, our impression is that as efficient algorithms are constructed for the original problem, this economization will become less necessary and less popular.

An authoritative reference for eigenvalue algorithms prior to about

1970 is [21]. Before describing a newer method, we want to review a very widely used algorithm in this class: the *inverse iteration* or *inverse power* method. In its simplest form, for the eigenvalue problem $Ax = \lambda x$, inverse iteration proceeds by solving a linear system at each step: $Ay_{n+1} = x_n$. Then the approximation to λ is $\lambda_{n+1} = 1/\|y_{n+1}\|$, and the new approximation to x is the normalized vector $x_{n+1} = \lambda_{n+1}y_{n+1}$. These would be exact if x_n were an eigenvector. If we imagine that the starting vector x_0 is expanded in terms of the true eigenvectors v_j, $x_0 = \sum c_j v_j$, then the effect of n inverse iterations is to amplify each component by $(\lambda_j)^{-n}$; x_n is proportional to $\sum c_j (\lambda_j)^{-n} v_j$. If λ_1 is distinctly smaller than the other eigenvalues, then the first component will become dominant, and x_n will approach the unit eigenvector v_1. The convergence is like that of a geometric series, depending on the fraction λ_1/λ_2; the error $x_n - x$ is of order $(\lambda_1/\lambda_2)^n$. Obviously the method becomes more effective when this ratio is small.

One technique for reducing this ratio is to shift the origin, replacing A at the nth step by $A - \lambda_n I$. This shifts all the eigenvalues of A by the same amount λ_n. If λ_n is close to a true eigenvalue λ, so that $\lambda - \lambda_n$ is small, then the corresponding component of the vector y_{n+1} is amplified by the large quantity $(\lambda - \lambda_n)^{-1}$. There is no numerical difficulty with this process, even though $A - \lambda_n I$ is nearly singular. In fact, it is useful to have a better approximation to the eigenvalue than $\lambda_n = 1/\|y_n\|$. For example, the Rayleigh quotient $\lambda_n = (Ay_n, y_n)/(y_n, y_n)$ will be much more accurate. This quotient is very flat in the neighborhood of a true eigenvalue (where it has a stationary point), and the algorithm with these improved shifts possesses cubic convergence: $\lambda_{n+1} - \lambda \sim (\lambda_n - \lambda)^3$. On the other hand, shifting A to $A - \lambda_n I$ means that Gauss elimination has to be redone at each iteration; the triangular factors of A cannot be stored and used over and over, as is done with the simple iteration $Ay_{n+1} = x_n$.

In the generalized eigenvalue problem, the simplest idea would be to solve $Ky_{n+1} = Mx_n$ at each stage and then obtain x_{n+1} by normalizing y_{n+1}. In this form, however, the iteration matrix $K^{-1}M$ will be nonsymmetric. Therefore, if the Cholesky factorization of K is BB^T, it is preferable to use $B^{-1}M(B^T)^{-1}$ as the (symmetric) iterating matrix. This appears to involve the nonsparse matrices we wished to avoid, but obviously the iteration does not begin by multiplying these matrices. Instead the approximation u_{n+1} is determined from $B^T v_{n+1} = u$, $Bw_{n+1} = Mv_{n+1}$, $u_{n+1} = $ normalized w_{n+1}. Convergence of the normalizing factors to λ_1, and of the v_n (not u_n!) to the eigenvector x_1, again depends on the ratio λ_2/λ_1.

A very useful variant of inverse iteration is the *block power method*, introduced by Bauer and improved by Rutishauser and by Peters and Wilkinson.† The idea is to find several eigenvalues at once, by carrying l approxi-

---

† It is known in the engineering literature as subspace iteration.

mate eigenvectors in the iterations. (Obviously they must be coupled as the algorithm proceeds, or the result would only be l different approximations to the same fundamental mode.) Convergence to λ_i occurs at the rate λ_i/λ_{l+1}, and repeated eigenvalues can be handled without difficulty.

Block iteration for $Ax = \lambda x$ proceeds as follows. Suppose the l starting vectors, assumed to be orthonormal, are the columns of an $N \times l$ matrix P_0. The first step is to solve the l equations $AZ_1 = P_0$. Then, rather than normalizing separately each column of Z_1, we orthonormalize all l columns. For this we form the $l \times l$ matrix $Z_1^T Z_1$ and find its eigenvalues μ_i^{-2} (which are the first approximation to $\lambda_1^{-2}, \ldots, \lambda_l^{-2}$) and corresponding eigenvectors w_i. The new approximation P_1 is the product of Z_1 and the matrix W with columns $w_1 \mu_1, \ldots, w_l \mu_l$. The columns of P_1 are orthonormal—they are the approximate eigenvectors of A—since $P_1^T P_1 = W^T Z_1^T Z_1 W = I$. At the next step $AZ_2 = P_1$, and so on.

This algorithm would reproduce exactly the eigenvectors $v_1, \ldots, v_l$ of A. More precisely, let each column of P_0 be a linear combination of $v_1, \ldots, v_l$—say $P_0 = VQ$, where Q is an $l \times l$ orthogonal matrix and V is the $N \times l$ matrix whose columns are $v_1, \ldots, v_l$. We note that $V^T V = I$ and $AV = V\Lambda$, if Λ is the diagonal matrix whose entries are $\lambda_1, \ldots, \lambda_l$.

Thus the first block power step gives $Z_1 = A^{-1} P_0 = A^{-1} VQ = V\Lambda^{-1}Q$, and $Z_1^T Z_1$ becomes $Q^T \Lambda^{-2} Q$. Since Q is orthogonal, it follows that the eigenvalues μ_i^{-2} of $Z_1^T Z_1$ are equal to the entries λ_i^{-2} of the diagonal matrix Λ^{-2}. In other words, if $P_0 = VQ$ then the first step of the method would produce $P_1 = V$, and the correct eigenvalues.

The changes involved in working with $AQ = \lambda MQ$ rather than $Ax = \lambda x$ are described in the final paragraph of this chapter. The block power algorithm as restated there is completely straightforward to program.

These inverse power methods are very simple and practical, especially when only modest accuracy is required. However, there is a new technique, based on a more subtle matrix theorem, which has also become a most highly recommended algorithm for the band eigenvalue problem $KQ = \lambda MQ$. It is due to Peters and Wilkinson [P1, P2], and depends on the following lovely matrix theorem: *The number of eigenvalues less than a given λ_0 can be determined just by counting the number of negative pivots when Gauss elimination is applied to $K - \lambda_0 M$.*

This suggests an algorithm based on bisection. Suppose it is determined that there are n_0 eigenvalues below the first guess λ_0. Then the Gaussian pivots for $K - (\lambda_0/2)M$ reveal the number n_1 which is below $\lambda_0/2$; and the remaining $n_0 - n_1$ eigenvalues must lie between $\lambda_0/2$ and λ_0. Repeated bisection will isolate any eigenvalue with tremendous accuracy, but the process requires a Gauss elimination at each step and is rather expensive. It must be speeded up by using the *values* of the pivots (or their product, the determinant $d(\lambda) = \det(K - \lambda M)$) rather than just their signs. Clough, Bathe, and

Parlett recommend an accelerated secant iteration

$$\lambda_{k+1} = \lambda_k - 2d(\lambda_k) \frac{\lambda_k - \lambda_{k-1}}{d(\lambda_k) - d(\lambda_{k-1})}.$$

This is a variant of the usual Newton's method, in which the derivative is replaced by a difference quotient; the factor 2 is for acceleration. Once λ_k is close to a true λ, the computer may cease further matrix factorizations, and turn to ordinary inverse iteration. We can confirm that the Peters–Wilkinson algorithm was very successful in the numerical experiments reported in Chapter 8.

We want to give the theoretical background for this algorithm, which can be based on a classical theorem known as *Sylvester's law of inertia: If two real symmetric matrices A and D are related by a* congruence transformation $A = BDB^T$, *where B is any nonsingular matrix, then they have the same number of negative eigenvalues, of positive eigenvalues, and of zero eigenvalues.*

The proof is especially neat when D is nonsingular; that is, it has no zero eigenvalues. Let B_θ be a family of nonsingular matrices which changes gradually from the identity matrix (at $\theta = 0$) to the matrix B (at $\theta = 1$). [We cannot be sure that the particular family $B_\theta = \theta B + (1 - \theta)I$ will be nonsingular, but the construction of a suitable B_θ is never difficult.] The matrices $B_\theta D B_\theta^T$ are always symmetric and always nonsingular. Therefore, their eigenvalues $\lambda(\theta)$ are real, change gradually with θ, and never cross zero. Consequently, the number of eigenvalues on each side of zero must be the same at $\theta = 1$ as at $\theta = 0$. In other words, A and D have the same number of negative eigenvalues and of positive eigenvalues. If D happens to be singular, we use the same argument for the matrices $D \pm \epsilon I$ and prove the theorem in the limit as $\epsilon \to 0$.

Now we apply this law of inertia to Gauss elimination. If the matrix A is factored into LDL^T (as in Section 1.5), where L is lower triangular with 1's on the diagonal and D is the diagonal matrix of pivots, then *the signs of the pivots determine the signs of the eigenvalues.* (In case two rows are exchanged during the elimination process, as will be required if one of the early pivots happens to be zero, the corresponding two columns must be exchanged to preserve the symmetry of the matrix. But such an exchange of both rows and columns is again a congruence transformation, brought about by a permutation matrix B. Therefore, the law of inertia still applies, and after such an exchange the Gauss elimination pivots still give the signs of the eigenvalues correctly.)

The application to the generalized eigenproblem $KQ = \lambda MQ$ is as follows. Let $A = K - \lambda_0 M$, and count the number of negative pivots in Gauss elimination. We claim that this equals the number n_0 of eigenvalues below λ_0 in the given problem $KQ = \lambda MQ$.

Proof. Rewrite this given problem as $M^{-1/2}KM^{-1/2}(M^{1/2}Q) = \lambda(M^{1/2}Q)$, so that n_0 is the number of eigenvalues of $M^{-1/2}KM^{-1/2}$ which lie below λ_0. This is the number of negative eigenvalues of $M^{-1/2}KM^{-1/2} - \lambda_0 I$. But by the law of inertia (again! Choose this last matrix as D, and $B = M^{1/2}$) this coincides with the number of negative eigenvalues of $A = K - \lambda_0 M$. Therefore, n_0 is easily determined by applying Gauss elimination to A. The eigenvectors follow in one or at most two steps of inverse iteration with the matrix $C = K - \lambda_{\text{computed}} M$. A good initial guess for the eigenvector is described in [P2].

Experiments are now going on (see the supplementary bibliography) to compare the performance of eigenvalue algorithms. Clough and Bathe have found the determinant search techniques, of which the Peters-Wilkinson method is one variant, to be most competitive when the bandwidth is not large; about five triangular factorizations were required for each eigenvalue, and the cost of every such iteration is proportional to the square of the bandwidth. The most generally successful technique was the inverse block power method, alias *subspace iteration*. They express the algorithm in the following equivalent form: starting from an eigenvector guess X_{n-1}, which is a matrix with l orthonormal columns, solve $KY_n = MX_{n-1}$. Then solve the l-dimensional eigenproblem

$$(55) \qquad (Y_n^T K Y_n)Q = \nu(Y_n^T M Y_n)Q.$$

ν_i are the approximate eigenvalues, and the new matrix X_n of approximate eigenvectors is the product of Y_n with a square matrix of order l, formed from the eigenvectors Q of (55). Two computational problems have been considered in detail by Bathe and Parlett—the solution of the small eigenproblem (55), for which they use a Jacobi-type sweep since the matrices become near to diagonal, and the selection of the starting X_0. There is also a compromise to be made on the choice of l; with a large l there are few iterations, but each one is expensive. They chose $l = \min(2p, p + 8)$ in computing the first p eigenvalues, and found that eight iteration steps gave excellent results. This technique is effective even on problems too large to be handled in core, and it may well become the accepted algorithm for finite element eigenvalue problems.

7 INITIAL-VALUE PROBLEMS

7.1. THE GALERKIN–CRANK–NICOLSON METHOD
FOR THE HEAT EQUATION

So far we have discussed only steady-state problems—elliptic equations and eigenvalue problems. Galerkin's principle is flexible enough to apply also to initial-value problems, and in this chapter we want to consider finite element approximations. They still have important advantages over finite differences, for general geometries and for problems which evolve comparatively slowly in time (low Reynolds numbers, in the case of fluid flow). For high-speed wave problems, we shall point out some disadvantages.

The main theorem for finite differences is well known: A formally consistent method is convergent if and only if it is stable. This applies verbatim to Galerkin approximations. In fact, the only difference is that stability and consistency are often easier to verify for finite elements than for finite differences. We show below how the Galerkin principle imitates the stability of the differential equation, and we repeat now the disguise which is assumed by consistency: *The Galerkin method is consistent if the subspaces S^h become dense in the admissible space.* This means that every admissible v is approximated more and more closely, as $h \longrightarrow 0$, by trial functions v^h. The approximation theorems of Chapter 3 establish exactly this property, and they give even the *degree of approximation*—which translates into the degree of consistency, or the *order of accuracy*, of the Galerkin equations.

A natural setting in which to illustrate these ideas is provided by the heat equation

$$(1) \qquad \frac{\partial u}{\partial t} - \frac{\partial^2 u}{\partial x^2} = f(x, t), \qquad 0 < x < \pi, \quad t > 0.$$

This parabolic differential equation represents heat conduction in a rod; $u = u(x, t)$ is the temperature at the point x and time $t > 0$, and $f(x, t)$ is a heat-source term. As in our previous examples, we impose a Dirichlet boundary condition at $x = 0$ and a natural condition at $x = \pi$:

$$(2) \qquad\qquad u(0, t) = \frac{\partial u}{\partial x}(\pi, t) = 0.$$

The first condition means physically that the temperature at the left end of the rod is held at 0. The Neumann condition, on the other hand, means that the right end of the rod is insulated; there is no temperature gradient across $x = \pi$. To complete the statement of the problem we specify the initial temperature

$$(3) \qquad\qquad u(x, 0) = u_0(x), \qquad 0 \le x \le \pi.$$

This classical formulation of heat conduction is fraught with difficulties. For example, (2) and (3) are contradictory if u_0 fails to vanish at $x = 0$, or if $\partial u_0/\partial x$ does not vanish at $x = \pi$. In addition, f could be a point source which is singular at some x_0, and so (1) is not literally true at that point. In all these cases the underlying physical problem still makes sense, to find the temperature distribution associated with a given initial temperature u_0 and heat source f. We therefore seek an alternative integral formulation as in the steady-state case.

Since there is no natural minimum energy principle—the equation is not self-adjoint—we turn to the weak formulation: At each time $t > 0$,

$$(4) \qquad\qquad \int_0^\pi (u_t - u_{xx} - f)v \, dx = 0.$$

In the steady state, with $u_t = 0$ and $f = f(x)$, this coincides with the earlier Galerkin formulation.

To achieve greater symmetry between trial function and test function, we integrate $-u_{xx}v$ by parts. The integrated term $u_x v$ vanishes naturally at $x = \pi$, and we are led back to the essential condition $v(0) = 0$, in other words to the space $\mathfrak{IC}_E^1$:

$$(5) \qquad\qquad \int_0^\pi (u_t v + u_x v_x - f v) \, dx = 0 \qquad \text{for all } v \text{ in } \mathfrak{IC}_E^1.$$

This is the starting point for the finite element approximation. Given an N-dimensional subspace S^h of $\mathfrak{IC}_E^1$, the Galerkin principle is to find a function $u^h(x, t)$ with the following property: At each $t > 0$, u^h lies in S^h and satisfies

$$(6) \qquad\qquad \int_0^\pi (u_t^h v^h + u_x^h v_x^h - f v^h) \, dx = 0 \qquad \text{for all } v^h \text{ in } S^h.$$

Notice that the time variable is still continuous: the Galerkin (or, more correctly, Faedo–Galerkin) formulation is discrete in the space variables and yields a system of ordinary differential equations in time. It is these equations which have to be solved numerically. To make this formulation operational, choose a basis $\varphi_1, \ldots, \varphi_N$ for the trial space S^h and expand the unknown solution as

$$u^h(x, t) = \sum_1^N Q_j(t)\varphi_j(x).$$

The optimal weights Q_j are determined by the Galerkin principle (6):

(7) $$\sum_j \int_0^\pi \left(\frac{\partial Q_j}{\partial t}\varphi_j\varphi_k + Q_j\frac{\partial \varphi_j}{\partial x}\frac{\partial \varphi_k}{\partial x} - f\varphi_k \right) dx = 0, \qquad k = 1, \ldots, N.$$

Since every v^h is a combination of these basis functions φ_k, it is enough to apply the principle only to the basis functions. The result is a system of N ordinary differential equations for the N unknowns $Q_1(t), \ldots, Q_N(t)$, and the boundary conditions are already incorporated in these equations. The initial condition $u = u_0$ is still to be accounted for, and here there are several possibilities. Mathematically, a natural choice of approximate initial condition u_0^h is the best least-squares approximation to u_0; u_0^h is in S^h and satisfies $(u_0^h, v^h) = (u_0, v^h)$ for all v^h, or in other words,

$$\sum Q_{j0} \int \varphi_j\varphi_k \, dx = \int u_0\varphi_k \, dx, \qquad k = 1, \ldots, N.$$

In practice, this means that the integrals on the right have to be computed, and the mass matrix on the left has to be inverted. Therefore, it is often more efficient to use the interpolate of u_0 as initial condition in the finite element system: $u_0^h = (u_0)_I$. The order of accuracy is not affected.

Now we put the Galerkin equations (7) into vector notation, with $Q = (Q_1, \ldots, Q_N)$. The striking point is that *they involve exactly the same matrices and load vector which appear in the steady-state problem*; the coefficient of Q' is the mass matrix M, and the coefficient of Q is the stiffness matrix:

(8) $$MQ' + KQ = F(t).$$

The components of the right hand side are $F_k = \int f(x, t)\varphi_k(x) \, dx$. If there were inhomogeneous boundary conditions, time-dependent or not, their effect would also appear in F.

It is natural to ask why finite elements are not used also in the time direction. This has certainly been attempted, but not with great success. Mathematically, it is perfectly reasonable to study the discretization in two steps— first to analyze the finite element error $u(x, t) - u^h(x, t)$, and then the error incurred in solving the ordinary differential equations for u^h. In the time variable there are no geometrical difficulties to overcome by finite elements, and in fact a straightforward application of the Galerkin principle may couple all the time levels, and destroy the crucial property of propagation *forward in time*. We see no reason to forego the extra flexibility of finite differences.

First we must mention the technique of *mode superposition*, a competitor of conventional finite differences. The idea is simply to analyze the initial u_0 and the forcing term f into the natural modes of the problem—the eigenfunctions u_l^h of Chapter 6. All the computational work is transferred to the eigenproblem. Only the lower eigenvalues are followed forward in time; Nickell suggests that fewer than 30 modes out of 1000 can produce very good results, unless f is unusually rich in the higher harmonics.

A conventional difference scheme in time, which couples all the modes, has to contend with the extreme *stiffness* of the equation (8); the condition number of $M^{-1}K$ can easily exceed 1000, so that the modes are decaying at radically different rates. A "trapezoidal rule" scheme (Crank–Nicolson, Neumark β) automatically filters out the useless high modes, if Δt is properly chosen. Of course these schemes are *implicit*, but so is Galerkin's differential equation; M cannot be inverted in (8) without destroying the band structure. (The alternative of lumping is discussed at the end of the chapter.) In one respect an implicit equation is not such a serious drawback for parabolic equations, such as the heat equation, since the stability requirements on explicit difference schemes are in any case severe: the time step must be limited to $\Delta t \leq Ch^2$ or there will be exponential instabilities in the difference equation. By contrast, an implicit scheme can be unconditionally stable; the size of Δt is restricted only by the demands of accuracy, and not by stability. This distinction between implicit and explicit methods is natural for the heat equation, where there is an *infinite speed of propagation*: the temperature Q^{n+1} at any point x_0 depends on the previous temperature Q^n at all points of the medium, however small the time step may be. This dependence is reflected in the fact that M^{-1} is not sparse in the Galerkin equation, and that $\Delta t \sim h^2$ is required for stability in an explicit difference equation.

We shall analyze the *Crank-Nicolson scheme*, which is centered at $(n + \frac{1}{2})\Delta t$ and therefore achieves second-order accuracy in time:

$$(9) \qquad M\frac{Q^{n+1} - Q^n}{\Delta t} + K\frac{Q^{n+1} + Q^n}{2} = \frac{f^{n+1} + f^n}{2}.$$

Rewritten, the approximation Q^{n+1} is determined by

$$\frac{M + K\,\Delta t}{2}Q^{n+1} = \frac{M - K\,\Delta t}{2}Q^n + \frac{(f^{n+1} + f^n)\,\Delta t}{2}.$$

In an actual computation, the matrix on the left can be factored by Gauss elimination into LL^T, where L is Cholesky's lower triangular matrix, and then Q^{n+1} would be computed at each step by two back substitutions:

$$LQ^{n+1/2} = \frac{M - K\,\Delta t}{2}Q^n + \frac{(f^{n+1} + f^n)\,\Delta t}{2}, \qquad L^T Q^{n+1} = Q^{n+1/2}.$$

If the coefficients in the problem are time-dependent (or nonlinear), then in the strict Galerkin theory the matrices M and K must be recomputed at each time step. It seems very likely that another variational crime will be found necessary, leading to some hybrid of finite elements and finite differences—in order to produce a stiffness matrix which is approximately correct without recomputing every integral. In large problems it may also be too expensive to solve exactly for Q^{n+1}; an iteration (possibly starting with Q^n as initial guess) may be more efficient. Douglas and Dupont [D6, D9] have introduced several iteration techniques for nonlinear problems in order to circumvent the solution of a large nonlinear system at every time step. Their analysis completely justifies these modifications of the pure Galerkin method.

The two remaining sections are occupied with verifying the stability and the expected rates of convergence for parabolic and hyperbolic equations. For the simple equation $u_t = u_x$ in the last section, something unexpected finally occurs.

7.2. STABILITY AND CONVERGENCE IN
PARABOLIC PROBLEMS

There are two approaches to the theory. One is more explicit and revealing; the other may be more general. In the first approach each eigenfunction is followed forward in time in all three equations—the partial differential equation for u, the ordinary differential equation for u^h, and the finite difference equation for Q^n. If the coefficients and boundary conditions in the equations are independent of time (the stationary case), then this simple approach is extremely successful; the previous chapter gave precise bounds on the eigenfunction errors, and the arguments become completely elementary.† In the nonstationary case the analysis is much more technical, but parabolic equations are so strongly dissipative that time-dependent (and even nonlinear) effects can ultimately be accounted for. In the second approach, which is based on energy inequalities at each time, this accounting becomes comparatively simple.

We consider a parabolic equation $u_t + Lu = f$, where L is exactly the kind of elliptic operator studied in the previous chapters. It is of order $2m$ ($m = 1$ is by far the most common) and its coefficients may depend on the

†This is also the approach to use in analyzing mode superposition.

position vector x. Suppose first that $f \equiv 0$, and that the initial function u_0 is expanded into the orthonormal eigenfunctions:

$$u_0 = \sum_1^\infty c_j u_j(x), \qquad c_j = \int_\Omega u_0(x) u_j(x) \, dx.$$

Each eigenfunction decays at its own rate in time, and the solution evolves according to

$$(10) \qquad\qquad u(t, x) = \sum c_j e^{-\lambda_j t} u_j(x).$$

For $t > 0$, this solution lies in the admissible space $\mathcal{H}_E^1$. Even if the initial function u_0 is discontinuous, it is easy to see that as time goes on u becomes increasingly smooth; the derivatives at any positive time satisfy

$$\left\| \left(\frac{\partial}{\partial t} \right)^k u \right\|_0^2 = \| \sum c_j (-\lambda_j)^k e^{-\lambda_j t} u_j \|_0^2 = \sum c_j^2 \lambda_j^{2k} e^{-2\lambda_j t}.$$

The exponential term makes this sum finite, and it does the same for the spatial derivatives. These norms are monotonically decreasing in t, in particular

$$(11) \qquad \sum c_j^2 e^{-2\lambda_j t} \leq e^{-2\lambda_1 t} \sum c_j^2 \quad \text{or} \quad \| u(t) \|_0 \leq e^{-\lambda_1 t} \| u_0 \|_0.$$

The fundamental frequency $\lambda_1 > 0$ gives the rate of decay of the solution.

The Galerkin equation $MQ' + KQ = 0$ is actually a little *more* stable than the equation it approximates. Suppose the initial Q_0 is expanded in terms of the discrete eigenvectors Q_j of $M^{-1}K$, or, equivalently, the initial u_0^h is expanded in terms of the approximate eigenfunctions u_j^h:

$$Q_0 = \sum_1^N d_j Q_j \quad \text{or} \quad u_0^h = \sum_1^N d_j u_j^h.$$

Then the solution at a later time t is

$$Q(t) = \sum d_j Q_j \exp(-\lambda_j^h t) \quad \text{or} \quad u^h(x, t) = \sum d_j u_j^h(x) \exp(-\lambda_j^h t).$$

Therefore, the rate of decay is λ_1^h, which is slightly larger than λ_1:

$$\| u^h(x, t) \|_0 \leq \| u_0^h \|_0 \exp(-\lambda_1^h t) \leq \| u_0^h \|_0 \exp(-\lambda_1 t).$$

Finally, the Crank–Nicolson scheme is also stable—with no restriction on the size of Δt. The difference operator at each step has the same eigenvectors as $M^{-1}K$, since it is given by

$$\frac{M + K \Delta t}{2} Q^{n+1} = \frac{M - K \Delta t}{2} Q^n$$

or

$$Q^{n+1} = \left(I + \frac{M^{-1}K\,\Delta t}{2}\right)^{-1}\left(I - \frac{M^{-1}K\,\Delta t}{2}\right)Q^n.$$

The solution after n steps is $Q^n = \sum d_j(\mu_j^h)^n Q_j$, where the *amplification factor* μ_j^h is

(12)
$$\mu_j^h = \frac{1 - \lambda_j^h\,\Delta t/2}{1 + \lambda_j^h\,\Delta t/2}.$$

Since every λ_j^h is nonnegative, it follows that every $|\mu_j^h| \leq 1$, and therefore the Crank–Nicolson scheme is automatically stable. The decay of the principal eigenfunction u_1^h is governed by its amplification factor μ_1^h, which is to be compared with the true amplification (or contraction) factor $\exp(-\lambda_1^h\,\Delta t)$ of the Galerkin equation over a single time step:

$$\mu_1^h \sim 1 - \lambda_1^h\,\Delta t + \tfrac{1}{2}(\lambda_1^h\,\Delta t)^2 - \tfrac{1}{4}(\lambda_1^h\,\Delta t)^3\ldots,$$
$$e^{-\lambda_1^h\Delta t} \sim 1 - \lambda_1^h\,\Delta t + \tfrac{1}{2}(\lambda_1^h\,\Delta t)^2 - \tfrac{1}{6}(\lambda_1^h\,\Delta t)^3\ldots.$$

Since μ_1^h is the smaller, this component of the solution decays slightly faster in the finite difference equation. The discrepancy is of order Δt^3, reflecting the second-order accuracy of the Crank–Nicolson scheme.

In the finite difference case, however, *the high-frequency components do not decay at faster and faster rates.* As $\lambda_j^h \to \infty$, the amplification factor μ_j^h converges to -1, and the weights attached to the high frequencies change sign at each time step. This does not occur in the Galerkin equation or in a fully implicit difference scheme. The latter would have $\mu_j^h = (1 + \lambda_j^h\,\Delta t)^{-1}$, and obviously $\mu_j^h \to 0$ as $\lambda_j^h \to \infty$. In the Crank–Nicolson case, however, μ_j^h will become negative and start to increase in magnitude at the frequency $\lambda_j^h = 2/\Delta t$. The highest frequency the mesh can hold is $\lambda_N^h \sim ch^{-2m}$, which normally exceeds $2/\Delta t$. Therefore the very high frequencies (presumably present only in small amounts) are actually damped less strongly than the moderate frequencies. If this should present any difficulty, then as in hyperbolic problems it is possible to add a simple dissipation term.

The rate of convergence is easy to determine from the eigenfunction expansions

(13)
$$u = \sum_1^\infty c_j e^{-\lambda_j t} u_j(x), \qquad u^h = \sum_1^N d_j e^{-\lambda_j^h t} u_j^h(x).$$

There are two sources of error: the *initial error* and the *evolution error*. Whether u_0^h is computed as a best approximation to the true u_0 or simply chosen to be its interpolate, the initial error $u_0 - u_0^h$ will be of order h^k—and like any other initial condition it will decay with time like $e^{-\lambda_1 t}$. The remaining error arises when u_0^h is used as initial condition in both equations. In the true equation it is expanded into the eigenfunctions u_j, and in Galerkin's equation it is expanded into the u_j^h—with a different evolution in time. We know from the previous chapter that

$$\lambda_j^h - \lambda_j \sim h^{2(k-m)}\lambda_j^{k/m}, \qquad \|u_j^h - u_j\|_0 \sim h^k\lambda_j^{k/2m}.$$

[In case $k < 2m$, which is not usual in practice, h^k should be replaced by $h^{2(k-m)}$.] These estimates show that the difference in the weights is only $c_j - d_j = \int u_0^h(u_j - u_j^h)\,dx \sim h^k$. Therefore, comparing u and u^h in (13), the evolution error is also of order h^k. The error in the derivatives will be of the usual order h^{k-s}, again decaying at the rate $e^{-\lambda_1 t}$.

We emphasize the simplicity of this technique for deriving error bounds. It would lead equally easily to the $O(\Delta t^2)$ errors in the Crank–Nicolson process. The simplicity would seem to require that the spatial part of the problem be self-adjoint, in order to apply the eigenvalue and eigenfunction estimates of the previous chapter, but this hypothesis is actually inessential. In fact, there is a simple formula which bypasses the eigenvalue theory entirely and relates the evolution error directly to the basic estimates for steady-state problems. With the same initial function u_0^h in both equations, the solutions differ at time t by

$$(14) \qquad\qquad u - u^h = \frac{1}{2\pi i}\int_C e^{zt}(u_z - u_z^h)\,dz.$$

Here z is a complex number, u_z solves the (non-self-adjoint) steady-state problem $(L + z)u_z = u_0^h$, and u_z^h is its Galerkin approximation. [In effect, u_z and u_z^h are the Laplace transforms of $u(t)$ and $u^h(t)$, respectively, and the integral (14) inverts the Laplace transform; the contour C runs along two rays $\arg z = \pm(\pi/2 + \varepsilon)$ in the left half-plane, so that the exponential e^{zt} produces a convergent integral.] With this formula—which reduces to the eigenfunction expansions in the self-adjoint, discrete spectrum case—the evolution error at time t follows directly from the steady-state errors of Theorem 2.1. The result is the expected order h^k, even for non-self-adjoint equations.

To complete the discussion of this first approach, we note that Duhamel's principle places the inhomogeneous case $f \neq 0$ into the same framework. According to this principle, the source term f entering at any time τ acts like an initial condition over the remaining time $t - \tau$. If the solution u is originally connected to u_0 by some rule $u(t) = E(t)u_0$, then in the nonhomogeneous case

$$u(t) = E(t)u_0 + \int_0^t E(t - \tau)f(\tau)\,d\tau.$$

In the eigenfunction expansions, this becomes

$$u(t) = \sum u_j(x)\left[c_j e^{-\lambda_j t} + \int_0^t f_j(\tau)e^{-\lambda_j(t-\tau)}\,d\tau\right], \qquad f_j(\tau) = \int_\Omega f(x, \tau)u_j\,dx.$$

The error $u - u^h$ will still be of order h^k, but if the source f acts at all times, then the decay factors $e^{-\lambda t}$ will no longer apply; there will be errors committed at times τ so near to t that their decay has not begun.

We try now a different idea—the second approach to parabolic problems mentioned at the beginning of the section—to estimate at each time t the *rate of change* of the error $u - u^h$. Rather than connecting u and u^h by eigenfunction expansions back to their data at $t = 0$, the error at time $t + dt$ (or $t + \Delta t$, in case of the difference equation) is determined from the error at time t. This technique has been pioneered by Douglas and Dupont [D6, D9], following early papers by Swartz and Wendroff [S11] and Price and Varga [P11]; Wheeler, Dendy, and others have made more recent contributions.

Suppose that the differential equation and its Galerkin approximation are written in variational form at each time t,

(15) $$(u_t, v) + a(u, v) = (f, v) \qquad \text{for all } v \text{ in } \mathfrak{IC}_E^1,$$

(16) $$(u_t^h, v^h) + a(u^h, v^h) = (f, v^h) \qquad \text{for all } v^h \text{ in } S^h.$$

In the stationary case, the energy inner product $a(v, w)$ is independent of time; it arises from (Lv, w). We retain the notation P for the Ritz projection of the admissible space $\mathfrak{IC}_E^1$ onto its subspace S^h, defined as in (6.38) by $a(u - Pu, v^h) = 0$ for all v^h.

Splitting the Galerkin error into $u - u^h = (u - Pu) + (Pu - u^h)$, the size of $u - Pu$ is known from approximation theory (6.39), and the energy inequalities required for the $Pu - u^h$ term are consequences of the following identity.

LEMMA 7.1

If $e(t) = Pu(t) - u^h(t)$, then

(17) $$(e_t, e) + a(e, e) = (Pu_t - u_t, e).$$

Proof. Since e lies in S^h, we may put $v = e$ in (15), $v^h = e$ in (16), and subtract:

$$(u_t - u_t^h, e) + a(u - u^h, e) = 0.$$

The second term is just $a(e, e)$, after applying the identity $a(u - Pu, e) = 0$. Rearranging, we have

$$(Pu_t - u_t^h, e) + a(e, e) = (Pu_t - u_t, e).$$

It remains to identity the first term as the inner product (e_t, e), in other words to know that Pu_t is the same as $(Pu)_t$: The Ritz projection P commutes with the differentiation $\partial/\partial t$. This is true only in the stationary case. If the energy inner product were to depend on t, the term $(Pu_t - (Pu)_t, e)$ would also appear in the identity and would have to be estimated. This difficulty is only technical (we omit the details) if the operator L depends smoothly on time. In the stationary case it is obvious that P is independent of time: Differentiating the identity $(u - Pu, v^h) = 0$, we have $(u_t - (Pu)_t, v^h) = 0$ for all v^h, so that $(Pu)_t$ must coincide with Pu_t. This completes the proof of the lemma.

From this identity the rate of change of the error is easy to find. The first term in (17) can be rewritten as

$$(e_t, e) = \frac{\partial}{\partial t} \frac{(e, e)}{2} = \|e\|_0 \frac{\partial}{\partial t} \|e\|_0.$$

The term $a(e, e)$ is at least as large as $\lambda_1 \|e\|_0^2$, since λ_1 is the minimum of the Rayleigh quotient. Finally, the right side of the identity is bounded by $\|u_t - Pu_t\|_0 \|e\|_0$. Cancelling the common factor $\|e\|_0$, the identity leads to

$$(18) \qquad \frac{\partial}{\partial t} \|e\|_0 + \lambda_1 \|e\|_0 \leq \|u_t - Pu_t\|_0.$$

Multiplying by $e^{\lambda_1 \tau}$, and integrating with respect to τ from 0 to t,

$$(19) \qquad e^{\lambda_1 t} \|e(t)\|_0 - \|e(0)\|_0 \leq \int_0^t e^{\lambda_1 \tau} \|u_t(\tau) - Pu_t(\tau)\|_0 \, d\tau.$$

The initial error $e(0) = Pu_0 - u_0^h$ is of order $Ch^k \|u_0\|_k$, whether u_0^h is computed by interpolation or least-squares approximation of u_0. The main theorem, giving the correct order of convergence although not necessarily the most precise estimates, follows immediately from (19).

THEOREM 7.1

Suppose that S^h is a finite element space of degree $k - 1$. Then the error in the Galerkin approximation satisfies

$$\|u(t) - u^h(t)\|_0 \leq \|u(t) - Pu(t)\|_0 + \|e(t)\|_0$$
$$(20)$$
$$\leq Ch^k \left[\|u(t)\|_k + e^{-\lambda_1 t} \|u_0\|_k + \int_0^t e^{\lambda_1(\tau - t)} \|u_t(\tau)\|_k d\tau \right].$$

Thus the error is of the same order h^k as in steady-state problems, and it decays as fast as the fundamental mode if there is no source term.

This theorem agrees with the h^k estimate derived earlier by means of eigenfunction expansions. An averaged error bound in energy follows easily by direct integration of the identity (17):

$$2 \int_0^t a(e(\tau), e(\tau)) \, d\tau \leq \|e(0)\|_0^2 - \|e(t)\|_0^2 + 2 \int_0^t |(Pu_t - u_t, e)|^2 \, d\tau.$$

According to the theorem just proved, the right side is of order h^{2k}. This suggests that in the energy norm e is negligible in comparison with $u - Pu$, and that the Galerkin error $u - u^h = u - Pu + e$ satisfies

$$(21) \qquad a(u - u^h, u - u^h) \sim a(u - Pu, u - Pu) \leq C^2 h^{2(k-m)} \| u \|_k^2.$$

This completes the technical error estimates for parabolic problems; there are no surprises in the results. Our impression is that just as in static problems, finite elements are particularly effective in *coarse mesh calculations*, with a large value of h. In this situation the physics is often more adequately represented by Galerkin's principle, on which the finite element method is based, than by supposing difference quotients to be close to the derivatives. Because of the integrals to be evaluated, however, there is a price to be paid in computing time. Perhaps there will ultimately be a satisfactory combination of finite elements and finite differences.

7.3. HYPERBOLIC EQUATIONS

It is natural to experiment with finite elements also for hyperbolic problems. In fact, the Galerkin principle can be formulated in such generality—the equation $M(u) = 0$ is replaced by $(M(u^h), v^h) = 0$ for all v^h—that we may expect finite elements to be tested in an increasing variety of applications. Some preliminary tests are already going on, but the conclusions are not yet in (and may never be, given the ambiguous nature of most numerical experiments). Maybe all that can be expected is agreement on some general guidelines.

Mathematically, one property that can be guaranteed† is that if energy is conserved in the true problem, then it is conserved in Galerkin's method, and that if it is decreasing with time in the true problem, then it is decreasing in the Galerkin approximation. For a first-order system $u_t + Lu = 0$ this is easy to see. The rate of change of the energy $(u, u) = \int u^2 \, dx$ can be computed by multiplying the equation by u and integrating:

$$(u_t, u) + (Lu, u) = \frac{\partial}{\partial t} \frac{(u, u)}{2} + (Lu, u) = 0.$$

The equation is *conservative*—(u, u) is constant in time—in case (Lu, u) is identically zero, and it is *dissipative*—the energy (u, u) is decreasing—if $(Lu, u) \geq 0$ for all possible states u. Parabolic equations are strongly dissipative, since (Lu, u) is in that case a positive-definite expression in the mth derivatives of u. Hyperbolic equations are either conservative or at best

---

†Provided the test space V^h coincides with the trial space S^h.

weakly dissipative; energy may leak out at the boundaries, but only very slowly. This makes their analysis much more delicate. In the Galerkin method, let Q denote the projection of $\mathcal{3C}^0$ onto the subspace S^h—Qu is the best least-squares approximation in S^h to u, where earlier Pu was the best approximation in the strain energy norm $a(v, v)$. Then the Galerkin approximation u^h is determined by projecting the differential equations onto the subspace:

$$Q(u_t^h + Lu^h) = 0.$$

Since u^h is required to lie in S^h, it is automatic that $u^h = Qu^h$, and the Galerkin equation can be written more symmetrically as

$$u_t^h + QLQu^h = 0, \qquad u_0^h \text{ in } S^h.$$

In other words, the true generator L is replaced by QLQ. But then a conservative equation remains conservative, $(QLQu, u) = (LQu, Qu) = 0$, and a dissipative equation remains dissipative: $(QLQu, u) = (LQu, Qu) \geq 0$. The corresponding nonlinear operators are called *monotone* (Section 2.4), and the same result holds.

It is interesting that the conservative property is not always desirable, particularly in *nonlinear* hyperbolic equations. The simplest example is the conservation law $u_t = (u^2)_x$. The solutions of these problems may develop spontaneous discontinuities (shocks) and the conservation of energy is lost, even though some other conservation laws of mass and momentum are retained. In the Galerkin equation these shocks apparently never quite appear, and the approximate equation remains conservative—from which it follows that *convergence to the correct solution is impossible*. The standard remedy for finite differences is to dissipate energy by means of artificial viscosity, and apparently that will also be necessary for finite elements.

There are two forms in which hyperbolic equations may appear—either as a first-order system in time, say $w_t + Lw = f$ with a vector unknown w, or as a second-order equation $u_{tt} + Lu = f$. We begin with the latter case, in which L is elliptic; a typical example is the wave equation $u_{tt} - c^2 u_{xx} = 0$. The weak form of such an equation is just

$$(22) \qquad (u_{tt}, v) + a(u, v) = (f, v) \qquad \text{for } v \text{ in } \mathcal{3C}_E^1, \quad t > 0.$$

In the Galerkin approximation u and v are replaced by u^h and v^h; this means that $u^h = \sum Q_j(t)\varphi_j(x)$ is determined by

$$(23) \quad (\sum Q_j''\varphi_j, \varphi_k) + a(\sum Q_j\varphi_j, \varphi_k) = (f, \varphi_k) \qquad \text{for } k = 1, \ldots, N, \quad t > 0.$$

This is again an ordinary differential equation in the time variable. In this case the same mass and stiffness matrices appear, but the equation is of second order:

(24) $$MQ'' + KQ = F(t).$$

The starting values are approximations from within S^h to the true initial displacement $u_0(x)$ and initial velocity $u_0'(x)$. The behavior of the solutions is completely at variance with the parabolic case, where Q' appears instead of Q''. In that case, with $F \equiv 0$, the solution decays very rapidly; each eigenvector is associated with an exponential $e^{-\lambda_j t}$, and discontinuities immediately disappear. In the hyperbolic case the exponent changes to $\pm i\lambda_j t$, and Q *oscillates rather than decays*. The solution is no smoother than its initial data, and discontinuities are propagated indefinitely in time.

For the one-dimensional wave equation and linear elements, Galerkin's approximation becomes

$$\frac{Q_{j+1}'' + 4Q_j'' + Q_{j-1}''}{6} = c^2 \frac{Q_{j+1} - 2Q_j + Q_{j-1}}{h^2}.$$

We remark again on the coupled form of these equations, which automatically leads to an implicit difference equation. It appears from the experiments of Clough and others, and the theoretical discussion by Fujii at the Second U.S.–Japan Seminar, that there will be *no loss in accuracy* if M is replaced (through a suitable lumping process) by a diagonal matrix. This would not be the case for elements of all degrees; lumping implicitly uses elements of low degree (generally piecewise constants) in dealing with terms which are not differentiated with respect to x, and this loses overall accuracy *once the other terms in the equation are treated with high accuracy*.

Fujii has also given a valuable stability analysis for difference approximations (in the time direction) of the finite element equation (24). Suppose for example that the terms Q'' are replaced by centered second differences $\Delta t^{-2}(Q^{n+1} - 2Q^n + Q^{n-1})$. Then, as is well known for finite differences, the step size Δt must be restricted or the computed approximations Q^n will explode exponentially with n. For the one-dimensional wave equation, his stability conditions are $c\Delta t \leq h/\sqrt{3}$ for the consistent mass matrix M, and $c\Delta t \leq h$ for the lumped mass case. (Tong [T5] also observed the added stability with lumping.) Fujii has studied other finite difference schemes and more general hyperbolic initial-boundary-value problems, including the equations of elasticity.

The Galerkin approximation has two important properties: *conservation of energy* (if $f = 0$) *and convergence*. To measure the energy in a second-order hyperbolic problem we add the kinetic and potential energies:

$$E(t) = \tfrac{1}{2}[(u_t, u_t) + a(u, u)].$$

In the wave equation this energy is $\tfrac{1}{2}\int (u_t^2 + c^2 u_x^2)\, dx$. The quantity E is independent of time, since with $v = u_t$ in the weak form (22),

$$(25) \qquad \frac{dE}{dt} = (u_{tt}, u_t) + a(u, u_t) = 0.$$

For the wave equation, this becomes simply

$$\frac{dE}{dt} = \int (u_t u_{tt} + c^2 u_x u_{xt})\, dx = \int u_t(u_{tt} - c^2 u_{xx})\, dx = 0.$$

Conservation of energy in the Galerkin equation can be verified in the same way:

$$E^h(t) = \tfrac{1}{2}[(u_t^h, u_t^h) + a(u^h, u^h)], \qquad \frac{dE^h}{dt} = (u_{tt}^h, u_t^h) + a(u^h, u_t^h) = 0.$$

Thus the approximation, like the true equation, is only neutrally stable.

We shall sketch the proof of convergence, which follows from an identity analogous to Lemma 7.1: With $e = Pu - u^h$,

$$(26) \qquad (e_{tt}, e_t) + a(e, e_t) = ([Pu - u]_{tt}, e_t).$$

The left side is the derivative of the energy $E(t, e)$ in the quantity e. This expression is not quite conserved, but the right side is less than

$$\| (Pu - u)_{tt} \|_0 \, \| e_t \|_0 \leq Ch^k \sqrt{E(t, e)}.$$

Thus $E' \leq Ch^k \sqrt{E}$. Integrating from 0 to t,

$$E^{1/2} \leq E_0^{1/2} + \frac{Ch^k t}{2}.$$

The initial error E_0 will be of order $h^{2(k-1)}$ and so will the energy in $u - Pu$. Therefore, *the energy in the Galerkin error $u - u^h = u - Pu + e$ is of the optimal order $h^{2(k-1)}$.* Provided the initial data are smooth, this continues to hold even for large times, $t \sim 1/h$.

We turn now to the trivial but interesting example $u_t = u_x$. The equation itself is certainly not very exciting; it describes a wave traveling to the left with unit velocity, $u(x, t) = u_0(x + t)$. There is no distortion in the wave, and $\int_{-\infty}^{\infty} u^2\, dx$ is obviously conserved; this is the energy in a first-order problem. The Galerkin approximation at each time is $(u_t^h, v^h) = (u_x^h, v^h)$, and it, too, conserves energy. With linear elements $u^h(t, x) = \sum u_j(t)\varphi_j(x)$, where φ_j is the roof function centered at the node jh, this approximation becomes

$$(27) \qquad \frac{u'_{j+1} + 4u'_j + u'_{j-1}}{6} = \frac{u_{j+1} - u_{j-1}}{2h}.$$

Evidently the equation is again implicit—a serious drawback for hyperbolic problems. The truncation error is found to be $O(h^2)$, the usual rate of convergence for linear elements.

Dupont [D10] happened to be computing the corresponding rate of convergence for cubic trial functions, and *the expected power h^4 simply would not appear*. Instead, the error $u - u^h$ turned out to be $O(h^3)$, which is an order of magnitude larger than the best approximation to u by a cubic. His calculations, which were greeted with surprise and perhaps even some disbelief, were carried out for the Hermite cubics with u and u_x as unknowns at each node. For cubic splines his computations *did* give $O(h^4)$. Therefore, the rate of convergence does not depend only on the degree of the finite element polynomials, and in fact, the larger space of Hermite cubics gives a worse approximation than its subspace of cubic splines. Some explanation is in order.

We propose to compute the truncation error in general, by substituting the true solution of $u_t + Lu = 0$ into the Galerkin equation $u_t^h + QLQu^h = 0$. In our case $L = -\partial/\partial x$, and Q is the projection onto the subspace S^h. The truncation error is

$$Lu - QLQu = (I - Q)Lu + QL(I - Q)u.$$

This first term on the right is the error in least-squares approximation of $Lu = -u_x$. If S^h is of degree $k - 1$ and u is smooth, then this error is of the usual order h^k. It is the other term $QL(I - Q)u$ which is decisive. Since $L(I - Q)u$ is the derivative of the least-squares error, it is an error in the $\mathcal{3C}^1$ norm and cannot be better than h^{k-1}. The question is whether or not this term $L(I - Q)u$ is annihilated by the final projection Q; we believe that this does happen for linear elements on a regular mesh but not for the usual cubics. Since such a cancellation must be regarded as exceptional, *the normal rate of convergence in a first-order hyperbolic system will be h^{k-1} rather than h^k*. This rate of convergence has been established by Lesaint for a wide class of hyperbolic systems.

To understand how this cancellation could occur, we apply $QL(I - Q)$ to the polynomial of lowest degree which is not identically present in the subspace—x^2 for linear elements and x^4 for cubics. In the linear case $(I - Q)x^2$ is the error function illustrated in Section 3.2 (Fig. 3.3). Its derivative is a multiple of the *sawtooth function*; $L(I - Q)x^2$ goes linearly from $+1$ to -1 over each subinterval. The best continuous piecewise linear approximation of such a function is identically zero: $QL(I - Q)x^2 = 0$, and cancellation has occurred. In the cubic case, $L(I - Q)x^4$ happens to be a Hermite cubic, and the final projection Q leaves it unchanged; there is no cancellation, and the Galerkin error is of order $||u - u^h||_0 \sim h^3$.

In one sense the exponent $k - 1$ might have been anticipated. If the wave equation $u_{tt} = c^2 u_{xx}$ is reduced to a first-order system, the vector unknown is made up of first derivatives u_t and cu_x:

$$\begin{pmatrix} u_t \\ cu_x \end{pmatrix}_t = \begin{pmatrix} 0 & c \\ c & 0 \end{pmatrix} \begin{pmatrix} u_t \\ cu_x \end{pmatrix}_x.$$

Therefore, the ordinary energy $\|u_t\|_0^2 + \|cu_x\|_0^2$ in the vector unknown is precisely twice the energy $E(t)$; it comes from the same sum of the kinetic and potential energies. Since the error in this energy was of order $h^{2(k-1)}$ for the single equation, the exponent $k - 1$ is exactly what we should expect for a system.

From a practical point of view, these error bounds as $h \to 0$ are subordinate to the problem of obtaining reasonable accuracy at reasonable expense. With hyperbolic equations, we are not sure that this is achieved most effectively by finite elements. The finite speed of propagation in the true solution means that explicit finite difference equations are possible, with time steps Δt of the same order as h, and it is known how artificial viscosity can be introduced to promote stability. For finite elements the difference equations will be implicit and almost too conservative. (They can be explicit only if we *lump* the mass matrix, or if, as Raviart has proposed, we choose the nodes as the evaluation points ξ_i for numerical integration. By conserving mass in the lumping process—or by using lower degree trial polynomials as in [T8] for the element mass matrices—we achieve a consistent difference equation, with a typical Courant condition for numerical stability. Stability is not unconditional, as it was for the implicit pure Galerkin processes described earlier in the chapter.)

The one important advantage of finite elements in hyperbolic problems—which must somehow be imitated by finite difference schemes in the future—is the systematic achievement of high accuracy, even at curved boundaries.

8 SINGULARITIES

8.1. CORNERS AND INTERFACES

Perhaps the most characteristic property of elliptic boundary-value problems like

$$(1) \qquad -\nabla \cdot (p \, \nabla u) + qu = f \text{ in } \Omega, \qquad u = 0 \text{ on } \Gamma,$$

is that the solution u is smooth as long as the boundary Γ and data p, q, and f are smooth. In fact, *Weyl's lemma* states that u is analytic in any subregion Ω_1 of Ω provided p, q, and f are analytic in Ω_1. Similar conclusions are valid "up to the boundary" provided the boundary of Ω itself is analytic.

Singularities can therefore occur only when the boundary or some part of the data is not smooth. Unfortunately, these cases often arise, for example in fracture mechanics problems, and in the presence of singularities it is completely unsatisfactory to proceed with finite elements on a regular mesh. As with difference approximations, local mesh refinement in the sense that we have discussed in earlier chapters has been a popular and effective method for dealing with singularities. However, a great deal is known about the nature of singularities which arise in elliptic problems, and the special form of the variational method invites one to use this information in the Ritz–Galerkin approximation. This chapter is devoted to this task, and we begin by establishing the analytical form of the singularities which can arise.

Starting first with the case of nonsmooth boundaries, consider the Laplace equation

$$(2) \qquad -\Delta u = f \text{ in } \Omega, \qquad u = 0 \text{ on } \Gamma$$

defined in a region Ω which has a corner (Fig. 8.1). To fix ideas let us assume that f is analytic in the closed region $\bar{\Omega}$ and that Γ is analytic except at P. Then Weyl's lemma states that u is analytic in Ω except at P, and we seek a

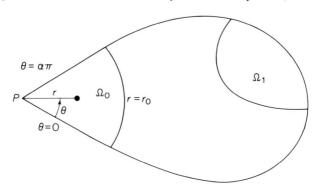

Fig. 8.1   A domain with corner angle $\alpha\pi$.

description of u near that point. In particular, we shall study the behavior of u in the sector

$$(3) \qquad \Omega_0 = \{(r, \theta) \mid 0 < r < r_0, \, 0 < \theta < \alpha\pi\} \subset \Omega,$$

where (r, θ) denotes polar coordinates at P. The weak form of (2) is

$$(4) \qquad \int_\Omega \nabla u \cdot \nabla v = \int_\Omega fv \qquad \text{for all } v \text{ in } \mathcal{3C}_E^1(\Omega).$$

If v is chosen to vanish near the corner and also outside the sector Ω_0, this reduces to

$$(5) \qquad 0 = \int_{\Omega_0} [\nabla u \cdot \nabla v - fv] = \int_0^{r_0} r \, dr \int_0^{\alpha\pi} \left[\frac{\partial u}{\partial r} \frac{\partial v}{\partial r} + r^{-2} \frac{\partial u}{\partial \theta} \frac{\partial v}{\partial \theta} - fv \right] d\theta.$$

Now, since u is analytic away from the corner and $u(r, 0) = u(r, \alpha\pi) = 0$, u may be expanded for each fixed $r > 0$ into

$$(6) \qquad u(r, \theta) = \sum_{j=1}^\infty u_j(r)\varphi_j(\theta),$$

where

$$(7) \qquad \varphi_j(\theta) = \sqrt{\frac{2}{\alpha\pi}} \sin v_j\theta, \qquad v_j = \frac{j}{\alpha}.$$

The Fourier coefficients $u_j(r)$ are determined by (5): Let $v = \psi(r)\varphi_j(\theta)$, and use the orthogonality properties

$$\int_0^{\alpha\pi} \frac{\partial\varphi_j}{\partial\theta}\frac{\partial\varphi_l}{\partial\theta}\,d\theta = v_i^2 \int_0^{\alpha\pi} \varphi_j(\theta)\varphi_l(\theta)\,d\theta = v_i^2\delta_{jl}.$$

Then (5) becomes, after the first term is integrated by parts in r,

(8)
$$\int_0^{r_0} dr\left[\frac{d}{dr}\left(r\frac{du_j}{dr}\right) - v_j^2 r^{-1}u_j - rf_j(r)\right]\psi(r) = 0,$$

$$f_j(r) = \int_0^{\alpha\pi} f\varphi_j(\theta)\,d\theta.$$

Since this holds for all ψ, the expression in brackets must vanish. This constitutes the basic differential equation for $u_j(r)$. Expanding $f_j(r)$ into $\sum f_{jl}r^l$, its general solution is

(9)
$$u_j(r) = \alpha_j r^{v_j} + \beta_j r^{-v_j} + \sum_{l=0}^{\infty} f_{jl}[(l+2)^2 - v_j^2]^{-1}r^{l+2},$$

where we agree to replace

(10)
$$[(l+2)^2 - v_j^2]^{-1}r^{l+2} \quad \text{by} \quad [2(v_j+1)]^{-1}r^{v_j}\ln r$$

whenever v_j^2 is of the form $(l+2)^2$. In addition, we reject the term r^{-v_j} and set $\beta_j = 0$, since otherwise that term has infinite energy. The other constant α_j is chosen so that at $r = r_0$, (9) is the correct Fourier coefficient of $u(r_0, \theta)$. Altogether, the solution near the corner takes the form (see Lehman [L2])

(11)
$$u(r, \theta) = \sum_{j=1}^{\infty} \alpha_j r^{v_j}\varphi_j(\theta) + \sum_{j=1}^{\infty}\sum_{l=0}^{\infty} f_{jl}[(l+2)^2 - v_j^2]^{-1}\varphi_j(\theta)r^{l+2}.$$

Suppose for the moment that $1/\alpha$ is not an integer. Then it follows from (11) that the leading term in the singularity of u is

$$r^{v_1}\sin v_1\theta = r^{1/\alpha}\sin\frac{\theta}{\alpha}.$$

Observe that this singularity becomes more pronounced as the angle $\alpha\pi$ increases, and if the corner at P is not convex, that is, $\alpha > 1$, even the first derivatives of u are unbounded. The worst case is a region with a crack running into it; this is one of the problems which we examine numerically in the last section. Around the point P at the head of the crack (Fig. 8.3) there is a full interior angle of 2π, and the solution behaves like $r^{1/2}\sin\theta/2$.

We can easily determine the degree of smoothness of u near any such singularity. By direct calculation, the function $r^v\sin v\theta$ falls just short of possessing $1 + \nu$ derivatives in the mean-square sense (and only ν in the pointwise sense—here we absolutely need the mean-square approximation theory to predict any reasonable convergence). Thus for any $\beta < \nu_1 = 1/\alpha$, the solution has $1 + \beta$ fractional derivatives. At a reentrant corner, where Ω is not convex and $\alpha > 1$, it follows that u lies in $\mathcal{3C}^1$ but not in $\mathcal{3C}^2$. Around a crack u is not quite in $\mathcal{3C}^{3/2}$.

When $1/\alpha$ is an integer the first sum in (11) is analytic. However, except for the case $\alpha = 1$ in which Γ is a straight line near P, we cannot conclude that u is analytic in Ω_0; *logarithms generally appear in the second sum*. For example, if $\alpha = \frac{1}{2}$, so that Γ makes a right angle at P, then $\nu_j^2 = (l + 2)^2 = 4$ in case $j = 1$ and $l = 0$. This is an instance in which the replacement (10) is required, and a term $(f_{10} r^2 \ln r \sin 2\theta)/6$ appears in the solution u. Observe that this term is in $\mathcal{3C}^2$ but it is not in $\mathcal{3C}^l$ for any $l > 2$.

Similar arguments give the behavior of the solution to other problems near corners. In the variable coefficient problem (1) a more complicated calculation shows that the singularity is still of the form (11). More generally, the singularity in a $2m$th-order problem is determined by the principal part of the operator. The leading term in u is of the form $r^\lambda \varphi_\lambda(\theta)$, where φ_λ is a smooth function of θ and λ is an eigenvalue of an auxiliary problem—both of which depend on the boundary conditions. We refer the reader to Kellogg's work [K2], which also includes the three-dimensional case, and to the Russian work [K3], as well as to standard engineering references [18, P6, W4, H5].

We now permit the second type of singularity, when the boundary is smooth yet one or more of the data is not smooth. Such a singularity typically arises in *interface problems*, and a simple example is provided by

$$(12) \qquad -\nabla \cdot (p \, \nabla u) = f \text{ in } \Omega, \qquad u = 0 \text{ on } \Gamma,$$

where Ω is the region shown in Fig. 8.2. We take the coefficient p to be

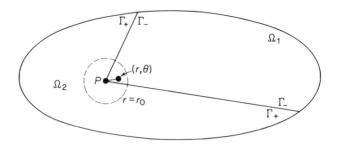

Fig. 8.2   An angular interface.

piecewise constant:

(13)
$$p = \begin{cases} p_1 & \text{in } \Omega_1, \\ p_2 & \text{in } \Omega_2. \end{cases}$$

The classical formulation of the problem is to require that the differential equation (12) hold separately in Ω_1 and Ω_2, with u and $p\,\partial u/\partial v$ being continuous across the interface Γ (v is the normal to Γ). Thus, referring to Fig. 8.2,

(14)
$$p_1 \frac{\partial u}{\partial v}\bigg|_{\Gamma_-} = p_2 \frac{\partial u}{\partial v}\bigg|_{\Gamma_+}.$$

As we noted in Section 1.3, the weak form of this equation is still

$$\int_\Omega p\, \nabla u \cdot \nabla v = \int_\Omega fv \qquad \text{for all } v \text{ in } \mathfrak{IC}^1_E,$$

and (14) is a *natural boundary condition*. It will *not* have to be satisfied by the trial functions in the Ritz method. Indeed, it would be very difficult to satisfy at a corner of Γ.

   Except when the interface is a straight line ($\alpha = 1$) or when $p_1 = p_2$, the solution u will be singular at the point P in Fig. 8.2, and we seek a description analogous to (11) for the behavior of u near this point. Following Birkhoff [B13] and Kellogg [K1], we introduce the periodic Sturm–Liouville system

(15)
$$-\frac{d}{d\theta}\left(p \frac{d\varphi}{d\theta}\right) = \lambda p \varphi, \qquad p = p(\theta) = \begin{cases} p_1 & \text{if } 0 < \theta < \alpha\pi, \\ p_2 & \text{if } \alpha\pi < \theta < 2\pi. \end{cases}$$

The eigenfunctions $\varphi(\theta)$ are required to be periodic, $\varphi(\theta) = \varphi(\theta + 2\pi)$, and to satisfy the interface conditions

(16)
$$\lim_{\theta \downarrow 0}\left[p_1 \frac{d\varphi}{d\theta}(\theta) \right] = \lim_{\theta \downarrow 0}\left[p_2 \frac{d\varphi}{d\theta}(-\theta) \right],$$

(17)
$$\lim_{\theta \downarrow 0}\left[p_1 \frac{d\varphi}{d\theta}(\alpha\pi - \theta) \right] = \lim_{\theta \downarrow 0}\left[p_2 \frac{d\varphi}{d\theta}(\alpha\pi + \theta) \right].$$

There is an infinite sequence of positive eigenvalues $\lambda_j = v_j^2$, and the associated eigenfunctions $\varphi_j(\theta)$ are orthogonal:

$$\int_0^{2\pi} p(\theta)\varphi_j'(\theta)\varphi_l'(\theta)\, d\theta = v_j^2 \int_0^{2\pi} p(\theta)\varphi_j(\theta)\varphi_l(\theta)\, d\theta = \delta_{jl} v_j^2.$$

For each fixed $r > 0$, the solution $u(r, \theta)$ satisfies the jump conditions

(16)–(17), and hence

(18) $$u(r, \theta) = \sum_{j=1}^{\infty} u_j(r) \varphi_j(\theta), \qquad u_j(r) = \int_0^{2\pi} p(\theta) u(r, \theta) \varphi_j(\theta) \, d\theta.$$

We now proceed in exactly the same way as in the case of boundary singularities. Substituting (18) into the differential equation (12) and using orthogonality, we obtain a differential equation for the Fourier coefficients $u_j(r)$. These equations can be solved exactly and we obtain an expression like (11) for u, except that in this case the exponents $\{v_j\}$ are the square roots of the eigenvalues of (15) and the $\{\varphi_j\}$ are the associated eigenfunctions.

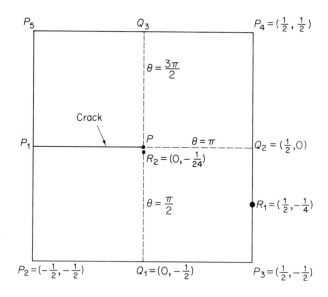

Fig. 8.3 A domain with a crack—an internal angle of 2π.

For this simple problem one can obtain exact formulas for the eigenfunctions and eigenvalues—in a more complicated problem numerical techniques will be required—and for illustration we consider the specific case $\alpha = \frac{1}{2}$. The eigenfunctions of (15) fall into two symmetry groups: those which are symmetric about $\theta = 0$ and those which are antisymmetric. In the former case the eigenfunctions have the form

(19) $$\varphi_v(\theta) = \begin{cases} \cos v\theta & \text{for } |\theta| < \pi/4, \\ \alpha_v \cos v(\pi - \theta) & \text{for } |\theta| > \pi/4. \end{cases}$$

The constant α_v is chosen so that the interface conditions (16)–(17) are satisfied:

$$\alpha_\nu = -\left[p_1 \sin \frac{\nu\pi}{4} \right] \Big/ \left[p_2 \sin \frac{3\nu\pi}{4} \right].$$

The eigenvalue ν is determined by substituting (19) into (15), and it can be shown that either $\tan \nu\pi/4 = 0$, that is, $\nu = 4n$, or $\nu = \pm\nu_1 \pm 4n$, where ν_1 is the smallest positive root of

(20) $$\left[3 - \tan^2 \frac{\nu\pi}{4} \right] \Big/ \left[1 - 3 \tan^2 \frac{\nu\pi}{4} \right] = -\frac{p_1}{p_2}.$$

Similar formulas are valid for the odd symmetry class.

The dominant singularity in the solution is therefore $r^\nu \varphi_\nu(\theta)$, where $\nu = \nu_1$ lies between 0 and 2. Note that $\nu_1 = 1$ in (20) if and only if $p_1 = p_2$. Otherwise u lies only in the fractional spaces $\mathfrak{IC}^{1+\beta}$, $\beta < \nu_1$, and in particular it is not in $\mathfrak{IC}^2$ when $\nu_1 < 1$.

There is no difficulty in extending this analysis to the case in which any number of interfaces meet at the point P. The coefficient $p(\theta)$ in (15) will have several discontinuities, with a jump condition of the form (16) at each one. Kellogg has observed that at a crossing of two straight interfaces, with suitably chosen coefficients p_i in each of the four quadrants, the leading exponent ν_1 can be arbitrarily small. Therefore, the singularity can be very severe; without some extra attention to the trial functions at all such exceptional points, the finite element method will give disappointing results.

8.2. SINGULAR FUNCTIONS

The expansions given in Section 8.1 suggest a modification of finite element spaces which will improve the approximation of singular solutions. Suppose that we can construct independent functions $\psi_1, \ldots, \psi_s$ such that $u - \sum c_i \psi_i$ is smooth, say in $\mathfrak{IC}^k$, for suitable (but unknown) constants $c_1, \ldots, c_s$. Then why not add $\psi_1, \ldots, \psi_s$ to the finite element space S^h? The idea is obviously to let the singular functions $\psi_1, \ldots, \psi_s$ approximate u near the singularity, with the conventional finite elements carrying the burden elsewhere. As a result, it is necessary to define the singular functions only locally near each singularity. We therefore choose, both for corners and for interfaces,

(21) $$\psi_j(r, \theta) = \begin{cases} r^{\nu_j} \sin \nu_j \theta & \text{for } 0 \le r \le r_0, \\ p_j(r) \sin \nu_j \theta & \text{for } r_0 \le r \le r_1, \\ 0 & \text{for } r_1 \le r. \end{cases}$$

The transition points r_0 and r_1 are fixed (independent of h), and the polynomials p_j are chosen so as to merge the coefficient r^{v_j} smoothly into zero. If we want the trial functions ψ_i to lie in $\mathcal{C}^{k-1}$ (and therefore in $\mathcal{X}^k$), then the polynomials are of degree $2k - 1$ and are determined by the Hermite conditions

$$(22) \qquad \frac{d^l}{dr^l}[p_j(r) - r^{v_j}]\big|_{r=r_0} = \frac{d^l}{dr^l}p_j(r)\big|_{r=r_1} = 0, \qquad l = 0, 1, \ldots, k - 1.\dagger$$

For example, suppose that we want to use cubic elements ($k = 4$) to solve Laplace's equation in a region with a crack. The exponents are $v_j = j - \frac{1}{2}$, and according to (11) there exist constants $\alpha_1, \ldots, \alpha_4$ such that

$$u - \sum_{j=1}^{3} \alpha_j \psi_j = \alpha_4 r^{7/2} \sin \frac{7\theta}{2} + \cdots + \text{analytic terms.}$$

The left side belongs to $\mathcal{X}^4$, and therefore it can be approximated to the optimal order h^4 by cubic elements. The best possible degree of approximation has been recovered by the inclusion of three singular functions ψ_j.

More generally, suppose that we have constructed $\psi_1, \ldots, \psi_s$ such that $u - \sum c_i \psi_i$ is in $\mathcal{X}^k$ for appropriate constants $c_1, \ldots, c_s$. In addition, let S^h be a standard finite element space of degree $k - 1$. Then if S_s^h is spanned by S^h together with $\psi_1, \ldots, \psi_s$, there is an approximation U^h in S_s^h to the singular solution u with error

$$(23) \qquad \|u - U^h\|_l \le Ch^{k-l}\|u - \sum_{i=1}^{s} c_i \psi_i\|_k.$$

This is clear: Since $u - \sum c_i \psi_i$ is in $\mathcal{X}^k$, we can approximate it by its interpolate v^h in the finite element space S^h. Then $U^h = v^h + \sum c_i \psi_i$.

This extra accuracy in approximation, obtained by adding singular functions to the trial space, will be reflected in extra accuracy of the Ritz–Galerkin approximation u^h. We shall postpone the analysis until the next section and concentrate here on the computational problems that accompany singular functions.

Two difficulties arise in applying these ideas to an actual computation, even after the forms of the singularities are known. The first is the evaluation of inner products involving singular functions, and the second is the inversion of the stiffness matrix. For the former there are a variety of tricks [F7, F9] which exploit the special form of the singular functions. Generally speaking, the most singular part—the radial dependence in the energy integrals $a(\psi_i, \psi_j)$—must be done analytically. The integration in θ, and if absolutely necessary also the integrals $a(\varphi_i, \psi_j)$ involving only one singular function, can be done

---

†A possible alternative is to multiply the singular function $r^v \sin v\theta$ directly by a polynomial $q(r)$ which merges smoothly into zero. This eliminates the condition at r_0 in favor of $q^{(l)}(0) = \delta_{0l}$.

by high-order numerical integration. A fixed quadrature rule of low accuracy is totally inappropriate.

The inversion of the stiffness matrix is a more serious problem. The addition of singular functions destroys the band structure of the matrix and may lead to extra operations in elimination ("fill in") as well as extra storage requirements. In addition, the conditioning of the stiffness matrix is very much in question. The singular functions $\psi_1, \ldots, \psi_s$ can be approximated by the other finite elements, and hence the basis for S_s^h will be "nearly linearly dependent."

In practice, both difficulties can be avoided by correctly ordering the unknowns. Let $\varphi_1, \ldots, \varphi_N$ be a basis for S^h and $\psi_1, \ldots, \psi_s$ be the singular functions. Then we order the unknowns so that the components of $\psi_1, \ldots, \psi_s$ appear last; that is, the vector of unknowns is $(Q_1 \ldots Q_N P_1 \ldots P_s) = (QP)$. The stiffness matrix for S_s^h is therefore

$$K_s = \begin{pmatrix} K_{11} & K_{12} \\ K_{21} & K_{22} \end{pmatrix},$$

where K_{11} is the standard stiffness matrix for S^h and the other blocks involve the singular functions. The entries of K_{22} are the energy inner products $a(\psi_i, \psi_j)$.

Faddeev and Faddeeva [6] then factor K_s into a product of triangular matrices, which we write in block form as

$$L = \begin{bmatrix} L_{11} & 0 \\ L_{21} & L_{22} \end{bmatrix}, \quad U = \begin{bmatrix} U_{11} & U_{12} \\ 0 & U_{22} \end{bmatrix}.$$

Therefore,

$$K_{11} = L_{11}U_{11}, \, K_{12} = L_{11}U_{12}, \, K_{21} = L_{21}U_{11}, \, K_{22} = L_{21}U_{12} + L_{22}U_{22}.$$

Obviously L_{11} and U_{11} are the factors of the usual stiffness matrix K_{11} associated with S^h. As a consequence, it is necessary only to store the bands of K_{11} and the much smaller matrices $K_{12}, K_{21},$ and K_{22}. (Note that $K_{12} = K_{21}^T$ in the symmetric case.) The additional storage required is only $sN + s^2$, which is orders of magnitude less than the storage requirement for K_{11}. In addition, the factorization and the calculation of the unknowns Q and P by back substitution represent only $O(w^2N)$ operations, where w is the band-width of K_{11}. This is the same as is required to solve $K_{11}Q = F_1$. In fact, the bulk of the work is in the factorization. The effects of numerical instabilities are isolated in the smaller matrices, and it is relatively easy to control the rounding errors. Typically the only real source of trouble is in the formation of $K_{22} - L_{21}U_{12}$, and it is often desirable to do the latter multiplication in high precision.

These ideas can also be used for eigenvalue problems. The bisection and (block) inverse power method require factorizations of the mass and stiffness matrix into LU, and if the Faddeev–Faddeeva bordering is used, no extra problems of storage and numerical stability are created by the addition of singular functions.

8.3. ERRORS IN THE PRESENCE OF SINGULARITIES

Let L be a $2m$th-order self-adjoint elliptic operator with homogeneous boundary data, and let $a(v, w)$ be the associated inner product on the energy space $\mathcal{3C}_E^m$. If the problem $Lu = f$ has interface or boundary singularities such as those described in Section 8.1, the error estimates derived in 3.4 are no longer valid. In this section we shall modify the earlier analysis to obtain the correct rates of convergence in the presence of singularities.

Using expansions analogous to those derived in Section 8.1, we can write the exact solution as a sum of singular functions plus a smooth function:

$$(24) \qquad\qquad u = \sum_{i=1}^{s} c_i \psi_i + w.$$

Each singular function ψ_i is in $\mathcal{3C}^\sigma$ for some $\sigma > m$, and is independent of the data f; it depends only on the geometry of Ω in the case of corners in the boundary, and on the coefficients of L for interface problems. We may use either $\psi_i = r^{\nu_i}\varphi_i(\theta)$, as in (11), or the function constructed in (21); all that matters is to keep the correct behavior near P. The smooth function w and the coefficients $c_1, \ldots, c_s$, on the other hand, do depend on f. According to the fundamental work of Kondrat'ev [K3], it is not only possible to ensure that w is in $\mathcal{3C}^k$ (by including sufficiently many ψ_i) but even to estimate its size:

$$(25) \qquad\qquad \| w \|_k \le C \| f \|_{k-2m}, \qquad \max_{1 \le i \le s} | c_i | \le C \| f \|_0.$$

Observe that if Ω and the coefficients of L are smooth, then the associated singular functions $\psi_1, \ldots, \psi_s$ are zero and hence (25) reduces to the usual bound for the solution in terms of the data.

As a first case let us derive the rate of convergence for the finite element method when *no special tricks*—mesh refinement or singular functions—*are used*. The error in strain energy presents no difficulties. As always, u^h is the closest trial function to u, and if u lies in $\mathcal{3C}^k$, then this error is of order $h^{2(k-m)}$. In general, however, u will lie only in some less smooth space $\mathcal{3C}^\sigma$, and the rate of convergence in energy is reduced to $h^{2(\sigma-m)}$. This error is likely to be unacceptably large.

Suppose that we attempt to carry out Nitsche's trick, as in Sections 1.6 and 3.4, to estimate the error in displacements without singular functions. As before, we take $u - u^h$ to be the data g in an auxiliary problem, whose solution is denoted by $z: Lz = u - u^h$. Then the argument can proceed unchanged, except at one crucial point: The estimate $\|z\|_{2m} \leq C\|u - u^h\|_0$ may no longer be true. Indeed, in light of (24), it may happen that z contains nonzero components of the singular terms $\psi_1, \ldots, \psi_s$, and hence we must be content to work with the weaker inequality

$$\|z\|_\sigma \leq C\|u - u^h\|_0,$$

which follows from (25) (replacing u with z). The effect of this weaker inequality is that a factor $h^{2m-\sigma}$ has to be sacrificed, and the optimal bound is

$$\|u - u^h\|_0 \leq C[h^{r+\sigma-2m} + h^{2(k-m)}]\|u\|_r, \qquad r = \min(k, \sigma).$$

For example, in the torsion experiment for a crack described in the next section, $\sigma = \frac{3}{2}$. *For any choice of element* this forces the error in the slopes to be $O(h^{1/2})$ and the error in displacement to be $O(h)$.

This is the error over the whole domain Ω. Away from the singularity it is reasonable to hope for something better, since elliptic equations always have a strong smoothing effect in the interior of Ω. In fact, if it were a question of ordinary least-squares approximation by piecewise polynomials, there is apparently no pollution from the singularity; if u has k derivatives in a subdomain Ω', then even without special tricks the best least-squares approximation over Ω is correct to order h^k in Ω' [N6]. For second-order equations this is no longer true, and some pollution does occur. However, the exponent still is better within Ω' than near the singularity. Suppose, for example, that the solution behaves like r^α near a corner in the domain. Then, according to Nitsche and Schatz, the error in the energy norm over Ω' which is attributable to the singularity is of order $h^{2\alpha}$. For the region with a crack this means an error in $\mathfrak{K}^1$ of order h away from the singularity, as compared with $h^{1/2}$ over the whole domain Ω.

Now we turn to the important question—the rate of convergence when singular functions are introduced into the trial space. For the energy norm we note that by construction the singular space S_s^h contains at least one function U^h satisfying (according to (23))

$$(26) \qquad a(u - U^h, u - U^h) \leq Ch^{2(k-m)} \left\| u - \sum_{i=1}^s c_i \psi_i \right\|_k^2.$$

Therefore, *the same bound holds for* $u - u^h$.

Proceeding to the $\mathfrak{K}^0$ estimate we again consider the auxiliary problem $Lz = u - u^h$. The crucial point of Kondrat'ev's theory is that z can be written as $\sum_{i=1}^{s} d_i \psi_i + v$, where v lies in $\mathcal{K}^k$ and

(27) $$\|v\|_{2m} \leq C\|u - u^h\|_0.$$

Since v is smooth, we can approximate it with a trial function v^h in S^h to order h^{k-m} in the energy norm. (We assume $k \geq 2m$, as is normal for finite elements.) On the other hand, the function $\sum d_i \psi_i$ is in our singular space S_s^h, and so is $z^h = \sum d_i \psi_i + v^h$. Therefore,

(28) $$\|z - z^h\|_m = \|v - v^h\|_m \leq C'h^m\|v\|_{2m} \leq C''h^m\|u - u^h\|_0.$$

Repeating the (rather complicated) argument of Section 4.4,

(29) $$\|u - u^h\|_0 \leq C'''h^k\|u - \sum_{i=1}^{s} c_i \psi_i\|_k \leq C^{(iv)}h^k\|f\|_{k-2m}.$$

This means that *by including enough singular functions, it is possible to obtain the same rate of convergence as for smooth problems.* A similar conclusion applies to mesh refinement, if the mesh size is taken to be an average $\bar{h} = N^{-1/2}$, based on the dimension N of the trial space. In this case the singular functions are not in the space, but according to the final comments in Section 3.2, a suitable mesh refinement permits their approximation to order $\bar{h}^k$. With this estimate the Nitsche argument goes as before.

It is obvious that all these theoretical predictions must be thoroughly tested. In complicated physical problems, it may be extremely difficult to identify the singularities and to incorporate them into special trial functions. Therefore, such a construction will be carried out only if the benefits are correspondingly great. Even mesh refinement introduces some complications, although it is usually much simpler than the construction of singular functions. All we can do here is to try each of these possibilities on a number of simplified physical problems and report the results. By glancing at the graphs in the following section, the reader can anticipate our conclusion: High accuracy has to be paid for, either by computer time on a simple method or by programming time with a more subtle technique. The prices vary with the problem. But almost certainly, from his own experience, the reader already knew that.

8.4. EXPERIMENTAL RESULTS

We conclude this chapter with three examples drawn from physics: (1) the computation of the rigidity and deformation of a cracked square elastic beam under torsion; (2) the criticality computation in an idealized square nuclear reactor consisting of a homogeneous square core surrounded by a square reflector, in the one-group diffusion approximation; and (3) the computation of the fundamental frequency of a vibrating L-shaped membrane.

In the torsion problem we give a numerical comparison of local mesh refinement as against the use of singular functions. For a given finite element their rates of convergence are the same, and efficiency hinges largely on the number of unknowns required. In the reactor problem, on the other hand, we shall be less concerned with singularities and concentrate attention on efficient methods for treating interfaces. Finally, the L-shaped membrane is included because of its long history as a model of an elliptic problem with a singularity. In fact, special methods developed for this problem have produced extremely accurate approximations to the vibrational frequencies. We shall compare these results with those obtained by the finite element method.

The differential equation governing the torsion problem (1) can be written in normalized form as

$$(30) \qquad\qquad -\Delta u = 1 \text{ in } \Omega, \qquad u = 0 \text{ on } \Gamma.$$

Ω is the region shown in Fig. 8.3 and Γ includes the crack $P_1 P$. Our expansions about P in Section 8.1 reduce to

$$u(r, \theta) = \sum_{j=1}^{\infty} c_j r^{v_j} \sin v_j \theta, \qquad v_j = \frac{2j-1}{2},$$

plus analytic terms. Thus the dominant term in the singularity at the point P is $r^{1/2} \sin (\theta/2)$. Its coefficient

$$(31) \qquad\qquad c_1 = \lim_{r \to 0} r^{-1/2}[u(r, \pi) - u(0, \pi)]$$

has great engineering significance; it is the commonly accepted measure of the torsion which the beam can endure before fracture occurs and is called the *stress intensity factor* [18]. We note that because of the factor $r^{1/2}$, u falls just short of $\frac{3}{2}$ derivatives in the mean-square sense.

In the problem as stated there are also singularities of the form $\rho^2 \ln \rho$ at each of the corners P_2, P_3, P_4, and P_5. Since these singularities are comparatively insignificant, we shall remove them by changing the boundary conditions to

$$(32) \quad u = 0 \text{ on } PP_1, P_2 P_3, \text{ and } P_4 P_5, \qquad \frac{\partial u}{\partial v} = 0 \text{ on } P_1 P_2, P_1 P_5, \text{ and } P_3 P_4,$$

where v denotes the normal to Γ. That the solution u of the new boundary-value problem is analytic away from P can be verified using the techniques of Section 8.1.†

We shall compute approximations from four different spaces; the first

---

†By the Saint-Venant principle this change of boundary conditions will not affect the singularity at P, a fact which is also evident from our analysis in Section 8.1.

three deal with a uniform mesh of length h and use singular functions constructed from $r^{\nu_j} \sin \nu_j \theta$, $\nu_j = (2j - 1)/2$, as in Section 8.2. These are the following:

1. S_L^h denotes the space of continuous functions on a uniform mesh which reduce to a bilinear polynomial $a + bx + cy + dxy$ on each subdivision. This space also includes the singular function ψ_1 constructed from $r^{1/2} \sin(\theta/2)$ so that we have the approximation property (26) with $k = 2$.

2. S_H^h denotes the *bicubic Hermite space* (Section 1.8) together with the singular functions ψ_1, ψ_2, and ψ_3, so that (26) holds with $k = 4$.

These examples fall into the category of *nodal finite element* spaces in the sense that we have discussed earlier, and so it is perhaps appropriate to include also an example of an *abstract finite element space*. The space of bicubic splines which are of class $\mathcal{C}^2$ in Ω is the most likely candidate; however, it is a difficult space to work with for this problem. The troubles are related to the essential boundary condition $u = 0$ on the crack PP_1. A bicubic spline which vanishes along this line will be so constrained at the point P that it cannot possibly match the true solution u, whose derivatives are all singular. To avoid this, we impose only simple continuity ($\mathcal{C}^0$) across the lines PQ_2, PQ_3, and PQ_4 and obtain what is usually called a *spline-Lagrange space:*

3. S_{SL}^h consists of piecewise bicubic polynomials which are $\mathcal{C}^2$ everywhere except at the lines PQ_2, PQ_3, and PQ_4. Across these lines the normal derivatives are permitted to be discontinuous. As with S_H^h, we include the singular functions ψ_1, ψ_2, and ψ_3 so that (26) holds with $k = 4$.

The advantage of the spline-Lagrange space is that it has only one unknown per node, except along the lines PQ_1, PQ_2, PQ_3 where there are three. Thus the dimension of S_{SL}^h is nearly four times less than that of the Hermite space S_H^h, and virtually the same as the dimension of the piecewise bilinear space S_L^h. Incidentally, a basis for S_{SL}^h is readily obtained from the standard spline formulas, by considering the lines PQ_1, PQ_2, and PQ_3 as a *triple confluence of nodal lines*.

Our final space consists of triangular elements and uses a graded mesh of maximum length h and minimum length δ.

4. $S_L^{h,\delta}$ denotes the space of continuous functions which reduce to *linear polynomials* on each triangle. We assume that $\delta = O(h^2)$, so that (26) holds with $k = 2$.†

The estimates in Section 8.3 imply that the square root of $\int |u - u^h|^2$

---

†In the calculations, the transition from meshsize h to $\delta = h^2$ was achieved by progressively halving the triangles.

is $O(h^4)$ for the cubic spaces, and $O(h^2)$ for S_L^h and $S_L^{h,\delta}$. We cannot measure this error directly, however, since the solution u is not known in closed form. Therefore, we also used a quintic element with several singular functions. More precisely, we computed with the quintic splines, of class $\mathcal{C}^4$ in Ω except across the lines PQ_1, PQ_2, and PQ_3—a quintic spline-Lagrange space, augmented by six singular functions. A very rapidly convergent approximation is obtained with this space; the rate of convergence is h^6, and the approximate solutions for $h = \frac{1}{6}, \frac{1}{8}$, and $\frac{1}{10}$ differed only in the seventh place. The first six figures were taken to be a correct value for u.

Figures 8.4 and 8.5 show the errors at the points R_1 and R_2 plotted against an average mesh length $\bar{h} = N^{-1/2}$.† This permits the selection of the most "efficient" space, meeting a given error tolerance with the least number of unknown parameters. In this regard S_L^h appears to be more efficient than the space $S_L^{h,\delta}$ of triangular elements using local mesh refinement. For the latter we used $h = \frac{1}{4}$, $\delta = \frac{1}{16}$ and $h = \frac{1}{8}$, $\delta = \frac{1}{64}$ according to the rule in 4. Thus many extra elements are needed to obtain the $h^{3/2}$ convergence, yet with S_L^h only one extra unknown—the coefficient of the singular function ψ_1—is required. The spaces S_H^h, S_{SL}^h of cubic elements were far superior to S_L^h, and the space of minimum dimension, namely the spline-Lagrange space S_{SL}^h, appears to be the most efficient of all.

The graphs also indicate the (larger) errors for finite element spaces *without singular functions*. There are three interesting things to note about these figures. The first, naturally, is the improvement obtained with singular functions—at R_1 and $h = \frac{1}{4}$, for example, the relative error without singular functions is about 40 per cent, and this drops to 0.1 per cent when singular functions are added. The second is that in the absence of singular functions, the pointwise errors are largest near P. In particular, it appears that these errors are of order $O(h^{1/2})$ near P and $O(h)$ everywhere else.‡ The third is that the standard Hermite cubics come out *worse* than the simplest linear elements. The cubics are too smooth to cope with the singularity.

Finally, in Fig. 8.6, we give approximations to the stress intensity factor (31). For the spaces S_L^h, S_H^h, and S_{SL}^h we use the coefficient c_1 of the singular function ψ_1, since

$$c_1 = \lim_{r \to 0} r^{-1/2}[u^h(r, \pi) - u^h(0, \pi)].$$

For the space $S_L^{h,\delta}$ this quantity is zero, and it is necessary to choose some

†While the error estimates cited above give no information concerning pointwise errors, the rates of convergence appear to be the same for this problem, as can be verified from the graphs.

‡An analysis similar to that in Section 8.3 shows that the mean-square error in this case is $O(h)$, hence the area of the region about P where the pointwise error is $O(h^{1/2})$ must be quite small and converge to zero as $h \to 0$. This explains the kink in the error with linear functions; R_2 is within the boundary layer, when h is small.

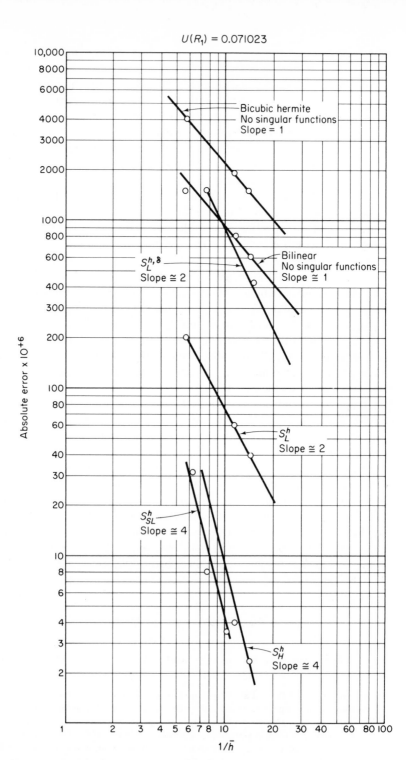

Fig. 8.4

$V(R_2) = 0.027425$

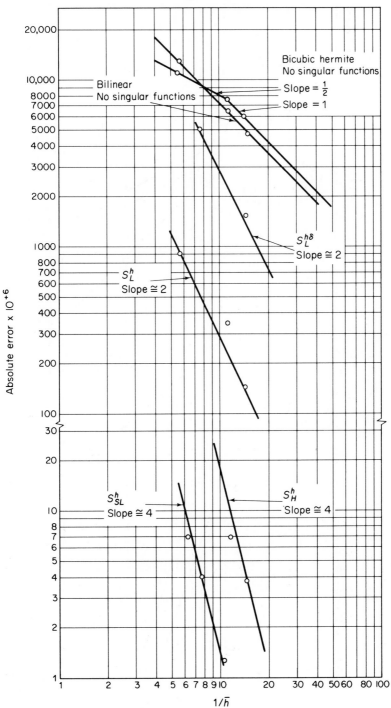

Fig. 8.5

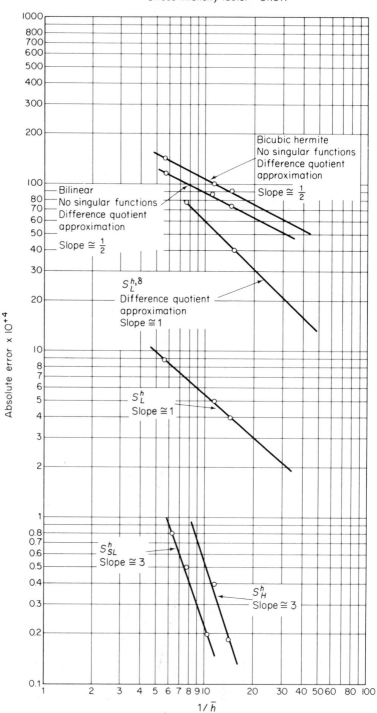

Fig. 8.6

The most important quantity to calculate is the lowest eigenvalue λ, which measures the criticality of the reactor.

Independent of any corner, the existence of interfaces fundamentally alters the best choice of an appropriate finite element space S^h. To use piecewise polynomials which are $\mathcal{C}^1$ across Γ would clearly lead to poor approximations, since u has discontinuous derivatives. Moreover, the use of trial functions satisfying the jump condition (35) leads to extra difficulties at the corners $P_j, j = 1, 2, 3, 4$. If we force the trial functions to satisfy the jump conditions along $P_1 P_2$, then they will still be influential on a portion of $Q_1 P_1$ where the solution is smooth.

Because of the convenient geometry, a spline-Lagrange space like that described in the torsion problem appears to be the most appropriate. The idea is to subdivide the region Ω into squares, with the interfaces lying on mesh lines. The trial space S^h consists of piecewise bicubic polynomials which are C^2 everywhere except on the lines $Q_1 Q_4$, $Q_2 Q_7$, $Q_3 Q_6$, and $Q_5 Q_8$, across which the elements are only continuous. We ignore the jump condition (35), which is a *natural boundary condition*, and allow the trial functions to have arbitrary jump discontinuities in their normal derivatives across the interfaces. Since the Galerkin method gives us in some sense a best approximation, it will presumably work out the jump discontinuities across the interfaces in a satisfactory way!

Approximations to the first eigenvalue using this space are given in Fig. 8.8 for the case

$$p_1 = 5, \qquad p_2 = 1, \qquad q = 0, \qquad p = 1.$$

The approximations λ^h are reasonably accurate, yet *the rate of their convergence to λ is very slow*. In particular,

$$(36) \qquad\qquad \lambda - \lambda^h = O(h^{2\nu}), \qquad \nu = 0.78.$$

The reason is that the eigenfunction u has unbounded first derivatives at the points P_1, P_2, P_3, and P_4. In fact, the analysis in Section 8.1 shows that the dominant term in the singularity is of the form $r^\nu \varphi_\nu(\theta)$, where φ_ν is a periodic function of θ and $\nu = 0.78$. Therefore, the eigenvalue error cannot be better than $h^{2\nu}$. By including $r^\nu \varphi_\nu$ in the trial space, the convergence rate can be increased to approximately $h^{6-2\nu}$. This is also confirmed by the numerical data.

Because of the singularity, it may seem surprising that λ^h, on a uniform mesh without singular functions, is reasonably accurate. This is in striking contrast to the situation for the torsion problem. The reason is that the coefficient of $r^\nu \varphi_\nu$ is quite small; computer plots of the eigenfunction u^h show that it is virtually a constant in the core Ω_1. We have checked this property (which is a consequence of the physics) by computing also the critical eigenvalue for

difference quotient

$$c_1 = \frac{u^h(\xi_h, \pi) - u^h(0, \pi)}{\xi_h^{1/2}}.$$

Since the errors in the finite element approximation without sing
tions grow quite rapidly as the point P is approached, it turns ou
choice $\xi_h = O(h)$ is as good as any. These are the values given in
The conclusions are the same as in the previous experiments, except t
refinement is even less competitive because of the extra error ma
difference quotient.

Our second example is the one-group, two-region reactor gov

(33) $-\nabla \cdot (p \, \nabla u) + qu = \lambda p u.$

This differential equation holds in the core Ω_1 and the reflector Ω_2 (
8.7), and we require the energy flux u to vanish on the outer bo

(34) $u = 0$ on Γ_2.

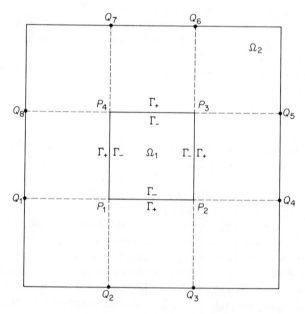

Fig. 8.7   Square core surrounded by a square reflector.

The coefficients p, q, and ρ are regionwise constants. Thus the curve Γ sep
rating Ω_1 and Ω_2 is an interface, and we require that u and $p \, \partial u / \partial v$ be continu
ous across Γ:

(35) $p \frac{\partial u}{\partial v}\bigg|_{\Gamma_+} = p \frac{\partial u}{\partial v}\bigg|_{\Gamma_-}, \qquad u\big|_{\Gamma_+} = u\big|_{\Gamma_-}.$

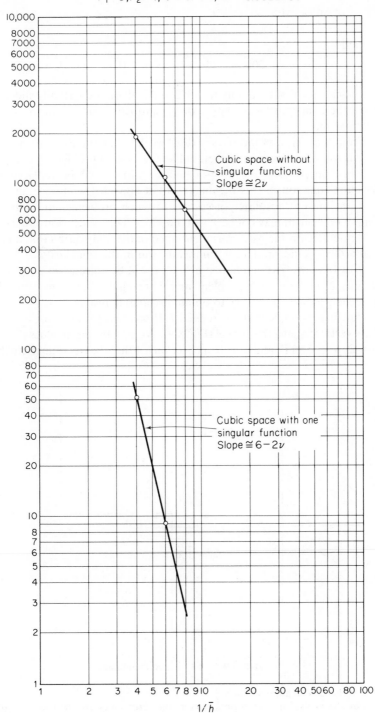

$P_1 = 5, P_2 = 1, \ \nu \cong 0.784, \ \lambda \cong 5.5822736$

Cubic space without
singular functions
Slope $\cong 2\nu$

Cubic space with one
singular function
Slope $\cong 6 - 2\nu$

Absolute error $\times 10^{+6}$

$1/\bar{h}$

Fig. 8.8

277

the case†

$$p_1 = 500, \qquad p_2 = 1, \qquad q = 0, \qquad p = 1.$$

This eigenvalue is shifted only to 5.582 and is therefore virtually independent of p_1; the contribution to the Rayleigh quotient from the inner square Ω_1 is almost zero.

From a physical point of view this "weak" coupling to a "strong" singularity is quite satisfactory. In fact, unlike the torsion problem, the singularity at P has no physical meaning—equation (33) should be replaced with a transport equation in this region. It would appear, therefore, that singular functions are not absolutely necessary for such problems, and a great deal of mesh refinement may not be worth the effort. This conclusion does *not* apply to all interface problems. It was mentioned in the first section that a crossing of interfaces could produce a dominant term in the singularity of any order $r^\epsilon \varphi_\epsilon(\theta)$.‡ It seems reasonable that a small ϵ will be reflected in poor approximations, as in the torsion problem, and that singular functions or local mesh refinement may be necessary to obtain acceptable results.

Our final example is the perennial L-shaped membrane—an overworked but nevertheless effective model (Fig. 8.9). We seek the eigenvalues of $-\Delta u = \lambda u$ in Ω, $u = 0$ on Γ, and note from Section 8.1 that the eigenfunctions behave

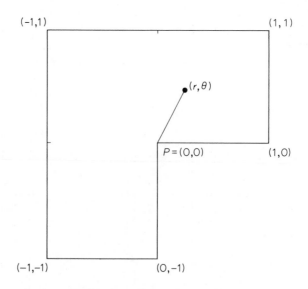

Fig. 8.9   An L-shaped membrane.

†v in this case in approximately $\frac{2}{3}$.

‡For the one-group, two-region reactor, the eigenvalues are the roots of (20), and $\epsilon \geq \frac{2}{3}$.

like

$$(37) \qquad u(r, \theta) = \sum_{j=1}^{\infty} \sum_{l=0}^{\infty} c_{jl} r^{v_j + 2l} \sin v_j \theta, \qquad v_j = \frac{2j}{3},$$

(plus analytic terms) near the reentrant corner P.

The most accurate approximations to date have been computed by Fox–Henrici–Moler [F10]. The idea is to take a linear combination of solutions to $\Delta u + \lambda u = 0$, in this case combinations of $K_v = J_v(\sqrt{\lambda}\, r) \sin v\theta$, $v = v_j$, and to determine the coefficients by minimizing the linear combination on the boundary Γ. This method is in a sense dual to the Galerkin method: It works with exact solutions and approximates the boundary conditions. The essential point, however, is the class of functions which are used. It is known that the eigenfunctions u can be very accurately approximated by linear combinations of the K_v, and therefore a Galerkin method would produce approximations with similar accuracy if the same class of functions were used. Strictly speaking, the functions K_v are not admissible, since they violate the essential boundary conditions. However, calculations with the closely related functions $(1 - x^2)(1 - y^2)r^{v+2l} \sin v\theta$ are reported in [F9] which are comparable with those obtained by Fox–Henrici–Moler.

These calculations reflect the remarkable accuracy which can be achieved by knowing exactly the right trial functions—which are not always piecewise polynomials! (Fourteen singular functions would be needed, together with bicubics, to improve on the results described above.) Our chief interest, however, is to obtain good accuracy with simple modifications of a standard finite element program. Therefore, we have computed the first and fourth eigenvalues, using a cubic spline space with and without singular functions. Figure 8.10 illustrates how the slow rate of convergence ($h^{4/3}$) in the latter case is greatly increased by the introduction of three singular trial functions. There is no question that the construction of these special functions makes the whole program more efficient.

The conclusion which we draw from all the numerical evidence is this: Even for coarse meshes and for singular problems, the rates of convergence which the theory predicts are clearly reproduced by the computations. The engineering literature contains a large number of numerical experiments, also leading to the same conclusion. This means that our goal, to analyze the steps of the finite element method and to explain its success, is largely achieved. We hope that this analysis will provide a theoretical basis for the future development of the method. The simplicity and convenience of polynomial elements was already clear, and now the accuracy which they achieve is mathematically confirmed.

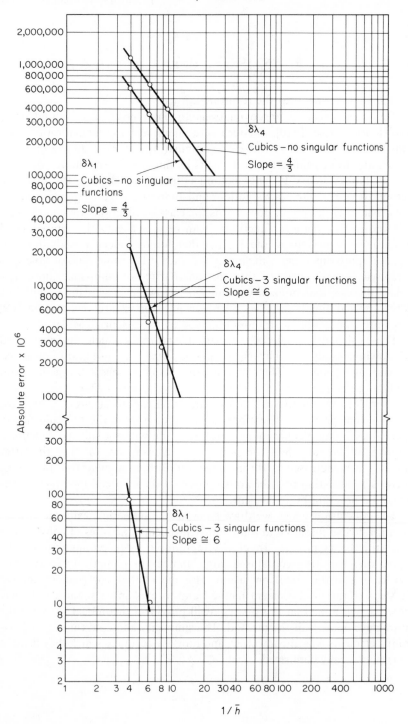

Fig. 8.10

BIBLIOGRAPHY

The number of publications on the finite element method is growing, within both the engineering and the numerical analysis literatures, at a rate which renders impossible any attempt at completeness. A bibliography which includes 170 titles prior to 1971 is contained in the excellent survey

ZIENKIEWICZ, O. C. (1970), The finite element method: from intuition to generality, *Appl. Mech. Rev.* 23, 249–256.

The following books have been referred to in preparing the present text, except for the three which appeared too recently (1972) for us to consult:

1. AGMON, S. (1965), *Lectures on Elliptic Boundary Value Problems*, Van Nostrand Reinhold, New York.

2. ARGYRIS, J. H. (1954–1955), *Energy Theorems and Structural Analysis*, Butterworth, London, 1960.

3. ARGYRIS, J. H. (1964), *Recent Advances in Matrix Methods of Structural Analysis*, Pergamon Press, Elmsford, N.Y.

4. AUBIN, J. P. (1972), *Approximation of Elliptic Boundary-Value Problems*, Wiley, New York.

5. DESAI, C., and J. ABEL (1972), *Introduction to the Finite Element Method*, Van Nostrand Reinhold, New York.

6. FADDEEV, D. K., and V. N. FADDEEVA (1963), *Computational Methods of Linear Algebra*, Dover, New York.

7. KRASNOSEL'SKII, M. A., G. M. VAINIKKO, et al. (1969), *Approximate Solution of Operator Equations*, Nauka, Moscow (in Russian).

8. HOLAND, I., and K. BELL, eds. (1969), *Finite Element Methods in Stress Analysis*, Tapir, Trondheim, Norway.

9. LIONS, J. L. (1969), *Quelques méthodes de résolution des problèmes aux limites non linéaires*, Dunod, Paris.

281

10. LIONS, J. L., and E. MAGENES (1968), *Problèmes aux limites non homogènes et applications*, Dunod, Paris.

11. MIKHLIN, S. G. (1964), *Variational Methods in Mathematical Physics*, Pergamon Press, Elmsford, N.Y.

12. MIKHLIN, S. G. (1965), *The Problem of the Minimum of a Quadratic Functional*, Holden-Day, San Francisco.

13. MIKHLIN, S. G. (1971), *The Numerical Performance of Variational Methods*, translated from the 1966 Russian edition, Wolters-Noordhoff, Groningen.

14. NEČAS, J. (1967), *Les méthodes directes en théorie des équations elliptiques*, Academia, Prague.

15. ODEN, J. T. (1972), *Finite Elements of Nonlinear Continua*, McGraw-Hill, New York.

16. PRZEMIENIECKI, J. S. (1968), *Theory of Matrix Structural Analysis*, McGraw-Hill, New York.

17. SYNGE, J. L. (1957), *The Hypercircle in Mathematical Physics*, Cambridge University Press, New York.

18. VAINBERG, M. M. (1964), *Variational Methods for the Study of Nonlinear Operators*, Holden-Day, San Francisco.

19. VISSER, M. (1968), *The Finite Element Method in Deformation and Heat Conduction Problems*, Delft, Holland.

20. WILKINSON, J. H. (1963), *Rounding Errors in Algebraic Processes*, Prentice-Hall, Englewood Cliffs, N.J.

21. WILKINSON, J. H., and C. REINSCH (1971), *Linear Algebra*, Springer-Verlag, Berlin.

22. ZIENKIEWICZ, O. C. (1971), *The Finite Element Method in Engineering Science*, 2nd ed. McGraw-Hill, New York.

23. ZIENKIEWICZ, O. C. and G. S. HOLISTER, eds. (1965), *Stress Analysis*, Wiley, New York.

It will also be useful to list some of the recent conferences and symposia which have concentrated on the theory and applications of finite elements. Their published proceedings contain a large number of valuable papers. Those to which we have referred specifically in the text will reappear in the list of individual papers below, with an indication such as "Wright-Patterson Conference II."

1. *Proceedings of the First Conference on Matrix Methods in Structural Mechanics*, Wright-Patterson AFB, Ohio, 1965.

2. *Proceedings of the Second Conference on Matrix Methods in Structural Mechanics*, Wright-Patterson AFB, Ohio, 1968.

3. *Proceedings of the Third Conference on Matrix Methods in Structural Mechanics*, Wright-Patterson AFB, Ohio, 1971 (to appear).

4. *Proceedings IUTAM Symposium*, High Speed Computing of Elastic Structures, Liège, Belgium, 1970.

5. *Numerical Solution of Partial Differential Equations* (SYNSPADE, University of Maryland), ed. B. Hubbard, Academic Press, New York, 1971.

6. *The Mathematical Foundations of the Finite Element Method.* (University of Maryland at Baltimore), Academic Press, New York, 1973.

7. *Numerical Solution of Field Problems in Continuum Physics*, ed. G. Birkhoff, Duke University, SIAM-AMS Proceedings, Vol. 2, 1970.

8. *SMD Symposium on Computer-Aided Engineering*, ed. G. L. M. Gladwell, University of Waterloo, May 1971.

9. *Finite Element Techniques in Structural Mechanics*, eds. H. Tottenham and C. Brebbia, Southampton University Press, 1970.

10. *Conference on Variational Methods in Engineering*, Southampton University, England, 1972.

11. *Conference on the Mathematics of Finite Elements and Applications*, Brunel University, England (to be published by Academic Press), 1972.

12. *Proceedings of the International Symposium on Numerical and Computer Methods in Engineering*, University of Illinois, 1971.

13. *Proceedings of the American Nuclear Society Meeting*, Boston, 1971.

14. *Proceedings of the First International Conference on Nuclear Reactor Structures*, Berlin, 1971.

15. *Proceedings of the First Symposium on Naval Structural Mechanics*, eds. J. H. Goodier and H. J. Hoff, Pergamon Press, 1960.

16. *Proceedings of the Symposium on Finite Element Techniques*, ISD, Stuttgart, 1969.

17. *Recent Advances in Matrix Methods of Structural Analysis and Design*, eds. J. T. Oden, R. H. Gallagher, and Y. Yamada, University of Alabama Press, 1971. (First Japan–U.S. seminar; the second will be held at Berkeley in 1972.)

18. *Application of Finite Element Methods to Stress Analysis Problems in Nuclear Engineering*, ISPRA, Italy, 1971.

19. *Conference on Computer Oriented Analysis of Shell Structures*, Lockheed Palo Alto Research Laboratories, Palo Alto, Calif., 1970 (papers to appear in the *Journal of Computers and Structures*).

20. *Symposium on the Application of Finite Element Methods in Civil Engineering*, Vanderbilt University, 1969.

21. *Symposium on Application of the Finite Element Method in Stress Analysis*, Swiss Society of Architects and Engineers, Zurich, 1970.

22. *National Symposium on Computerized Structural Analysis and Design*, George Washington University, 1972.

23. *On General Purpose Finite Element Computer Programs*, ed. P. V. Marcal, Amer. Society of Mechanical Engineers.

24. *Proceedings of the NATO Advanced Study Institute*, Lisbon, 1971.

25. *Computational Approaches in Applied Mechanics*, ASME Joint Computer Conference, Chicago, 1969.

The following bibliography contains those papers which are commented on in the text, together with many others which we have found to be valuable references. At this point especially, it must be repeated that a search of the literature would yield a much larger number of important titles. This is particularly true of engineering papers. The list below does indicate those journals in which the finite element method is strongly represented, and together with the conference proceedings it should be a reasonable guide to the significant analytical and theoretical work on the method.

A1.   AHMAD, S., B. M. IRONS, and O. C. ZIENKIEWICZ (1968), Curved thick shell and membrane elements with particular reference to axisymmetric problems, *Wright-Patterson II*.

A2.   ALLMAN, D. J. (1971), Finite element analysis of plate buckling using a mixed variational principle, *Wright-Patterson III*.

A3.   ANDERHEGGEN, E. (1970), A conforming triangular finite element plate bending solution, *Int. J. for Num. Meth. in Eng.* 2, 259–264.

A4.   ARGYRIS, J. H., and S. KELSEY (1963), *Modern Fuselage Analysis and the Elastic Aircraft*, Butterworth, London.

A5.   ARGYRIS, J. H., I. FRIED, and D. W. SCHARPF (1968), The Hermes eight element for the matrix displacement method, *J. Royal Aero. Soc.*, 613–617.

A6.   ARGYRIS, J. H., and I. FRIED (1968), The LUMINA element for the matrix displacement method, *J. Royal Aero Soc.*, 514–517.

A7.   ARGYRIS, J. H., O. E. BRÖNLUND, I. GRIEGER, and M. SÖRENSEN (1971), A survey of the application of finite element methods to stress analysis problems with particular emphasis on their application to nuclear engineering problems, *ISPRA Conference*.

A8.   AUBIN, J. P. (1967), Approximation des espaces de distributions et des opérateurs différentiels, *Bull. Soc. Math. France, Mémoire 12.*

A9.   AUBIN, J. P. (1968), Evaluation des erreurs de troncature des approximations des espaces de Sobolev, *J. Math. Anal. Appl.* 21, 356–368.

A10. AUBIN, J. P., and H. BURCHARD (1971), Some aspects of the method of the hypercircle applied to elliptic variational problems, *SYNSPADE*, 1–67.

B1.   BABUSKA, I. (1961), Stability of the domain of definition . . . (in Russian), *Czech. Math. J.* 11(86), 76–105, 165–203.

B2.   BABUSKA, I. (1970), Approximation by hill functions, *Tech. Note 648, Univ. Maryland.*

B3.   BABUSKA, I. (1970), Finite element method for domains with corners, *Computing* 6, 264–273.

B4.   BABUSKA, I. (1971), Error bounds for the finite element method, *Numer. Math.* 16, 322–333.

B5.   BABUSKA, I. (1971), Finite element method with penalty, *Rept. BN-710, Univ. Maryland.*

B6.   BABUSKA, I. (1973), The finite element method with Lagrangian multipliers, *Numer. Math.* 20, 179–192.

B7.   BÄCKLUND, J. (1971), *Mixed Finite Element Analysis of Elastic and Elastoplastic Plates in Bending,* Chalmers Institute of Technology, Göteborg, Sweden.

B8.   BAUER, F. L. (1963), Optimally scaled matrices, *Numer. Math.* 5, 73–87.

B9.   BAZELEY, G. P., Y. K. CHEUNG, B. M. IRONS, and O. C. ZIENKIEWICZ (1965), Triangular elements in plate bending—conforming and nonconforming solutions, *Wright-Patterson I.*

B10. BERGER, A., R. SCOTT, and G. STRANG (1972), Approximate boundary conditions in the finite element method, *Symposium on Numerical Analysis,* Istituto Nazionale di Alta Matematica, Rome; to appear in *Symposia Mathematica,* Academic Press, New York.

B11. BIRKHOFF, G. (1969), Piecewise bicubic interpolation and approximation in polygons, *Approximation with Special Emphasis on Spline Functions,* ed. I. Schoenberg, Academic Press, New York, 185–221.

B12. BIRKHOFF, G. (1971), Numerical solutions of elliptic equations, *SIAM Regional Conference Series,* Vol. 1.

B13. BIRKHOFF, G. (1972), Angular singularities of elliptic problems, *J. Approx. Th.* 6, 215–230.

B14. BIRKHOFF, G., and C. de BOOR (1964), Error bounds for spline interpolation, *J. of Math. and Mech.* 13, 827–836.

B15. BIRKHOFF, G., and C. DE BOOR (1965), Piecewise polynomial interpolation and approximation, *Approximation of Functions,* ed. H. L. Garabedian, Elsevier, Amsterdam.

B16. BIRKHOFF, G., C. DE BOOR, B. SWARTZ, and B. WENDROFF (1966), Rayleigh–Ritz approximation by piecewise cubic polynomials, *SIAM J. Num. Anal.* 13, 188–203.

B17. BIRKHOFF, G., and G. FIX (1967), Rayleigh–Ritz approximation by trigonometric polynomials, *Indian J. of Math.* 9, 269–277.

B18. BIRKHOFF, G., and G. FIX (1970), Accurate eigenvalue computations for elliptic problems, *Duke University SIAM–AMS Symposium.*

B19. BIRKHOFF, G., M. H. SCHULTZ, and R. VARGA (1968), Piecewise Hermite interpolation in one and two variables with applications to partial differential equations, *Numer. Math.* 11, 232–256.

B20. BLAIR, J. J. (1971), Bounds for the change in the solutions of second order elliptic PDEs when the boundary is perturbed, to appear.

B21. DE BOOR, C. (1968), The method of projections as applied to the numerical solution of two-point boundary value problems using cubic splines, Thesis, University of Michigan.

B22.   DE BOOR, C. (1968), On local spline approximation by moments, *J. of Math. and Mech.* 17, 729–736.

B23.   DE BOOR, C., and G. FIX (1972), Spline approximation by quasi-interpolants, to appear in *J. of Approx. Theory.*

B24.   DE BOOR, C., and B. SWARTZ (1972), Collocation at Gaussian points, *Los Alamos Rept. 72–65.*

B25.   BRAMBLE, J. H. (1971), Variational methods for the numerical solution of elliptic problems, Lecture notes, Chalmers Institute of Technology, Göteborg, Sweden.

B26.   BRAMBLE, J. H., T. DUPONT, and V. THOMÉE (1972), Projection methods for Dirichlet's problem in approximating polygonal domains with boundary value corrections, *MRC Tech. Rept. 1213, Univ. Wisconsin.*

B27.   BRAMBLE, J. H., and S. R. HILBERT (1970), Estimation of linear functionals on Sobolev spaces with application to Fourier transforms and spline interpolation, *SIAM J. Num. Anal.* 7, 113–124.

B28.   BRAMBLE, J. H., and S. R. HILBERT (1971), Bounds for a class of linear functionals with applications to Hermite interpolation, *Numer. Math.* 16, 362–369.

B29.   BRAMBLE, J. H., and J. OSBORN (1972), Rate of convergence estimates for nonselfadjoint eigenvalue approximations, *MRC Tech. Rept. 1232, Univ. Wisconsin.*

B30.   BRAMBLE, J. H., and A. H. SCHATZ (1970), Rayleigh–Ritz–Galerkin methods for Dirichlet's problem using subspaces without boundary conditions, *Comm. Pure Appl. Math.* 23, 653–675.

B31.   BRAMBLE, J. H., and A. H. SCHATZ (1971), On the numerical solution of elliptic boundary value problems by least square approximation of the data, *SYNSPADE*, 107–133.

B32.   BRAMBLE, J. H., and M. ZLAMAL (1970), Triangular elements in the finite element method, *Math. of Comp.* 24, 809–821.

C1.   CARLSON, R. E., and C. A. HALL (1971), Ritz approximations to two-dimensional boundary value problems, *Numer. Math.* 18, 171–181.

C2.   CÉA, J. (1964), Approximation variationelle des problèmes aux limites, *Ann. Inst. Fourier,* 14, 345–444.

C3.   CHERNUKA, M. W., G. R. COWPER, G. M. LINDBERG, and M. D. OLSON (1972), Finite element analysis of plates with curved edges, *Int. J. for Num. Methods in Eng.* 4, 49–65.

C4.   CIARLET, P. G., M. H. SCHULTZ, and R. S. VARGA (1967), Numerical methods of higher order accuracy for nonlinear boundary value problems, *Numer. Math.* 9, 394–430; *Numer. Math.* 13, 51–77.

C5.   CIARLET, P. G., and C. WAGSCHAL (1971), Multipoint Taylor formulas and applications to the finite element method, *Numer. Math.* 17, 84–100.

C6. CIARLET, P. G., and P. A. RAVIART (1972), General Lagrange and Hermite interpolation in R^n with applications to the finite element method, *Arch. Rat. Mech. Anal.* 46, 177–199.

C7. CIARLET, P. G., and P. A. RAVIART (1972), Interpolation theory over curved elements, with applications to finite element methods, *Computer Methods in Appl. Mech. and Eng.* 1, 217–249.

C8. CLOUGH, R. W., and J. L. TOCHER (1965), Finite element stiffness matrices for analysis of plates in bending, *Wright-Patterson I.*

C9. CLOUGH, R. W., and C. A. FELIPPA (1968), A refined quadrilateral element for analysis of plate bending, *Wright-Patterson II.*

C10. CLOUGH, R. W. (1969), Comparison of three-dimensional elements, *Vanderbilt Symposium.*

C11. COURANT, R. (1943), Variational methods for the solution of problems of equilibrium and vibrations, *Bull. Amer. Math. Soc.* 49, 1–23.

C12. COWPER, G. R. (1972), CURSHL: *A High-Precision Finite Element for Shells of Arbitrary Shape*, National Research Council of Canada Report.

C13. COWPER, G. R. (1972), Gaussian quadrature formulas for triangles, manuscript.

C14. COWPER, G. R., E. KOSKO, G. M. LINDBERG, and M. D. OLSON (1969), Static and dynamic applications of a high-precision triangular plate bending element, *AIAA J.* 7, 1957–1965.

D1. DEMJANOVIC, J. K. (1964), The net method for some problems in mathematical physics, *Dokl. Akad. Nauk SSSR* 159, *Soviet Math. Dokl.* 5.

D2. DEMJANOVIC, J. K. (1966), Approximation and convergence of the net method in elliptic problems, *Dokl. Akad. Nauk SSSR* 170, *Soviet Math. Dokl.* 7.

D3. DENDY, J. (1971), Thesis, Rice University.

D4. DESCLOUX, J. (1972), On finite element matrices, *SIAM J. Num. Anal.* 9, 260–265.

D5. DESCLOUX, J. (1970), On the numerical integration of the heat equation, *Numer. Math.* 15, 371–381.

D6. DOUGLAS, J., and T. DUPONT (1970), Galerkin methods for parabolic problems, *SIAM J. Numer. Anal.* 4, 575–626.

D7. DOUGLAS, J., and T. DUPONT (1972), A finite element collocation method for quasilinear parabolic equations, manuscript.

D8. DOUGLAS, J., and T. DUPONT (1972), manuscript on interpolation of coefficients, unpublished.

D9. DOUGLAS, J., and T. DUPONT (1973), Galerkin methods for parabolic equations with nonlinear boundary conditions, *Numer. Math.*, 20, 213–237.

D10. DUPONT, T. (1973), Galerkin methods for first-order hyperbolics: an example, *SIAM J. Numer. Anal.*, to appear.

D11. Dupuis, G., and J. J. Goël (1969), A curved element for thin elastic shells, *Tech. Rept., Brown Univ.*

D12. Dupuis, G., and J. J. Goël (1969), Eléments finis raffinés en élasticité bidimensionelle, *ZAMP* 20, 858–881.

D13. Dupuis, G., and J. J. Goël (1970), Finite element with high degree of regularity, *Int. J. for Num. Meth. in Eng.* 2, 563–577.

E1. Ergatoudis, I., B. Irons, and O. C. Zienkiewicz (1968), Curved isoparametric quadrilateral elements for finite element analysis, *Int. J. Solids Struct.* 4, 31–42.

F1. Felippa, C. A. (1966), Refined finite element analysis of linear and nonlinear two-dimensional structures, *Rept., Univ. California at Berkeley.*

F2. Felippa, C. A. (1969), Analysis of plate bending problems by the finite element method, *SESM Rept.* (Dept. Civil Eng.), *Univ. California at Berkeley.*

F3. Felippa, C. A., and R. W. Clough (1970), The finite element method in solid mechanics, *Duke University SIAM–AMS Symposium*, 210–252.

F4. Fix, G. (1968), Orders of convergence of the Rayleigh–Ritz and Weinstein–Bazley methods, *Proc. Nat. Acad. Sci.* 61, 1219–1223.

F5. Fix, G. (1969), Higher-order Rayleigh–Ritz approximations, *J. Math. Mech.* 18, 645–658.

F6. Fix, G., and G. Strang (1969), Fourier analysis of the finite element method in Ritz–Galerkin theory, *Studies in Appl. Math.* 48, 265–273.

F7. Fix, G., and S. Gulati (1971), Computational problems arising from the use of singular functions, *Rept., Harvard Univ.*

F8. Fix, G., and N. Nassif (1972), On finite element approximations to time-dependent problems, *Numer. Math.* 19, 127–135.

F9. Fix, G., S. Gulati, and G. I. Wakoff (1972), On the use of singular functions with the finite element method, *J. Comp. Physics*, to appear.

F10. Fox, L., P. Henrici, and C. Moler (1967), Approximation and bounds for eigenvalues of elliptic operators, *SIAM J. Numer. Anal.* 4, 89–102.

F11. Fraeijs de Veubeke, B. (1965), Displacement and equilibrium models in the finite element method, Chap. 9 of *Stress Analysis*, eds. O. C. Zienkiewicz and G. S. Holister, Wiley, New York.

F12. Fraeijs de Veubeke, B. (1968), A conforming finite element for plate bending, *Int. J. Solids Structures* 4, 96–108.

F13. Frederickson, P. O. (1971), Generalized triangular splines, *Math. Rept.* 7, *Lakehead Univ.*, Canada.

F14. Fried, I. (1971), Condition of finite element matrices generated from nonuniform meshes, *AIAA J.* 10, 219–221.

F15. Fried, I. (1971), Accuracy of finite element eigenproblems, *J. of Sound and Vibration* 18, 289–295.

F16.   FRIED, I. (1971), Basic computational problems in the finite element analysis of shells, *Int. J. Solids Struct.* 7, 1705–1715.

F17.   FRIED, I. (1971), Discretization and computational errors in high-order finite elements, *AIAA J.* 9, 2071–2073.

F18.   FRIED, I. (1972), The l_2 and l_∞ condition numbers . . . , *Conference at Brunel University.*

F19.   FRIEDRICHS, K. (1928), Die Randwert- und Eigenwertprobleme aus der Theorie der elastischen Platten, *Math. Ann.* 98, 205–247.

F20.   FRIEDRICHS, K. O., and H. B. KELLER (1966), A finite difference scheme for generalized Neumann problems, in *Numerical Solutions of Partial Differential Equations*, ed. J. Bramble, Academic Press, New York.

G1.   GALLAGHER, R. H., and A. K. DHALLA (1971), Direct flexibility finite element elastoplastic analysis, *Berlin Symposium* (see also papers by Gallagher in Vanderbilt and Japan–U.S. Symposia).

G2.   GEORGE, A. (1971), Computer implementation of the finite element method, Thesis, Stanford University.

G3.   GEORGE, A. (1971), Block elimination of finite element systems of equations, manuscript.

G4.   GOËL, J. J. (1968), Construction of basic functions for numerical utilization of Ritz's method, *Numer. Math.* 12, 435–447.

G5.   DI GUGLIELMO, F. (1969), Construction d'approximations des espaces de Sobolev sur des réseaux en simplexes, *Calcolo* 6, 279–331.

H1.   HANNA, M. S., and K. T. SMITH (1967), Some remarks on the Dirichlet problem in piecewise smooth domains, *Comm. Pure Appl. Math.* 20, 575–593.

H2.   HARRICK, I. I. (1955), On the approximation of functions vanishing on the boundary of a region by functions of a special form, *Mat. Sbornik*, N.S. 37(79), 353–384.

H3.   HERBOLD, R. J., M. H. SCHULTZ, and R. S. VARGA (1969), Quadrature schemes for the numerical solution of boundary value problems by variational techniques, *Aequ. Math.* 3, 96–119.

H4.   HERRMANN, L. R. (1967), Finite-element bending analysis for plates, *J. Eng. Mech. Div. ASCE* 94, 13–25.

H5.   HILTON, P. D., and J. HUTCHINSON (1971), Plastic intensity factors for cracked plates, *Eng. Fract. Mech.* 3, 435–451.

H6.   HULME, B. L. (1968), Interpolation by Ritz approximation, *J. Math. Mech.* 18, 337–342.

I1.   IRONS, B. M. (1966), Engineering applications of numerical integration in stiffness methods, *AIAA J.* 4, 2035–2037.

I2.   IRONS, B. M. (1968), Roundoff criteria in direct stiffness solutions, *AIAA J.* 6, 1308–12.

I3.   IRONS, B. M. (1969), Economical computer techniques for numerically integrated finite elements, *Int. J. for Num. Meth. in Eng.* 1, 201–203.

I4.   IRONS, B. M. (1970), A frontal solution program for finite element analyses, *Int. J. for Num. Meth. in Eng.* 2, 5–32.

I5.   IRONS, B. M. (1971), Quadrature rules for brick based finite elements, *AIAA J.* 9, 293–294.

I6.   IRONS, B. M., and A. RAZZAQUE (1971), A new formulation for plate bending elements, manuscript.

I7.   IRONS, B. M., E. A. DE OLIVEIRA, and O. C. ZIENKIEWICZ (1970), Comments on the paper: Theoretical foundations of the finite element method, *Int. J. Solids Struct.* 6, 695–697.

I8.   IRWIN, G. R. (1960), Fracture mechanics, *Symposium on Naval Structural Mechanics.*

K1.   KELLOGG, B. (1970), On the Poisson equation with intersecting interfaces, *Tech. Note BN-643, Univ. Maryland.*

K2.   KELLOGG, B. (1971), Singularities in interface problems, *SYNSPADE*, 351–400.

K3.   KONDRAT'EV, V. A. (1968), Boundary problems for elliptic equations with conical or angular points, *Trans. Moscow Math. Soc.* 17.

K4.   KOUKAL, S. (1970), Piecewise polynomial interpolations and their applications to partial differential equations, *Czech. Sbornik VAAZ, Brno*, 29–38.

K5.   KRASNOSELSKII, M. A. (1950), The convergence of the Galerkin method for nonlinear equations, *Dokl. Akad. Nauk SSSR* 73, 1121–1124.

K6.   KRATOCHVIL, J., A. ZENISEK, and M. ZLAMAL (1971), A simple algorithm for the stiffness matrix of triangular plate bending finite elements, *Int. J. for Num. Meth. in Eng.* 3, 553–563.

K7.   KREISS, H. O. (1971), *Difference Approximations for Ordinary Differential Equations*, Computer Science Department, Uppsala University.

L1.   LAASONEN, P. (1967), On the discretization error of the Dirichlet problems in a plane region with corners, *Ann. Acad. Scient. Fennicae* 408, 3–15.

L2.   LEHMAN, R. S. (1959), Developments near an analytic corner of solutions of elliptic partial differential equations, *J. Math. Mech.* 8, 727–760.

L3.   LINDBERG, G. M., and M. D. OLSON (1970), Convergence studies of eigenvalue solutions using two finite plate bending elements, *Int. J. for Num. Meth. in Eng.* 2, 99–116.

M1.   MARCAL, P. V. (1971), Finite element analysis with nonlinearities—theory and practice, *First Japan-U.S. Seminar.*

M2.   MARTIN, H. C. (1971), Finite elements and the analysis of geometrically nonlinear problems, *First Japan-U.S. Seminar.*

M3.   McCARTHY, C., and G. STRANG (1973), Optimal conditioning of matrices, *SIAM J. Num. Anal.*, to appear.

M4.   McLay, R. W. (1968), Completeness and convergence properties of finite element displacement functions—a general treatment, *AIAA Paper 67–143, 5th Aerospace Science Meeting.*

M5.   McLay, R. W. (1971), On certain approximations in the finite-element method, *Trans. of the ASME* 58–61.

M6.   Melosh, R. J. (1966), Basis for derivation of matrices for the direct stiffness method, *AIAA J.* 34, 153–170.

M7.   Mikhlin, S. G. (1960), The stability of the Ritz method, *Soviet Math. Dokl.* 1, 1230–1233.

M8.   Miller, C. (1971), Thesis, M.I.T.

M9.   Mitchell, A. R., G. Phillips, and R. Wachpress (1971), Forbidden shapes in the finite element method, *J. Inst. Maths. Appl.* 8.

M10. Morley, L. S. D. (1969), A modification of the Rayleigh–Ritz method for stress concentration problems in elastostatics, *J. of Mech. and Phys. of Solids* 17, 73–82.

N1.   Nitsche, J. (1968), Ein Kriterium für die Quasi-Optimalität des Ritzschen Verfahrens, *Numer. Math.* 11, 346–348.

N2.   Nitsche, J. (1968), Bemerkungen zur Approximationsgüte bei projektiven Verfahren, *Math. Zeit.* 106, 327–331.

N3.   Nitsche, J. (1970), Über ein Variationsprinzip zur Lösung von Dirichlet Problemen bei Verwendung von Teilräumen, die keinen Randbedingungen unterworfen sind, *Abh. Math. Sem. Univ. Hamburg* 36.

N4.   Nitsche, J. (1970), Lineare Spline-Funktionen und die Methoden von Ritz für elliptische Randwertprobleme, *Arch. Rat. Mech. Anal.* 36, 348–355.

N5.   Nitsche, J. (1971), A projection method for Dirichlet problems using subspaces with almost zero boundary conditions, manuscript.

N6.   Nitsche, J., and A. Schatz (1972), On local approximation of L_2-projections on spline-subspaces, Applicable Analysis 2, 161–168.

O1.   Oganesjan, L. A. (1966), Convergence of variational difference schemes under improved approximation to the boundary, *Soviet Math. Dokl.* 7, 1146–1150.

O2.   Oganesjan, L. A., and L. A. Rukhovets (1969), Study of the rate of convergence of variational difference schemes for second-order elliptic equations in two-dimensional regions with smooth boundaries, *Zh. Vychisl. Matem.* 9, 1102–1120.

O3.   Oliveira, E. A. de (1968), Theoretical foundations of the finite element method, *Int. J. Solids Struct.* 4, 929–952.

O4.   Olson, M. D., and G. M. Lindberg (1971), Dynamic analysis of shallow shells with a doubly-curved triangular finite element, *J. Sound Vibration* 19, 299–318.

P1.   Peters, G., and J. H. Wilkinson (1970), $Ax = \lambda Bx$ and the generalized eigenproblem, *SIAM J. Num. Anal.* 7, 479–492.

P2.   PETERS, G., and J. H. WILKINSON (1971), Eigenvalues of $Ax = \lambda Bx$ with band symmetric A and B, *Comput. J.* 14.

P3.   PHILLIPS, Z., and D. V. PHILLIPS (1971), An automatic generation scheme for plane and curved surfaces by isoparametric coordinates, *Int. J. for Num. Meth. in Eng.* 3, 519–528.

P4.   PIAN, T. H. H., and P. TONG (1969), Basis of finite element methods for solid continua, *Int. J. for Num. Meth. in Eng.* 1, 3–28.

P5.   PIAN, T. H. H. (1970), Finite element stiffness methods by different variational principles in elasticity, *Duke University SIAM-AMS Symposium*, 253–271.

P6.   PIAN, T. H. H., P. TONG, and C. H. LUK (1971), Elastic crack analysis by a finite element hybrid method, *Wright-Patterson III*.

P7.   PIERCE, J. G., and R. S. VARGA (1972), Higher order convergence results for the Rayleigh–Ritz method applied to eigenvalue problems I, *SIAM J. Num. Anal.* 9, 137–151.

P8.   PÓLYA, G. (1952), Sur une interprétation de la méthode des différences finies qui peut fournir des bornes supérieures ou inférieures, *Comptes Rendus* 235, 995–997.

P9.   PRAGER, W., and J. L. SYNGE (1947), Approximations in elasticity based on the concept of function space, *Quart. Appl. Math.* 5, 241–269.

P10.   PRAGER, W. (1968), Variational principles for elastic plates with relaxed continuity requirements, *Int. J. Solids Struct.* 4, 837–844.

P11.   PRICE, H. S., and R. S. VARGA (1970), Error bounds for semidiscrete Galerkin approximations of parabolic problems, *Duke University SIAM–AMS Symposium*, 74–94.

R1.   RAI, A. K., and K. RAJAIAH (1967), Polygon-circle paradox of simply supported thin plates under uniform pressure, *AIAA J.* 6, 155–156.

R2.   REID, J. K. (1972), On the construction and convergence of a finite-element solution of Laplace's equation, *J. Inst. Maths. Appl.* 9, 1–13.

S1.   SANDER, G. (1970), Application of the dual analysis principle, *IUTAM Symposium*.

S2.   SCHOENBERG, I. J. (1946), Contributions to the problem of approximation of equidistant data by analytic functions, *Quart. Appl. Math.* 4, 45–99, 112–141.

S3.   SCHULTZ, M. H. (1969), Rayleigh–Ritz methods for multidimensional problems, *SIAM J. Num. Anal.* 6, 523–538.

S4.   SCHULTZ, M. H. (1971), L^2 error bounds for the Rayleigh–Ritz–Galerkin method, *SIAM J. Num. Anal.* 8, 737–748.

S5.   STRANG, G. (1971), The finite element method and approximation theory, *SYNSPADE*, 547–584.

S6. STRANG, G. (1972), Approximation in the finite element method, *Numer. Math.* 19, 81–98.

S7. STRANG, G. (1972), Variational crimes in the finite element method, *Maryland Symposium*.

S8. STRANG, G. (1973), Piecewise polynomials and the finite element method, *Bull. Amer. Math. Soc.* 79, 1128–1137.

S9. STRANG, G., and A. E. BERGER (1971), The change in solution due to change in domain, *Proc. AMS Symposium on Partial Differential Equations, Berkeley*.

S10. STRANG, G., and G. FIX (1971), A Fourier analysis of the finite element method, *Proc. CIME Summer School, Italy*, to appear.

S11. SWARTZ, B., and B. WENDROFF (1969), Generalized finite difference schemes, *Math. of Comp.* 23, 37–50.

T1. TAYLOR, R. L. (1972), On completeness of shape functions for finite element analysis, *Int. J. for Num. Meth. in Eng.* 4, 17–22.

T2. THOMÉE, V. (1964), Elliptic difference operators and Dirichlet's problem, *Diff. Eqns.* 3, 301–324.

T3. THOMÉE, V. (1971), Polygonal domain approximation in Dirichlet's problem, *MRC Tech. Rept. 1188, Univ. Wisconsin*.

T4. TONG, P. (1969), Exact solution of certain problems by finite-element method, *AIAA J.* 7, 178–180.

T5. TONG, P. (1971), On the numerical problems of the finite element methods, *Waterloo Conference*.

T6. TONG, P., and T. H. H. PIAN, The convergence of finite element method in solving linear elastic problems, *Int. J. Solids Struct.* 3, 865–879.

T7. TONG, P., and T. H. H. PIAN (1970), Bounds to the influence coefficients by the assumed stress method, *Int. J. Solids Struct.* 6, 1429–1432.

T8. TONG, P., T. H. H. PIAN, and L. L. BUCCIARELLI (1971), Mode shapes and frequencies by the finite element method using consistent and lumped masses, *J. Comp. Struct.* 1, 623–638.

T9. TREFFTZ, E. (1926), Ein Gegenstück zum Ritzschen Verfahren, *Second Congress Applied Mechanics, Zurich*.

T10. TURNER, M. J., R. W. CLOUGH, H. C. MARTIN, and L. J. TOPP (1956), Stiffness and deflection analysis of complex structures, *J. Aero. Sciences* 23.

V1. VAINIKKO, G. M. (1964), Asymptotic error estimates for projective methods in the eigenvalue problem, *Zh. Vychisl. Mat.* 4, 404–425.

V2. VAINIKKO, G. M. (1967), On the speed of convergence of approximate methods in the eigenvalue problem, *USSR Comp. Math. and Math. Phys.* 7, 18–32.

V3. VAN DER SLUIS, A. (1970), Condition, equilibration, and pivoting in linear algebraic systems, *Numer. Math.* 15, 74–86.

V4.   VARGA, R. S. (1965), Hermite interpolation and Ritz-type methods for two-point boundary value problems, in *Numerical Solutions of Partial Differential Equations*, ed. J. H. Bramble, Academic Press, New York.

V5.   VARGA, R. S. (1970), Functional analysis and approximation theory in numerical analysis, *SIAM Regional Conference Series*, Vol. 3.

V6.   VISSER, W. (1969), A refined mixed-type plate bending element, *AIAA J.* 7, 1801–1803.

W1.   WAIT, R., and A. R. MITCHELL (1971), Corner singularities in elliptic problems by finite element methods, *J. Comp. Physics* 8, 45–52.

W2.   WEINBERGER, H. F. (1961), *Variational Methods in Boundary Value Problems*, University of Minnesota.

W3.   WIDLUND, O. B. (1971), Some recent applications of asymptotic error expansions to finite-difference schemes, *Proc. Roy. Soc. Lond.* A323, 167–177.

W4.   WILLIAMS, M. L. (1952), Stress singularities resulting from various boundary conditions in angular corners of plates in extension, *J. Appl. Mech.* 526–527.

W5.   WILSON, E. L., R. L. TAYLOR, W. P. DOHERTY, and J. GHABOUSSI (1971), Incompatible displacement models, *University of Ilinois Symposium*.

Y1.   YAMAMOTO, Y., and N. TOKUDA (1971), A note on convergence of finite element solutions, *Int. J. for Num. Meth. in Eng.* 3, 485–493.

Z1.   ZENISEK, A. (1970), Polynomial approximations on tetrahedrons in the finite element method, manuscript.

Z2.   ZENISEK, A. (1970), Higher degree tetrahedral finite elements, manuscript.

Z3.   ZENISEK, A. (1970), Interpolation polynomials on the triangle, *Numer. Math.* 15, 283–296.

Z4.   ZIENKIEWICZ, O. C. (1971), Isoparametric and allied numerically integrated elements—a review, *University of Illinois Symposium*.

Z5.   ZIENKIEWICZ, O. C., B. M. IRONS, et al. (1969), Isoparametric and associated element families for two and three-dimensional analysis, Holand and Bell [8].

Z6.   ZIENKIEWICZ, O. C., R. L. TAYLOR, and J. M. TOO (1971), Reduced integration technique in general analysis of plates and shells, *Int. J. for Num. Meth. in Eng.* 3, 275–290.

Z7.   ZLAMAL, M. (1968), On the finite element method, *Numer. Math.* 12, 394–409.

Z8.   ZLAMAL, M. (1970), A finite element procedure of the second order of accuracy, *Numer. Math.* 14, 394–402.

Z9.   ZLAMAL, M. (1972), Curved elements in the finite element method I, *SIAM J. Num. Anal.*, to appear.

Z10. ZLAMAL, M. (1972), The finite element method in domains with curved boundaries, *Int. J. for Num. Meth. in Eng.*, to appear.

SUPPLEMENTARY BIBLIOGRAPHY

S1. FINLAYSON, T. (1972), *The Method of Weighted Residuals and Variational Principles*, Academic Press, New York.

S2. Conference on Numerical Analysis, Royal Irish Academy, Dublin, 1972.

S3. Applications of the finite element method in geotechnical engineering, U. S. Army Engineers Symposium at Vicksburg, Mississippi, 1972.

S4. BATHE, K.-J. (1971), Solution methods for large generalized eigenvalue problems in structural engineering, Thesis, Univ. of Calif. at Berkeley.

S5. BATHE, K.-J., and E. L. WILSON (1972), Stability and accuracy analysis of direct integration methods, to appear.

S6. BIRMAN, M. S., and M. Z. SOLOMZAK (1967), Piecewise-polynomial approximation of functions of the classes W_p^α, Math. USSR-Sbornik 2, 295–317.

S7. CLOUGH, R. W., and K.-J. BATHE (1972), Finite element analysis of dynamic response, Second U.S.–Japan Seminar.

S8. FRIED, I., and S. K. YANG (1972), Best finite elements distribution around a singularity, *AIAA J.* 10, 1244–1246.

S9. FRIED, I., and S. K. YANG (1972), Triangular, 9 degrees of freedom, $\mathcal{C}^0$ plate bending element of quadratic accuracy, *Q. Appl. Math.*, to appear.

S10. FUJII, H. (1972), Finite element schemes: stability and convergence, Second U.S.–Japan Seminar.

S11. GIRAULT, V. (1972), A finite difference method on irregular networks, *SIAM J. Numer. Anal.*, to appear.

S12. GOLUB, G. H., R. UNDERWOOD, and J. H. WILKINSON (1972), The Lanczos algorithm for the symmetric $Ax = \lambda Bx$ problem, to appear.

S13. JEROME, J. W. (1973), Topics in multivariate approximation theory, Symposium on Approximation at Austin, Texas.

S14. JOHNSON, C. (1972), On the convergence of a mixed finite-element method for plate bending problems, *Numer. Math.*, to appear.

S15. JOHNSON, C. (1972), Convergence of another mixed finite-element method for plate bending problems, unpublished.

S16. KORNEEV, V. G. (1970), The construction of variational difference schemes of a high order of accuracy, *Vestnik Leningrad Univ.* 25, 28–40.

S17. McLEOD, R., and A. R. MITCHELL (1972), The construction of basis functions for curved elements in the finite element method, *J. Inst. Maths. Applics.* 10, 382–393.

S18. MOTE, C. D. (1971), Global–local finite element, *Int. J. for Numer. Meth. in Eng.* 3, 565–574.

S19. PIAN, T. H. H., and P. TONG (1972), Finite element methods in continuum mechanics, *Adv. in Appl. Mech.* 12, 1–58.

S20. VANDERGRAFT, J. S. (1971), Generalized Rayleigh methods with applications to finding eigenvalues of large matrices, *Lin. Alg. and Applics.* 4, 353–368.

S21. WEAVER, W., JR. (1971), The eigenvalue problem for banded matrices, *Computers and Structures* 1, 651–664.

INDEX OF NOTATIONS

This will be a little more than an index. We shall try to summarize in a convenient way the ideas which are essential in understanding three of the main themes of this book:

    I. Norms, function spaces, and boundary conditions
    II. Energy inner products, ellipticity, and the Ritz projection
    III. Finite element spaces and the patch test.

The definitions in I and II are more or less standard; those in III are special to the subject of finite elements.

I. NORMS, FUNCTION SPACES, AND BOUNDARY CONDITIONS

A *norm* is a measure of the size of a function ($\|u\|$) or of the distance between two functions ($\|u - v\|$). It satisfies the condition $\|cu\| = |c| \|u\|$ and the triangle inequality $\|u + v\| \le \|u\| + \|v\|$. Furthermore, unlike a *seminorm*, the norm is zero only if u is the zero function. It follows that if all solutions u to a linear problem are bounded in terms of their inhomogeneous terms ($\|u\| \le C \|f\|$) then solutions are *unique*: $u = 0$ if $f = 0$, or (by superposition) $u_1 - u_2 = 0$ if u_1 and u_2 share the same f.

Some familiar norms are

    i) the maximum norm = "sup norm" = L_∞ norm = $\sup\limits_{x \text{ in } \Omega} |u(x)|$;

    ii) the L_2 norm = $\mathcal{H}^0$ norm = $(\int_\Omega |u(x)|^2 \, dx_1 \cdots dx_n)^{1/2}$.

In the discrete case, for vectors $u = (u_1, u_2, \ldots)$ instead of functions $u(x)$, the integrals are replaced by corresponding sums. One generalization is to L_p norms; the exponents 2 and $1/2$ are replaced by p and $1/p$. The triangle inequality is satisfied (and we have a norm) for $p \geq 1$. These spaces become interesting and valuable in nonlinear problems; we have not found them essential to the linear theory.

A second generalization, absolutely basic, is to include derivatives as well as function values of u in computing a norm (pages 5, 143). The $\mathcal{3C}^s$ norm combines the $\mathcal{3C}^0$ (or L_2) norms of all partial derivatives

$$D^\alpha u = \left(\frac{\partial}{\partial x_1}\right)^{\alpha_1} \cdots \left(\frac{\partial}{\partial x_n}\right)^{\alpha_n} u \quad \text{of order} \quad |\alpha| = \alpha_1 + \cdots + \alpha_n \leq s:$$

(1)
$$\|u\|^2_{\mathcal{3C}^s} = \sum_{|\alpha| \leq s} \int |D^\alpha u|^2 \, dx_1 \cdots dx_n.$$

The seminorm $|u|_s$ includes only those terms of order exactly $|\alpha| = s$; it is zero if u is a polynomial of degree $s - 1$. The squares are introduced in equation (1) in order to have an inner product structure, in other words, to make $\mathcal{3C}$ a Hilbert space (see II below). *Fractional derivatives* (non-integer s) also have important applications (pages 144–5, 260), but their definition is rather technical [1, 10, 14]—except when Ω is the whole of n-space, when we use Fourier transforms,

$$\|u\|^2_s = \int_{-\infty}^{\infty} |\hat{u}(\xi)|^2 (1 + |\xi|^2)^s \, d\xi_1 \cdots d\xi_n.$$

Negative norms (the index s is negative, not the norm itself!) are defined by *duality* (pages 16, 73, 167):

$$\|u\|_{-s} = \max_{v \text{ in } \mathcal{3C}^s} \frac{|\int uv|}{\|v\|_s}.$$

The functions in these "Sobolev spaces"—as distinct from their norms—are usually defined by starting with a set of comparatively simple functions and then *completing* the space (page 11). The result is a *Banach space*; it contains the limit point of any sequence for which $|u_N - u_M| \to 0$ as $N, M \to \infty$. Intuitively, the holes are filled in.

The completed space will depend on the original set of simple functions. If we start with the set $\mathcal{C}^0$ of all continuous functions, then in the maximum norm, this set contains all its limit functions and there are no holes to fill in. If the original set includes also the *piecewise continuous* functions, the final space L_∞ is much larger (and more difficult to describe). This point reappears in connection with boundary conditions: if none are applied to the original set of all infinitely differentiable functions, then its completion is the whole space $\mathcal{3C}^s(\Omega)$; this is the admissible space for the *Neumann problem*, with free edges. If each member of the original set is required to vanish in a strip near the boundary Γ—the strip may be smaller for one function than another—then the completed space *in the same norm* is $\mathcal{3C}_0^s$, with derivatives of order less than s vanishing on Γ (page 67). This is the admissible space for the *Dirichlet problem*.

We mention two groups of important but technical theorems about these spaces. One group is typified by the *Sobolev inequality* (pages 73, 142), and answers the question: if the derivatives of order s_1 (integer or not) are in L_{p_1}, are those of order s_2 in L_{p_2}? In other words, which function spaces contain which others? (Sobolev:

$\mathfrak{K}^s$ contains $\mathbb{C}^q$ if and only if $s - q > n/2$.) The other technical results are the group of *trace theorems:* suppose u is in $\mathfrak{K}^s$, how smooth are its boundary values, considered as a function on Γ? Rough answer: they are in $\mathfrak{K}^{s-(1/2)}(\Gamma)$. Therefore $\mathfrak{K}^{m-(1/2)}$ is a suitable "data space" for the inhomogeneous boundary condition $u = g$; it matches the solution space $\mathfrak{K}_E^m$ for u.

The central problem of partial differential equations (pages 4–6) is to match a space of data to a space of solutions. Such a match is not automatic. For the example $-\Delta u = f$ in the L_∞ norm, it is not true that $\|u_{xx}\| + \|u_{yy}\| \leq C \|f\|$— and the pointwise theory suffers from it. The $\mathfrak{K}^s$ norms *are* well matched for elliptic problems of any order $2m$: $\|u\|_s \leq C \|f\|_{s-2m}$. For the Euler equation $s = 2m$: we take f in $\mathfrak{K}^0$ and look for u in $\mathfrak{K}_B^{2m}$, satisfying all m boundary conditions. For the variational problem $s = m$: u is found in the admissible space $\mathfrak{K}_E^m$, restricted only by the essential boundary conditions.

II. ENERGY INNER PRODUCTS, ELLIPTICITY, AND THE RITZ PROJECTION

Linear variational problems are posed in terms of quadratic functionals $I(v)$. The classical case is to minimize a *potential energy* $I(v) = a(v, v) - 2(f, v)$— here we separate the terms of second and first degree—over a space of admissible solutions v. The second-degree term is the *strain energy*, and it is associated with an *energy inner product*

$$(2) \qquad a(v, w) = \tfrac{1}{4}(a(v + w, v + w) - a(v - w, v - w)).$$

In terms of this inner product, the condition that u minimize $I(v)$ is the vanishing of the first variation, alias the *equation of virtual work:*

$$(3) \qquad a(u, v) = (f, v) \qquad \text{for all admissible } v.$$

Integration by parts alters this *weak form* (or *Galerkin form*) of the problem into the Euler differential equation for u—of order $2m$, without the presence of v, and with *jump conditions* appearing as integrated terms at any discontinuities.

NOTE: An inner product is bilinear,

$$a(u + v, w + z) = a(u, w) + a(v, w) + a(u, z) + a(v, z),$$

and it is only when the strain energy has a favorable form that this property holds. If this energy depended on the point of maximum strain, $a(v, v) = \max |\operatorname{grad} v|^2$, it would fail. Among L_p norms only the case $p = 2$ yields, through equation (2), such an inner product; it is therefore the only *Hilbert space*. The inner product property extends to the spaces $\mathfrak{K}^s$, and also (thank God) to the strain energies in linear elasticity and other applications.

Problems which are not self-adjoint begin directly with the equation (3) of virtual work, not with a minimization. The bilinear form $a(u, v)$ is no longer symmetric, $a(u, v) \neq a(v, u)$, and complex-valued functions must be admitted. Nevertheless, if the *real part* of $a(v, v)$ is elliptic (see below), then the results of the Galerkin theory (page 119) completely parallel those of the Ritz theory. They coincide, when $a(u, v)$ is symmetric.

The solvability of the fundamental variational equation (3) is guaranteed if the form is *elliptic*: $\text{Re } a(v, v) \geq \sigma \|v\|_m^2$. (So is the solvability of the corresponding *parabolic* equation.) In the case of *systems* of equations, with a vector of unknowns $u = (u_1(x), \ldots, u_r(x))$—as is typical in applications—there appear several varieties of ellipticity. One possibility [1, 10, 14] asks that the eigenvalues of certain matrices of order r have positive real parts; this is too weak to guarantee success when the Galerkin method is applied on a subspace. Strong ellipticity is a condition, not on the eigenvalues, but on the matrices themselves—and a still stronger condition is the familiar one on $\text{Re } a(v, v)$, which applies as successfully to systems as to single equations; v becomes an admissible vector of functions. Also for boundary conditions there is a catalogue of possibilities, clearly described by Kellogg in the Baltimore Symposium volume [6]; for applications the central step is still to distinguish the essential conditions, and thereby the admissible space—and to require the definiteness of $\text{Re } a(v, v)$ over that space.

The Ritz method is to minimize the functional $I(v)$ over a sequence of subspaces S^h. The fundamental theorem (page 39) establishes that the minimizing u^h is the *projection* of u onto S^h, in other words, u^h is the *closest function to u* in the strain energy norm $a(v, v)$. Therefore if each subspace S^h is contained in the next—as was supposed in the classical Ritz method, and usually occurs for finite elements when new elements are formed by subdividing the old ones—the convergence in strain energy is *monotonic* as $h \rightarrow 0$. So is the convergence of eigenvalues. This may be useful, but it is not critical to the Ritz theory; monotonicity of the S^h is an extra hypothesis, and monotonicity of convergence is an extra conclusion.

III. FINITE ELEMENT SPACES AND THE PATCH TEST

The usual description of a finite element specifies the form of the shape function (trial polynomial), and the location as well as the parameter (function value v, or some derivative $D_j v$) assigned to each node. Section 1.9 contains a number of examples. This is enough information to compute the element matrices, and to assemble them into the global stiffness and mass matrices K and M.

In Section 2.1, for mathematical reasons, we took one more step in describing the trial space S^h: we defined a set of basis functions $\varphi_1, \ldots, \varphi_N$ for the space. The function φ_j was directly associated with the node z_j, and with the particular nodal parameter $D_j v$. If the shape functions are uniquely defined by the values of the nodal parameters (which they must be!), then there is a unique trial function φ_j for which $D_j \varphi_j(z_j) = 1$, and all its other nodal parameters $D_i \varphi_j(z_i)$ are zero. These functions form a basis, because any trial function can be expanded in terms of its nodal parameters as

$$v^h = \sum q_j \varphi_j, \text{ where the weight } q_j = D_j v^h(z_j).$$

This leads immediately to the definition of the *interpolate* u_I of any given function

u. The interpolate assumes the same nodal parameters as u, but inside each element it is one of the trial polynomials: $u_I = \sum q_j \varphi_j$, where the weight $q_j = D_j u(z_j)$. The *approximation theorems* of Chapter 3 establish that u_I is close to u, in the norms described earlier. The error $u - u_I$ depends on the element sizes h_i, and on the degree $k - 1$ to which the shape functions are complete (page 136).

The *dimension* of the trial space S^h is N, the number of basis functions φ_j and free parameters q_j. Obviously N depends on the number of elements. A critical quantity is the number M of *parameters per vertex*; this permits the size of the assembled matrix K to be compared for two competing elements. Let d be the number of *degrees of freedom* (shape function coefficients) within each element; M is smaller, depending on the continuity constraints imposed between elements. (Our conjecture for the trial space described on page 84, complete through degree $k - 1$ and of continuity class $\mathcal{C}^q$ on triangles, is $M = (k - 1 - q)(k - 1 - 2q)$.)

We come now to the *patch test*. This test has had very little publicity, and until recently it was hardly known (at least by that name) even to the experts. It was created in the appendix to [B9], in order to explain why the Zienkiewicz triangle was convergent in one configuration but not in another (page 175). The test resurfaced, under its official name, in a brief comment [I9] on an earlier paper. The full story was told more recently by Irons at the Baltimore Symposium, but there he is still assailed by doubts that the test is sufficient for convergence.

We are convinced that, under reasonable hypotheses, it is. The test is described on page 174, and is very simple to conduct. We recall the equivalent form on page 177: if the strain energy involves derivatives $D^m v$ of order m, then all the integrals $\iint D^m \varphi_j$ should be calculated correctly—even though interelement boundary terms are ignored in the nonconforming case, or numerical quadrature is applied. The *higher-order patch test* asks that $\iint P_{n-m} D^m \varphi_j$ be correct for all polynomials of degree $n - m$; this generalization was made by the first author, and produces convergence in strain energy of order $h^{2(n-m+1)}$.

A word about the proof. On pages 179 and 186, the problem is reduced to estimating the inner product error $a_*(u, v^h) - a(u, v^h)$. (For numerical integration the linear terms involving f also appear.) Suppose we stay with the model problem $-\Delta u = f$, for which the energy inner product is $a(u, v) = \iint u_x v_x + u_y v_y$. Success in the patch test means that for any linear polynomial P,

$$(4) \qquad a_*(u, \varphi_j) - a(u, \varphi_j) = a_*(u - P, \varphi_j) - a(u - P, \varphi_j).$$

Choosing P close to u over the set E_j on which φ_j is non-zero, and renormalizing the basis by $a(\varphi_j, \varphi_j) = 1$, (4) is less than $ch\|u\|_{2, E_j}$. Therefore if we expand any v^h as $\sum q_j \varphi_j$,

$$(5) \qquad \begin{aligned} |a_*(u, v^h) - a(u, v^h)| &\leq ch \sum \|u\|_{2, E_j} |q_j| \\ &\leq ch \left(\sum \|u\|_{2, E_j}^2\right)^{1/2} \left(\sum q_j^2\right)^{1/2} \\ &\leq c'h \|u\|_2 \left(\sum q_j^2\right)^{1/2}. \end{aligned}$$

If it were true that

$$(6) \qquad \sum q_j^2 \leq C^2 a_*\left(\sum q_j \varphi_j, \sum q_j \varphi_j\right),$$

then the convergence proof would be complete. The expression Δ is less than $c'Ch\|u\|_2$, and according to the estimate at the top of page 179, the strains are in error by $O(h)$.

The result is correct, but regrettably the inequality (6) is false. It amounts to asking that the condition number of the stiffness matrix K be bounded—or that the φ_j be uniformly independent in the energy norm. This is true of the mass matrix (page 212), but not of K. There is, however, a way out. Any *conforming* φ_j can be ignored in our calculations; the difference in (4) is identically zero. Therefore if the trial space can be regarded as a conforming space *to which some uniformly independent nonconforming elements are added*, the proof succeeds. This was obvious in Wilson's case, since he began with the standard bilinear elements, and superimposed two nonconforming quadratics within each square. (Such internal degrees of freedom are called *nodeless variables*.) Because the squares never overlap, uniform independence was automatic.

Crouzeix and Raviart have recently given a beautiful treatment of nonconforming elements for divergence-free fluids. Their technique applies to all elements which pass the test in the following way: over each interelement edge or face, the integral of the nonconforming jump is zero. Their technique for deducing convergence has been formalized, and applied to plate elements, by both Ciarlet and Lascaux.

The increasing application of finite elements — to the Navier-Stokes equations, control problems, earthquake prediction, nonlinear elasticity and plasticity in soils as well as metals, and the design of tankers and reactors — promises a happy future for both the analyst and the engineer.

INDEX

DATE DUE